Natural Landscapes of Maine

A Guide to Natural Communities and Ecosystems

by Susan Gawler and Andrew Cutko

ISBN 0-615-34739-4

To cite this document:

Gawler, S. and A. Cutko. 2010. Natural Landscapes of Maine: A Guide to Natural Communities and Ecosystems. Maine Natural Areas Program, Maine Department of Conservation, Augusta, Maine.

Cover photo: Circumneutral Riverside Seep on the St. John River, Maine

Printed and bound in Maine using recycled, chlorine-free paper

Contents

Contents (continued)

Acknowledgements

This classification reflects the work of dozens of people over nearly two decades. Numerous field ecologists and naturalists offered data or advice as the classification developed, and several also reviewed draft versions. Contributors and reviewers include Dennis Anderson, Rob Bryan, Charles Cogbill, Ronald Davis, Norm Famous, Linda Gregory, Arthur Haines, Jim and Pat Hinds, Pat Ledlie, Julie Lundgren, Mark McCollough, Janet McMahon, John McPhedran, Don Mansius, Glenn Mittlehauser, Morten Moesswilde, Betsy Newcomer, Sally Rooney, Josh Royte, Nancy Sferra, Eric Sorensen, Marcia Spencer-Famous, Dan Sperduto, Lauren Stockwell, Lillie Vitelli, Jill Weber, Alan White, and Art Wilder. In addition, this work benefitted from natural community classifications at the regional and national level by Mark Anderson, Rick Enser, Jean Fike, Denny Grossman, David Hunt, Julie Lundgren, Ken Metzler, Lesley Sneddon, Eric Sorenson, Daniel Sperduto, and Pat Swain.

Linkages of natural communities to animal species were drafted by Mark Ward and reviewed by Mark McCollough, Phillip deMaynadier, and Tom Hodgman. Others providing useful comments on the wildlife information include Mac Hunter, Bob Houston, Peter Vickery, and Andy Whitman.

Numerous staff at the Maine Natural Areas Program have been involved with the project. Most notably, Lisa St. Hilaire took what had been a languishing project and kicked it into high gear, providing countless hours of editorial and organizational oversight to bring the project to publication. Don Cameron and Kristen Puryear provided review comments, suggested new types based on field work, and acted as sounding boards throughout the process. Chris Cabot, Brian Carlson, Emily Chase, Dan Coker, Sarah Demers, Janet Gannon, Toni Pied, and Matt Waterhouse helped with various technical and administrative components of the project. Program coordinator Molly Docherty was essential in securing the funding for this project and in overseeing its progress.

Most of the photos are from current or former Maine Natural Areas Program staff. Notable exceptions are photos from Ron Butler, Phillip deMaynadier, Keith Kanoti, Ron Logan, Jonathan Mays, Erik Nielsen, Barbara Spencer, and Beth Swartz. Layout and design work were patiently provided by Cory Courtois of Pages Plus.

Financial support for this project was provided by the Maine Outdoor Heritage Fund, Maine Forest Service, Sweet Water Trust, The Nature Conservancy, and the US Environmental Protection Agency.

Finally, we are grateful to our teachers, our colleagues, and our families, who have supported our passion for Maine's natural places.

Foreword

Of fir and fen: exploring ecological patterns

You may not want to read this book. Are you a person who sees trees and flowers rather than hemlocks and trilliums? Do you call all the birds you see swimming in a pond "ducks"? Do you think of those slimy plants growing in the intertidal zone as just "seaweed"? If you answered "yes" to one or more of these questions you may not be a good candidate to read this book. Perhaps you are one of those people who say things like "I love nature, but I just don't want my enjoyment of it cluttered up with a lot of names that I can't bother to remember" or perhaps you picked up this book thinking that it was a compilation of landscape paintings.

On the other hand, if you answered "no" to those questions there is a good chance that you are a naturalist with enough sophistication, depth, and curiosity to appreciate what this book has to offer. You are probably ready to take the next step toward fully savoring the myriad species that share our planet, ready to understand a critical aspect of their ecology, the patterns by which they are organized into natural communities.

These patterns are driven by the processes through which organisms interact with their physical environment (for example soil, water, and climate) and one another (for example, as predators or competitors). At a very basic level you already know these patterns--you certainly would be shocked to see a hemlock growing in a stand of cattails because you understand that a forest community is profoundly different from a marsh community.

At a much higher level of complexity these patterns and processes are so complex that no one can fully understand them, not even professional ecologists who devote their lives to studying a single ecosystem. As Frank Egler wrote, "Ecosystems are not just more complicated than we think they are; they are more complicated that we can think." We can get a feel for this complexity with a simple math exercise. Imagine a community with a typical number of species (1000 would be a rough ballpark estimate that will make the math easy). Next multiply that number times all the other species in the community with which each species interacts (999 in our example because indirect interactions count too). You also have to account for a few dozen environmental attributes by multiplying them by your number of species; let's say 50. So, for our example 1000 x (999 + 50) = 1,049,000. Think about the critical role of innumerable microorganisms that are too small to see readily, many of which have not even been identified by scientists, and the task of sorting out all this complexity is truly overwhelming.

The good news is that this book operates at a level that is readily accessible and practical. It does this first of all by operating at a spatial scale that is easy to grasp, one in which natural communities correspond to those patches of vegetation easily discerned from any overlook, patches that we would typically measure in acres. Conceptually, natural communities can exist at any scale; the invertebrates and microbes inhabiting a single acorn decomposing on the forest floor have been described as a community and all life on Earth constitutes the biggest community, the biosphere. The scale used in this book fits most human enterprises for it corresponds reasonably well to the scales at which we divide up the land for different uses and ownerships.

This book is also a convenient introduction to understanding ecological patterns because it is largely based on the conspicuous plant species that dominate and usually define natural communities. You will not have to discriminate among dozens of different sedges or measure soil pH or net butterflies to identify the natural communities defined here. Knowing a few dozen plants, many of them trees and other well known species, will take you a long way down the path of natural community classification.

The practical nature of this natural community classification also makes it an important tool for conservationists. Beginning in the 1980s conservationists started to look beyond the habitat needs of some rare birds, mammals, and a few plants to think about conserving all species. This idea was conceptualized under the banner of biological diversity, soon truncated to "biodiversity." Now every species, even fungi and worms yet to be described by science, is recognized as having value and meriting action to assure that it does not become extinct. But how do you undertake conservation action for huge numbers of species ... actually millions of species on a global scale?

One efficient approach is to protect entire communities on the assumption that this process will capture most of the species that constitute the community. Conserve a wetland and you protect habitat for hundreds of species, the majority of which are insects and other small creatures that you know almost nothing about. This is often called "coarse filter" conservation, "coarse" because the pores in this filter are large enough that some species will fall through and require individual attention. For example, conserving a wetland would not be sufficient to protect the peregrine falcons that occasionally fly by looking for an injured duck because peregrines have huge home ranges and, at least in Maine, are probably more limited by pesticide contamination than by a scarcity of wetlands. Thus peregrine conservation requires a fine-filter approach.

Employing a coarse-filter strategy depends on understanding how species are organized into predictable patterns and expressing that understanding in a community classification scheme. Such classifications are never perfect: some species are such habitat generalists that they seem almost ubiquitous across the landscape; some are so rare that they are seldom present even in their characteristic community. Nevertheless, if conservationists start with a good classification and protect a representative array of communities across the landscape this will go a very long way toward conserving a region's biodiversity.

So if you are a naturalist seeking to understand ecological patterns or a conservationist trying to use those patterns to undertake efficient conservation then this book has much to offer. And chances are, if you have read this far, you are just such a person.

Malcolm L. Hunter, Jr.
Libra Professor of Conservation Biology
Department of Wildlife Ecology
University of Maine
Orono, Maine

Introduction

Conservation of natural areas, or particularly of the vegetated habitats that comprise them, requires information on community composition, structure, distribution, and conservation status. This classification names and describes the 104 terrestrial natural communities of Maine and describes how different communities occur together as 24 larger-scale ecosystems over the Maine landscape.

Identifying and mapping natural communities on the landscape provides important baseline data for current and future conservation efforts. These efforts may include setting priorities for conservation action, developing management plans for specific areas, and local land use planning, among others. Consequently, the purpose of this classification is to provide a framework by which the Maine Natural Areas Program (MNAP) and others in applied conservation can identify natural communities and ecosystems, as these are important components of the state's ecological diversity. The classification is intended for a variety of users, ranging from professional ecologists to foresters, natural resource planners, land trust members, educators, and interested citizens.

Natural Community Concepts

Natural Communities and Ecosystems: What Are They?

A natural community is an assemblage of interacting plants and animals and their common environment, recurring across the landscape, in which the effects of human intervention are minimal. A natural community includes all of the organisms (plant, animal, etc.) in a particular physical setting, as well as the physical setting itself[1]. Careful study of natural communities allows us to generalize and group similar habitats into natural community types that repeat across the landscape where similar environmental conditions occur. For example, Oak - Pine Forests on rocky hillsides in York County and Androscoggin County should be recognizable as the same community type. A "natural community type" is thus a conceptual pigeonhole used to organize and make some sense of natural community variation.

Natural communities are the primary "coarse filter" for capturing and protecting biological diversity; that is, protecting a natural community conserves all of its component species, even ones we may not know are there. Recognizing that conservation often involves landscape processes, this classification also addresses how natural communities occur together on the landscape at broader scales. Aggregations of natural communities on the landscape that share similar environmental settings or ecological processes can be thought of as ecological systems (Poiani and Richter 1999) and are here termed ecosystems[2]. This classification adopts the convention that an ecosystem is a group of communities and their environment, occurring together over a particular portion of the landscape and held together by some common physical or biotic feature.

The different scales (natural communities and ecosystems) incorporated into this classification acknowledge different levels of ecological complexity. Just as species are integrated into natural communities, so are communities aggregated into ecosystems, and ecosystems aggregated into landscapes. In conservation work, the appropriate ecological scale depends on the conservation goal; in many cases it may be more instructive to consider the ecosystem scale unit, rather than an individual natural community or species. Conservation of a large peatland, for example, is

---

[1] The concept of a community as used here differs from the definition found in most ecology textbooks. A "community", in the ecological sense, is usually described as a particular group of organisms which interact and share a common environment, but the physical environment is not part of the community itself. One can refer to a "plant community", a "bird community", a "soil fauna community", etc.

[2] In the world of biology and ecology, "ecosystem" can actually have many different meanings. Ecosystems are defined at different scales, from "the ecosystem" of a rotting log in a forest to "the ecosystem" of the entire biosphere, depending on the context for the definition. Whatever the scale, an ecosystem includes all of the biotic (living) components, the abiotic (non-living) components, and their interactions.

better served by considering the entire ecosystem (e.g., a Domed Bog Ecosystem) rather than the individual natural communities (e.g., Sheep Laurel Dwarf Shrub Bog, Sedge – Leatherleaf Fen Lawn, Spruce – Larch Wooded Bog, and others) embedded within it. We hope that including both scales will make this classification more useful to those working to understand and conserve Maine's biodiversity.

Ecological Variation

Ecological systems and natural communities are both temporally and spatially variable. A Red Spruce – Mixed Conifer Woodland on a rocky hilltop is likely to have patches of very open areas, where lichens and a few vascular plants dominate, and patches that are more heavily wooded: this is spatial variability. The small open patches do not have the 25 – 60% tree cover characteristic of that community type, but the area overall does, and the spatial variability can be integrated into an overall view of the community.

When do patches represent variability within a community type, and when might they be natural communities on their own? That depends both on the level of detail in the classification and on how the patches occur within the landscape. In this classification, small patches that occur (often repeatedly) within a larger community type and that obviously bear a similarity to the general community type are treated as spatial variation or smaller scale inclusions (as in the example above). Conversely, a distinct area that occurs under different environmental conditions and has different flora than the surrounding or adjacent community could be considered its own type. For example, the bald summit of a coastal hill may only be a few acres, but is clearly more open than the adjacent Pitch Pine Woodland on the upper slopes. The open bald is also environmentally distinct (summit vs. slope), and is characterized by different species in the herb layer (smooth sandwort and three-toothed cinquefoil are abundant on the bald and do not occur in the woodland). Therefore, the open bald could be called a Three-toothed Cinquefoil – Blueberry Low Summit Bald. Field experience and familiarity with this classification will bring a sense of when an area should be considered a distinct community and when it is part of the larger overall community.

Succession and natural disturbance introduce temporal variability. A Circumneutral Riverside Seep will be dominated by herbaceous plants or very low shrubs for a few years following heavy ice scour, but it can become densely shrubby if several years elapse without much ice. Spruce – Fir forests in the portions of Acadia National Park that burned in 1947 are now recognizable as either Maritime Spruce – Fir Forests or Spruce – Fir – Broom-moss Forests even though they may have a sizeable component of aspen and paper birch remaining from the early successional years. This classification describes mid- to late- successional forests, and stands with a more recent harvesting or disturbance history may differ in overstory dominants from those described here. In these cases, the forest type can often be inferred from the later-successional trees present and by the understory and site characters. One early successional forest type is included, the Aspen – Birch Woodland/Forest Complex, to account for typical early successional forests originating after wildfire.

Patterns of Formation

The landscape pattern of a natural community describes the scale of a typical occurrence of this community in Maine. Some communities tend to extend over large areas, while others are characteristically confined to only a small acreage. We use the three categories developed by The Nature Conservancy, referred to as "small patch", "large patch", and "matrix-forming" (Anderson 1999).

- *Small patch communities* are the smallest landscape elements and occur under particular environmental conditions (e.g., Crowberry – Bilberry Summit Bald). Up to about 50 acres in size, they collectively occupy only a very small portion (less than 1%) of the total landscape. They are often strongly influenced by the ecological dynamics in the surrounding large patch and matrix communities. Many of the rare community types in Maine are small patch types.

- *Large patch communities* are associated with more particular environmental conditions than are matrix-forming communities (e.g., Northern White Cedar Swamp), and cover fairly large but discretely defined areas of the landscape. Natural occurrences are typically in the 50 – 1000 acre size range, occasionally larger, and collectively occupy an estimated 10% of the landscape.

- *Matrix-forming communities* are those that cover the majority of the natural landscape of the region. In Maine, these are forest communities, such as a Beech – Birch Maple Forest. Individual occurrences typically range from hundreds to several thousand acres. Together, the matrix-forming types form more or less contiguous cover over millions of acres, or roughly 90% of the landscape. Large patch and small patch communities are embedded within the matrix-forming types.

Rarity

To reflect the varying abundance of natural communities across the state, this classification assigns a rarity rank to each natural community type. Rarity ranks range from S1, the rarest, to S5, the most common (Appendix A). These same ranks are applied to plant species and animal species tracked by the state. In applying the rank criteria to each natural community, the absolute number of occurrences may be secondary to other factors like a limited geographic distribution or threats from competing uses. Readers may wonder why many common forest types (for example, Beech – Birch – Maple Forests) are given a rank of S4 even though the type may be abundant statewide. MNAP interprets the criterion for an S5 rank of "demonstrably secure" to mean that high-quality examples are secure and sufficiently distributed throughout the state. Common types that are in working forestland and not set aside as reserved lands are not considered secure high-quality examples for ranking purposes.

MNAP tracks the location and status of each occurrence of a natural community that is considered to be:

1. any example of a community type that is rare statewide (i.e., ranked S1, S2, or S3), or

2. an outstanding example of a more common community type (ranked S4 or S5).

Beyond our statewide view, it is also important to consider global rarity. Several natural community types are not only rare in Maine but also rare throughout North America. Other types that are rare statewide are more extensive elsewhere, such as White Oak – Red Oak Forests that are more common south of Maine. Assessing the global rarity of a community type is more complicated than developing state ranks, but continuing efforts (e.g., Grossman et al. 1994 and NatureServe 2001) are highlighting those community types especially in need of conservation attention wherever they occur. Appendix B lists the National Vegetation Classification types (defined by NatureServe) that correspond to the Maine types. Note that Maine's natural communities may be more broadly defined than the global types.

Globally Rare Natural Community Types in Maine	
Hudsonia River Beach	G1
Bilberry – Mountain-heath Alpine Snowbank	G2G3
Bluebell – Balsam Ragwort Shoreline Outcrop	G3
Diapensia Alpine Ridge	G2G3
Dwarf Heath Graminoid Alpine Ridge, in part	G2G3
Pitch Pine – Scrub Oak Barren	G2
Pitch Pine Dune Woodland	G2
Pitch Pine Woodland	G2
Freshwater Tidal Marsh, in part	G2
Three-way Sedge – Goldenrod Outwash Plain Pondshore	G2G3

Natural Communities and Climate Change

How should natural community concepts be considered in the context of a changing climate? Historic pollen data indicate that Maine's natural communities have been dynamic rather than static, responding to continuous climate shifts for thousands of years. Following the retreat of the last glaciers, for example, the northeastern U.S. became as much as 2°C (4°F) warmer and the air considerably drier than it is today. White pine was widespread and abundant from frequent fires, spruce was scarce, and both white pine and hemlock grew at higher elevations than they do now (Davis et al. 1980).

Looking forward, a changing climate will ensure that transitions in natural community distribution and composition continue. Recent projections suggest that Maine will experience warmer year round temperatures, increases in precipitation and storm intensity, and rising sea levels (University of Maine 2009). In light of these projections, the University of Maine recently suggested that "climate change will almost certainly lead to significant changes in Maine's overall assemblage of plants and animals." Among other changes, within the next few centuries red spruce and balsam fir are expected to be replaced by red oak, red maple, and white pine (US Forest Service 2009 http://www.nrs.fs.fed.us/atlas/. Moreover, if historic patterns foreshadow future trends, entire natural communities will not just pack up and shift north as a unit. Rather, the ranges of individual species will change in different and unpredictable ways, potentially resulting in combinations of species that do not occur together today.

Do seemingly inevitable changes diminish the importance of conserving natural communities that now exist on the landscape? On the contrary, a key adaptive strategy for climate change should be to maintain functional, resilient natural systems and viable landscape scale connections between them. As Maine's Beginning with Habitat program noted, "One thing we can control is how permeable our built landscape will be. By planning for undeveloped 'green belts', or habitat corridors that connect conserved habitat blocks, we can better support gradual species range shifts through and around developed areas" (Walker 2009). Habitat conservation may also help give plants and animals that are already rare the best chance to remain viable as the climate changes. Last and perhaps most importantly, habitat conservation offers a way to ensure that an adequate and representative sample of Maine's underlying physical features (including geology, soils, landform, and elevation) are safeguarded for the long term. These underlying features, together with climatic factors, will ultimately influence what species will occur where. With sufficient natural communities and physical features conserved, Maine's landscape will be most favorably positioned to accommodate whatever vegetation changes a shifting climate will impose.

About This Classification

How This Classification Was Developed

This classification is a comprehensive revision of MNAP's Natural Landscapes of Maine: a Classification of Ecosystems and Natural Communities (Gawler 1991). Additional information and field survey data amassed since that time provide the basis for refining the community types to better represent the diversity of Maine's vegetated habitats.

A draft set of revised types was derived based on previous analyses (McMahon 1998, Anderson and Davis 1997, Gawler 2000), the author's personal experience, comments from other users, and discussions with natural heritage program ecologists in New Hampshire, Vermont, Massachusetts, and New York. These draft types were cross-referenced to the existing types and to the National Vegetation Classification (NVC) developed by The Nature Conservancy (TNC) and the NatureServe. Quantitative analyses of over 1,000 natural community plot samples were conducted as the first approach to evaluating the proposed types. Additional details on the quantitative analysis used are in Appendix E.

The quantitative analyses were augmented with qualitative information from MNAP files. Hundreds of field forms with notes describing specific examples of natural communities in Maine, as well as journal articles and reports in the primary scientific literature, were reviewed as supplementary materials for the community type delineation.

Relationships to Other Classifications

A classification is a tool designed for specific purposes. While a forest classification for timber management would be different from an ecological classification, both are legitimate tools to serve their need. This natural community classification is designed to aid conservation of natural diversity in Maine and does not have a one-to-one relationship with other vegetation or ecosystem classifications. Refer to Appendix B for a comparison of Maine's natural community types to other classification systems.

SAF types: This vegetation classification, limited to forests, was developed by the Society of American Foresters and most recently revised in 1980. Types are described based almost entirely on overstory tree species. Some SAF types apply to more than one natural community type (e.g., SAF type 37, Northern White Cedar, is applied to four Maine types).

Other Natural Heritage and Conservation Data Centre (CDC) Classifications: The New Hampshire Natural Heritage Inventory has identified over 200 types for that state, so New Hampshire types are more finely split than Maine types.[3] Natural community classifications in the Maritimes and Quebec are currently under development but are also more finely split. All natural heritage and CDC classifications are cross-referenced to the National Vegetation Classification.

NVC types: The National Vegetation Classification (NVC) system was developed by NatureServe, including its constituent natural heritage programs in the United States. This system, also called International Vegetation Associations, is a comprehensive classification of vegetation types arranged in a physiognomic hierarchy. The basic unit of classification in the NVC is the association, which is roughly analogous to the present classification's natural community. Two-thirds of the natural community types in the Maine classification relate 1:1 with NVC associations; the remaining third of the Maine types are broader than the NVC types, i.e., more than one NVC type corresponds to a single Maine type. There is a difference in concept between Maine's natural communities and NVC associations. Maine natural community types are somewhat broader units that include a particular environmental setting and may include patches of different vegetation types. Some associations are more finely split – they are strictly vegetation-based and are theoretically more homogeneous in their structure and dominant species. For example, Maine's Three-way Sedge – Goldenrod Outwash Plain Pondshore natural community type includes several NVC plant associations in concentrically banded zones. While the individual associations differ in their dominant plants, the collection of associations is bound together by the basin hydrology, responds to the same environmental factors and from a practical standpoint makes up one conservation entity. It matters less what the unit(s) are called than that the appropriate target area is chosen for conservation actions.

International Ecological Systems: Recognizing the limitations of the fine-scaled NVC classification for landscape-scale mapping and conservation planning, NatureServe worked with a number of federal partners to develop an international classification of coarser-scaled ecological systems. Ecological systems are currently being mapped at regional, national, and international levels through federal projects such as the Landfire Project GAP project. Each ecological system typically encompasses a few to more than 10 NVC types. The ecosystem types in Maine's classification are closely related to the national ecological systems and may someday be replaced by them as a classification unit.

---

[3] For more information on New Hampshire community types, contact the New Hampshire Natural Heritage Inventory at 603-271-3623 or on the web at http://www.nhdfl.org/about-forests-and-lands/bureaus/natural-heritage-bureau/

The scaled relationships of the various classification types, ranging from fine scale (NVC) to coarse scale (ME Wildlife Action Plan Habitat Types) are shown below:

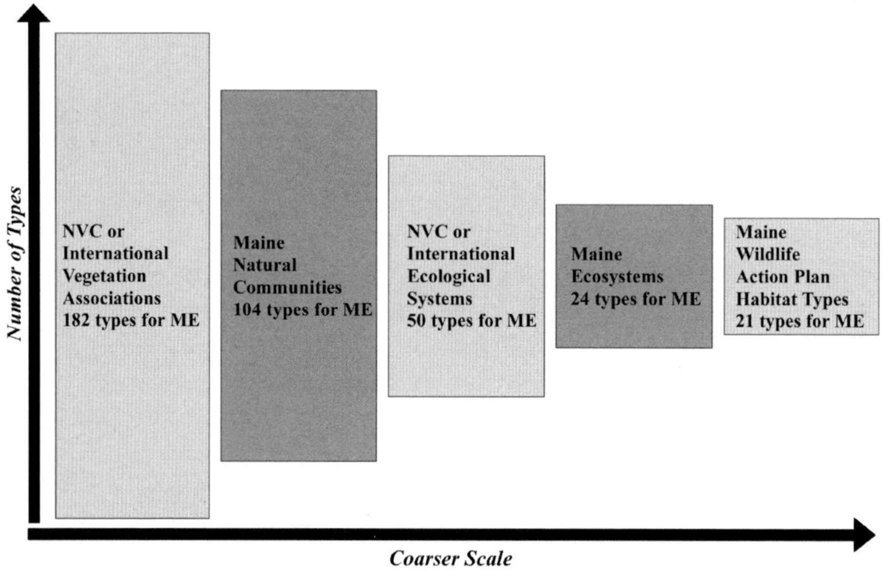

NWI types: The National Wetlands Inventory (NWI) classifies wetland vegetation according to physiognomy (structure) and hydrology. In almost all cases, the NWI types are broader than the natural community types. For example, the "PEM1" (Palustrine Emergent Marsh - Persistent) NWI type applies to 13 different natural community types. NWI types may indicate certain ecological functions relevant for regulatory purposes, but they are poor predictors of biodiversity. Also, the NWI mapping tends to be less accurate for forested wetlands than for open ones because of the difficulty of distinguishing some wetland forest types on aerial photos.

Ecoregions: Ecoregional classifications are fundamentally different from but complementary to vegetation classifications. Rather than focusing on particular vegetation or landform types, ecoregional classifications delineate broad geographic patterns incorporating major vegetation types, climate, and geology. Ecoregional classifications provide a useful way of stratifying large landscapes into discrete units for conservation planning efforts. The ecoregional classification used in this publication is the 'section' level (intermediate between coarser-scaled 'provinces' and finer-scaled 'subsections') and draws from various national and state classification efforts including Bailey (1995), Keys et al. (1995), and McMahon (1990). The

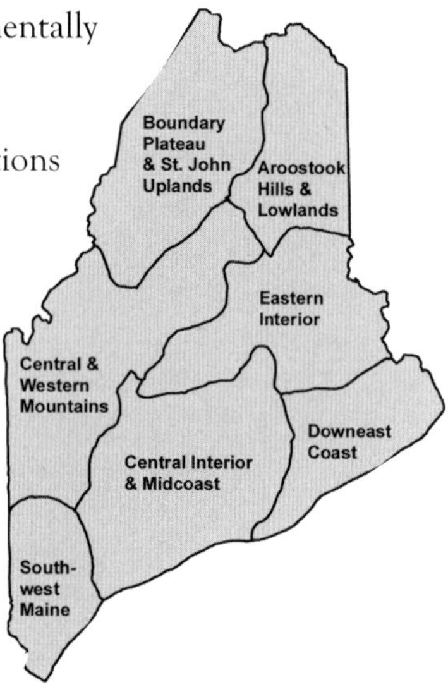

seven sections are used in the natural community profiles to depict the geographic distribution or range of communities throughout the state.

Ecological Land Types: Another commonly used approach to categorizing landscape-scale diversity is the use of Ecological Land Types (ELTs), also known as Ecological Land Units (ELUs). This is a GIS-based remote classification approach in which different combinations of physical landscape variables are arranged into groupings and the groupings mapped. Mapping ELUs gives some representation of the physical diversity of an area, which can be useful in stratifying areas or in selecting areas with particular characteristics for inventory. For example, using GIS layers of bedrock geology and landscape position (elevation, slope and aspect), Anderson et al. (1998) developed ELUs for the Connecticut River watershed. These ELUs were then used as a predictive model to identify natural communities on the landscape and select a set of priority areas (each >8000 acres) that together would encompass the major ecological features present in the watershed. However, using ELUs to select specific conservation targets has two major drawbacks. First, geographic information is not always available at the level of resolution at which it is needed. For example, soil surveys are available for southern Maine, but not for much of northern Maine; similarly, the bedrock of Maine is mapped at a decent resolution for part of the state, but the northern and especially northwestern portions are only generally mapped. The second drawback is that the relationship between ELUs and natural communities is not necessarily straightforward, so ELUs cannot be thought of as a surrogate for natural communities. ELUs have considerable promise for informing conservation efforts and our understanding of landscape diversity, but at this point the applications are not sufficiently developed to be a substitute for on-the-ground natural community inventories.

Associated Rare and Common Plant and Wildlife Species

Common names for plants and animals are used in this classification. See Appendix C for a cross reference to scientific names for plants and Appendix D for a cross reference for rare animals.

The list of associated rare plants notes species that are documented from that community type within the state. For space reasons, not all rare plants that occur in a type are necessarily listed. Rarely, plants not currently known from Maine but documented in that community type from an adjacent state/province are included (e.g., wild lupine in Pitch Pine – Scrub Oak Barrens). The Maine Natural Areas Program can provide lists and additional information on plants considered rare in Maine.

The list of associated rare animals includes those that are listed as endangered, threatened, or rare in Maine. The list of associated rare animals represents a systematic attempt to match rare animal species with natural community types based on the known habitat uses and preferences of the animals and the particular compositional, structural, and hydrological characteristics of the community. Animal species' geographic ranges and expected distributions of natural community types were also considered. The process involved a careful examination of the literature and review by wildlife biologists with expertise in specific taxonomic groups.

Because of the broad ecological amplitude and mobility of many animal species, their allegiance to particular substrate or compositional characteristic is not as strong as that for many plant species. Therefore, a conservative approach was taken in assigning animal species to natural community types and focused on species that showed a strong preference for one or a few community types (especially those communities that play a critical role in one of the life stages of the animal–i.e., breeding, feeding or overwintering). This approach was intended to avoid including habitat generalists or species that use habitats heavily influenced by human disturbance. The species list is intended as a guidepost and is not exhaustive.

Limitations

An important limitation of this classification is that it applies primarily to vegetated communities in a relatively natural condition. Aquatic communities that are not characterized by vascular plants are not covered, including the non-vegetated Lacustrine, Riverine, Estuarine, and Marine types in Maine's 1991 natural community classification. State and regional efforts to develop classifications of aquatic systems by The Nature Conservancy and others have more recently focused on the physical characteristics of ponds, lakes, streams, and rivers. Even in the vegetated realm, some gaps remain, for instance in the coverage of submerged aquatic vegetation.

Communities initiated and maintained by human activities (agricultural fields, lawns, roadsides, forest planta¬tions, etc.) are not considered natural communities and are not included in this classification.

Furthermore, it is important to realize that while conserving natural community diversity is an essential component of biological conservation, it will not protect all of the biological diversity in any given area. Additional attention must be provided to certain "fine filter" species (e.g., rare species found in small populations or in very particular habitats) as well as to wide-ranging animal species, two groups that are not adequately addressed by the coarse filter approach. A combination of coarse-filter and fine-filter approaches is most likely to be successful in conserving biological diversity in any given area.

This classification is dynamic and intended to be periodically updated. Our concepts of plant associations and natural communities can shift as we learn more about Maine's flora and fauna. We lack good descriptions of some communities, and, as inventory work continues, our understanding of the relative rarity of the different community types may change. The MNAP welcomes information about natural communities and ecosystems from field biologists and ecologists.

Table of Natural Community Types

Natural community types are arranged by class, sub-class, and group within each of four major divisions: wooded uplands, wooded wetlands, open uplands, and open wetlands. Natural community types ranked S1, S2, or S3 are considered rare in Maine.

Wooded Uplands

Class	Sub-class	Natural Community Type	State Rank	Page
Forests	Conifer	Fir – Heart-leaved Birch Subalpine Forest	S3	118
		Hemlock Forest	S4	76
		Jack Pine Forest	S1	78
		Maritime Spruce – Fir Forest	S4	84
		Red Pine – White Pine Forest	S3	108
		Spruce – Fir – Broom-moss Forest	S4	82
		Spruce – Fir – Wood-sorrel – Feather-moss Forest	S4	86
		White Pine – Mixed Conifer Forest	S4	126
	Deciduous	Beech – Birch – Maple Forest	S4	88
		Maple – Basswood – Ash Forest	S3	74
		Oak – Hickory Forest	S1	92
		Semi-rich Northern Hardwood Forest	S4	120
		White Oak – Red Oak Forest	S3	124
	Mixed	Oak – Pine Forest	S4	96
		Red Oak – Northern Hardwoods – White Pine Forest	S4	94
		Spruce – Northern Hardwoods Forest	S4	112
Woodlands	Conifer	Black Spruce Woodland	S3	68
		Jack Pine Woodland	S3	80
		Pitch Pine Dune Woodland	S1	100
		Pitch Pine Woodland	S3	106
		Pitch Pine – Heath Barren	S1	102
		Pitch Pine – Scrub Oak Barren	S1	104
		Red Pine Woodland	S3	110
		Red Spruce – Mixed Conifer Woodland	S4	114
		Spruce – Heath Barren	S2	66
		Spruce Talus Woodland	S4	116
		White Cedar Woodland	S2	122
	Deciduous	Aspen – Birch Woodland/Forest Complex	S5	72
		Birch – Oak Talus Woodland	S3	64
		Chestnut Oak Woodland	S1	70
		Ironwood – Oak – Ash Woodland	S2	90
	Mixed	Oak – Pine Woodland	S4	98

Wooded Wetlands

Class	Sub-class	Natural Community Type	State Rank	Page
Forests	Conifer	Atlantic White Cedar Swamp	S2	132
		Cedar – Spruce Seepage Forest	S4	140
		Northern White Cedar Swamp	S4	144
		Spruce – Fir – Cinnamon Fern Forest	S4	158
	Deciduous	Balsam Poplar Floodplain Forest	S1	134
		Hardwood River Terrace Forest	S2	160
		Silver Maple Floodplain Forest	S3	156
	Mixed	Black Ash Swamp	S4	136
		Hardwood Seepage Forest	S3	142
		Hemlock – Hardwood Pocket Swamp	S2	150
Woodlands	Conifer	Atlantic White Cedar Bog	S1	130
		Northern White Cedar Woodland Fen	S4	146
		Pitch Pine Bog	S2	148
		Spruce – Larch Wooded Bog	S4	138
	Deciduous	Red Maple Wooded Fen	S4	152
		Red Maple – Sensitive Fern Swamp	S4	154

Open Uplands

Class	Sub-class	Natural Community Type	State Rank	Page
Shrublands	Conifer	Spruce – Fir – Birch Krummholz	S3	196
	Deciduous	Mountain Alder – Bush-honeysuckle Subalpine Meadow	S2	198
		Rose – Bayberry Maritime Shrubland	S4	192
Dwarf Shrubland	Deciduous	Blueberry – Lichen Barren	S2	170
		Crowberry – Bilberry Summit Bald	S3	184
		Dwarf Heath – Graminoid Alpine Ridge	S3	180
	Evergreen	Bilberry – Mountain Heath Alpine Snowbank	S1	168
		Crowberry – Bayberry Headland	S3	176
		Diapensia Alpine Ridge	S1	200
		Labrador Tea Talus Dwarf-shrubland	S2	174
		Rocky Summit Heath	S4	190
Herbaceous	Forb	Alpine Cliff	S1	166
		Boreal Circumneutral Open Outcrop	S2	172
		Three-toothed Cinquefoil – Blueberry Low Summit Bald	S3	182
	Graminoid	Dune Grassland	S2	178
		Little Bluestem – Blueberry Sandplain Grassland	S1	194
	Consolidated Rock	Acidic Cliff – Gorge	S4	164
		Bluebell – Balsam Ragwort Shoreline Outcrop	S3	188
		Seaside Goldenrod – Goosetongue Open Headland	S4	186

Open Wetlands

Class	Sub-class	Natural Community Type	State Rank	Page
Shrublands	Deciduous	Alder Floodplain	S4	204
		Alder Shrub Thicket	S5	206
		Dogwood – Willow Shoreline Thicket	S2	254
		Mountain Holly – Alder Woodland Fen	S4	276
		Sweetgale Mixed Shrub Fen	S4	268
Dwarf Shrublands	Evergreen	Heath – Crowberry Maritime Slope Bog	S2	242
		Heath – Lichen Subalpine Slope Bog	S1	270
		Huckleberry – Crowberry Bog	S3	240
		Hudsonia River Beach	S1	258
		Leatherleaf Boggy Fen	S4	236
		Sheep Laurel Dwarf Shrub Bog	S4	226
Herbaceous	Forb	Twisted Sedge Cobble Rivershore	S4	224
		Circumneutral Riverside Seep	S2	256
		Lakeshore Sand/Cobble Beach	S4	232
		Pickerelweed – Macrophyte Aquatic Bed	S5	252
		Sand Cherry – Tufted Hairgrass River Beach	S2	234
		Three-way Sedge – Goldenrod Outwash Plain Pondshore	S1	250
	Graminoid	Bluejoint Meadow	S3	272
		Brackish Tidal Marsh	S3	210
		Bulrush Bed	S4	212
		Cattail Marsh	S5	214
		Cotton-grass – Heath Alpine Bog	S2	208
		Deer-hair Sedge Bog Lawn	S3	222
		Freshwater Tidal Marsh	S2	228
		Low Sedge – Buckbean Fen Lawn	S3	238
		Mixed Graminoid – Forb Saltmarsh	S4	244
		Mixed Graminoid – Shrub Marsh	S5	230
		Mixed Tall Sedge Fen	S5	274
		Sedge – Leatherleaf Fen Lawn	S4	264
		Shrubby Cinquefoil – Sedge Circumneutral Fen	S2	216
		Spartina Saltmarsh	S3	260
		Tussock Sedge Meadow	S3	266
	Hydro-Morphic	Circumneutral – Alkaline Water Macrophyte Suite	S2	218
		Pipewort – Water Lobelia Aquatic Bed	S5	262
		Water-lily – Macrophyte Aquatic Bed	S5	248
Sparse	Bryophyte	Bog Moss Lawn	S4	246
	Sand	Beach Strand	S4	220

Ecosystems

An ecosystem, as the term is used in this classification, is a group of communities and their environment, occurring together over a particular portion of the landscape, and held together by some common physical or biotic feature. Ecosystem types for Maine, with the natural communities that occur in each type, are listed and described below. The natural communities listed for each ecosystem will not all occur in one particular location of the ecosystem; rather, each ecosystem will be made up of some assortment of the communities listed. As noted previously, a number of federal agencies and conservation groups have recently developed a national ecological systems classification (Comer et al. 2003). The ecosystem types in Maine's classification are closely related to the national ecological systems and may eventually be replaced by them.

Group	Ecosystem Type	State Rank
Forests	Appalachian – Acadian Basin Swamp Ecosystem	S4
	Central Hardwoods Oak Forest Ecosystem	S3
	Coastal Plain Basin Swamp Ecosystem	S3
	Maritime Forest Ecosystem	S4
	Spruce – Fir – Northern Hardwood Forest Ecosystem	S4
	White Pine – Mixed Hardwood Ecosystem	S4
Open Uplands	Alpine Ecosystem	S2
	Pine Barrens Ecosystem	S1
	Rock Outcrop Ecosystem	S4
Peatlands	Coastal Plateau Bog Ecosystem	S3
	Domed Bog Ecosystem	S3
	Eccentric Bog Ecosystem	S3
	Kettlehole Bog Pond Ecosystem	S4
	Patterned Fen Ecosystem	S3
	Raised Level Bog Ecosystem	S4
	Unpatterned Fen Ecosystem	S4
Freshwater Shorelines	Appalachian – Acadian Rivershore Ecosystem	S4
	Coastal Plain Pondshore Ecosystem	S1
	Coastal Plain Rivershore Ecosystem	S2
	Lakeshore Ecosystem	S5
	Streamshore Ecosystem	S4
Tidal	Coastal Dune – Marsh Ecosystem	S3
	Coastal Headlands Ecosystem	S4
	Tidal Marsh Estuary Ecosystem	S3

Forested Ecosystems

The forested ecosystems are defined by macroclimate and the dominant types of trees. Maine's diversity of forest types stems from its position in the transition zone between the Eastern Deciduous Forest to our south and the Boreal Forest to our north. With at least 90% of its land tree-covered, Maine's history and economy are largely based on its forests.

Appalachian – Acadian Basin Swamp Ecosystem S4

These are topographic basins with only a small to mid-sized stream as an outlet, mostly to entirely forested, and with minimal peat accumulation. Swamps that occur immediately associated with a lake or larger stream are included under lakeshore and streamshore ecosystems; forested basins with closed drainage and peat accumulation are included under peatland ecosystems. Distributed statewide except for extreme southern Maine, where basin swamps are distinguished as a separate type (see following) characterized by assemblages with more southerly affinities. First-order or second-order streams may drain the swamp. Soils are hydric and usually saturated. Pockets without trees (shrub swamp or short graminoid marsh) may occur within the basin swamp.

Alder Shrub Thicket
Black Ash Swamp
Bluejoint Meadow
Cedar – Spruce Seepage Forest
Mixed Graminoid – Shrub Marsh
Northern White Cedar Swamp
Red Maple – Sensitive Fern Swamp
Spruce – Fir – Cinnamon Fern Forest
Tussock Sedge Meadow

The Appalachian – Acadian Basin Swamp Ecosystem in Maine corresponds to the following national Ecological Systems: Acadian – Appalachian Conifer Seepage Forest, Boreal – Laurentian Conifer Acidic Swamp, Laurentian – Acadian Alkaline Conifer – Hardwood Swamp, and Northern Appalachian – Acadian Conifer – Hardwood Acidic Swamp.

Central Hardwoods Oak Forest Ecosystem S3

These are forested areas in extreme southern Maine (York County), an area also known as the "Central Hardwoods – Hemlock – White Pine zone". These forests are typified by so-called "central hardwoods" (black oak, white oak, shagbark hickory, bitternut hickory) and associated species, species common to our south but with a restricted distribution in Maine. The particular community(ies) in any area will depend upon microclimate, aspect, slope position, soil type, etc. More exposed habitats may have woodlands dominated by oaks mixed with other species; more protected habitats will have continuous forest cover. This ecosystem is almost entirely limited to terrestrial communities, although it may include vernal pools and intermittent streams.

Chestnut Oak Woodland
Hemlock – Hardwood Pocket Swamp
Ironwood – Oak – Ash Woodland
Oak – Hickory Forest
Oak – Pine Forest
Oak – Pine Woodland
White Oak – Red Oak Forest

The Central Hardwoods Oak Forest Ecosystem in Maine corresponds to the following national Ecological Systems: Appalachian (Hemlock) – Northern Hardwood Forest, Central Appalachian Dry Oak – Pine Forest, Central Appalachian Pine – Oak Rocky Woodland, and Northern Atlantic Coastal Plain Hardwood Forest.

Coastal Plain Basin Swamp Ecosystem S3

This ecosystem is similar to the Appalachian – Acadian Basin Swamp Ecosystem type, but restricted to extreme southern Maine, where climatic influences and species affinities are allied with areas southward along the Atlantic coastal plain. These are topographic basins with only a small to mid-sized stream as an outlet, mostly to entirely forested, and with minimal peat accumulation. First-order or second-order streams may drain the swamp. Soils are hydric and usually saturated. Pockets without trees (shrub swamp or short graminoid marsh) may occur within the basin swamp.

Alder Shrub Thicket
Atlantic White Cedar Swamp
Hemlock – Hardwood Pocket Swamp
Mixed Graminoid – Shrub Marsh
Pitch Pine Bog
Red Maple – Sensitive Fern Swamp
Spruce – Fir – Cinnamon Fern Forest
Tussock Sedge Meadow

The Coastal Plain Basin Swamp Ecosystem in Maine corresponds to the Northern Appalachian – Acadian Conifer – Hardwood Acidic Swamp national Ecological System.

Maritime Forest Ecosystem S4

This ecosystem is the narrow belt of coniferous forest along the immediate coastline. The salt spray, frequent fogs, and temperate influence of the ocean make the maritime forests distinct from those only a few miles inland. Soils are generally thin, and trees often wind-flagged due to exposure. Where bedrock is at the surface, the typical spruce-fir forest may be replaced by a more open canopy woodland.

Aspen – Birch Woodland/Forest Complex
Black Spruce Woodland
Jack Pine Woodland
Maritime Spruce – Fir Forest
Pitch Pine Woodland
White Cedar Woodland

The Maritime Forest Ecosystem in Maine corresponds to the Acadian Low – Elevation Spruce – Fir – Hardwood Forest national Ecological System.

Spruce - Fir - Northern Hardwoods Forest Ecosystem S4

This broadly encompassing forest ecosystem can be found over most of the northern half of the state, and is often referred to by foresters as the "Spruce – Fir – Northern Hardwoods" zone. Red spruce and balsam fir are the most widespread conifers. Conifer-dominated and hardwood-dominated forests intergrade with one another: conifers are typically found in the lowlands or at the highest elevations, while northern hardwood forests are found at mid-elevation hillsides or along mid-elevation ridgelines. This ecosystem is almost entirely limited to terrestrial communities, although it may include vernal pools and intermittent streams.

Aspen – Birch Woodland/Forest Complex
Beech – Birch Maple Forest
Cedar Spruce Seepage Forest
Fir – Heart-leaved Birch Subalpine Forest
Hemlock Forest
Jack Pine Forest
Jack Pine Woodland
Labrador Tea Talus Dwarf Shrubland
Maple – Basswood – Ash Forest
Red Pine – White Pine Forest
Red Pine Woodland
Red Spruce – Mixed Conifer Woodland
Semi-rich Northern Hardwoods Forest
Spruce – Fir – Broom-moss Forest
Spruce – Fir – Cinnamon Fern Forest

Spruce – Fir – Wood-sorrel – Feather-moss Forest
Spruce – Heath Barren
Spruce – Northern Hardwoods Forest
Spruce Talus Woodland
White Pine – Mixed Conifer Forest

The Spruce – Fir – Northern Hardwoods Forest Ecosystem in Maine corresponds to the following national Ecological Systems: Acadian Low – Elevation Spruce – Fir – Hardwood Forest, Acadian Near – Boreal Spruce Barrens, Acadian Near – Boreal Spruce Flat, Acadian – Appalachian Montane Spruce – Fir Forest, Appalachian (Hemlock) – Northern Hardwood Forest, Boreal Jack Pine – Black Spruce Forest, Laurentian – Acadian Northern Hardwoods Forest, Laurentian – Acadian Northern Pine – (Oak) Forest, and Laurentian – Acadian Pine – Hemlock – Hardwood Forest.

White Pine - Mixed Hardwoods Forest Ecosystem S4

This ecosystem lies between the deciduous-dominated forests of southern Maine and the spruce-fir or northern hardwood forests of northern Maine, a belt often referred to as "Transitional Forest". It is characterized by white pine and hemlock as the most widespread conifers; important hardwood species include red oak, beech, sugar maple, and red maple. This ecosystem is almost entirely limited to terrestrial communities, although it may include vernal pools and intermittent streams.

Aspen – Birch Woodland/Forest Complex
Hardwood Seepage Forest
Hemlock – Hardwood Pocket Swamp
Hemlock Forest
Oak – Pine Forest
Oak – Pine Woodland
Pitch Pine Woodland
Red Oak – Northern Hardwoods – White Pine Forest
Red Pine – White Pine Forest
Red Pine Woodland
White Pine – Mixed Conifer Forest

The White Pine – Mixed Hardwoods Forest Ecosystem in Maine corresponds to the Appalachian (Hemlock) – Northern Hardwood Forest national Ecological System.

Open Upland Ecosystems

Include upland areas where tree cover is sparse or absent. Most communities in these ecosystems are terrestrial, but the ecosystems may include small pockets of palustrine communities.

Alpine Ecosystem S2

Alpine Ecosystems occur above treeline where elevation and exposure create extremely harsh conditions; restricted in Maine to mountains above 3500', although not all mountains above 3500' have alpine vegetation. Alpine ecosystems have low and often sparse vegetation due to the harsh environment. Certain tree species may be present, but grow only as krummholz, not erect.

Alpine Cliff
Cotton-grass – Heath Alpine Bog
Crowberry – Bilberry Summit Bald
Diapensia Alpine Ridge
Dwarf Heath – Graminoid Alpine Ridge
Heath – Lichen Subalpine Slope Bog
Mountain Alder – Bush Honeysuckle Subalpine Meadow
Spruce – Fir – Birch Krummholz
The Alpine Ecosystem in Maine corresponds to the following national Ecological Systems: Acadian – Appalachian Alpine Tundra and Acadian – Appalachian Subalpine Woodland and Heath – Krummholz.

Pine Barrens Ecosystem S1

Pine Barrens Ecosystems are sandy glacial outwash or till areas with nutrient-poor, excessively well-drained soils, flat to undulating topography, and a patchy canopy dominated by pine (generally a single species, either pitch pine, jack pine, or red pine), or, in more boreal settings, spruce (usually black spruce). Historically, at least, pine barrens burned frequently due to the dry conditions and flammability of the vegetation; species able to withstand fire or regenerate quickly after a fire (as well as being tolerant of low nutrient conditions and dry soil) are therefore typical. More recently, many pine barrens have been maintained in an early successional stage (represented below by the Little Bluestem – Blueberry Sandplain Grassland and Blueberry – Lichen Barren communities) through biennial burning for blueberry production. The actual vegetation of a barren depends on how recently or frequently it burned. Many pine barrens have a layer of heath shrubs (< 0.5 m tall) below the canopy, with occasional grassy openings among the shrubs. The overstory dominants of Maine barrens range from pitch pine in the southwestern Maine barrens, to red or white pine in the eastern Maine blueberry barrens, to spruce in far northwestern Maine.

Blueberry – Lichen Barren
Little Bluestem – Blueberry Sandplain Grassland
Pitch Pine – Heath Barren
Pitch Pine – Scrub Oak Barren
Spruce – Heath Barren

The Pine Barrens Ecosystem in Maine corresponds to the Northeastern Interior Pine Barrens national Ecological System.

Rock Outcrop Ecosystem S4

Rock Outcrop Ecosystems are areas (not including alpine areas) where bedrock, with thin soil pockets in places, makes up the ground surface and vegetation is consequently sparse, with few if any trees. Vegetation is dominated by low shrubs and herbs which can tolerate the exposed, usually dry, and low nutrient conditions. Scattered trees (i.e., canopy cover totaling less than 30%) may be present, but will often be stunted or wind-flagged. Species composition will vary depending on the nature of the bedrock; the distinction made here is between circumneutral to calcareous rock versus acidic rock. Rock outcrop ecosystems generally grade into woodlands (rocky areas with tree canopies of 30 - 60%) or forests (tree canopies covering > 60%). This is a geographically broad type, and some of the communities listed are unlikely to be found together (e.g., Oak - Birch Talus Woodland and Rocky Summit Heath). Which communities make up any occurrence of this ecosystem will depend on geographic location, substrate pH, and probably other factors.

Acidic – Cliff Gorge
Birch – Oak Talus Woodland
Boreal Circumneutral Open Outcrop
Crowberry – Bilberry Summit Bald
Rocky Summit Heath
Spruce Talus Woodland
Three-toothed Cinquefoil – Blueberry Low Summit Bald

The Rock Outcrop Ecosystem in Maine corresponds to the following national Ecological Systems: Laurentian – Acadian Acidic Cliff and Talus, Laurentian – Acadian Calcareous Cliff and Talus, Laurentian – Acadian Calcareous Rocky Outcrop, and Northern Appalachian – Acadian Rocky Heath Outcrop.

Peatland Ecosystems

Peatland ecosystems occur in closed or open basins, where conditions allow accumulation of a well-developed layer of *Sphagnum* and/or sedge peat (most commonly the former). Different researchers have classified Maine peatlands in different ways: by basin morphology, by vegetation, by hydrology, or by a combination of those factors. No classification is "right" or "wrong"; their utility depends on their purpose. A classification developed to evaluate peatlands for mining potential would clearly use different criteria than one used to evaluate peatland biota or to elucidate developmental processes in peatlands. The classification used here is derived from Davis and Anderson's work (Davis and Anderson 1991, 1999, 2001), which includes both morphological and vegetational criteria. This approach recognizes that peatland conservation should address both whole systems and individual collections of species. Thus, while a kettlehole bog and a domed bog may have nearly identical species complements, they are very different landscape units. Likewise, patterned fens and unpatterned fens are floristically almost identical, but the patterning is an interesting morphological character that separates that ecosystem type. Peatland complexes containing more than one of the above ecosystems also exist. These are treated in this classification as a collection of ecosystem units.

Coastal Plateau Bog Ecosystem S3

Coastal Plateau Bog Ecosystems are peatlands in east coastal Maine in which the surface is raised above the surrounding terrain, with the bog perimeter sloping sharply to mineral soil. The raised surface is flat or undulating, generally with few to no trees, and usually features extensive lawns of deer-hair sedge. Black crowberry and baked-apple berry are also characteristic. Some coastal plateau bogs support the rare crowberry blue butterfly.

Bog Moss Lawn
Deer-hair Sedge Bog Lawn
Huckleberry – Crowberry Bog
Mountain Holly – Alder Woodland Fen
Sheep Laurel Dwarf Shrub Bog
Spruce – Larch Wooded Bog

The Coastal Plateau Bog Ecosystem in Maine corresponds to the Acadian Maritime Bog national Ecological System.

Domed Bog Ecosystem S3

Domed Bog Ecosystems are a type of raised bog; these are large inland peatlands, usually more than 500 meters in diameter, with convex surfaces that rise several meters above the surrounding terrain and that display concentric patterning.

At least in the center, peat accumulation is sufficient to maintain a perched water table. Consequently, most water available for plant growth comes from precipitation and is nutrient poor. Most domed bogs show a vegetation zonation reflecting the nutrient gradient, where more nutrient-demanding (minerotrophic) vegetation occurs around the perimeter of the peatland (where water from surrounding uplands or draining from the center of the peatland flows) and low-nutrient vegetation occupies the raised portions of the bog. The peatland surface is characterized by hummocks and hollows. Patterned domed bogs have small, usually crescent-shaped pools near the highest point; unpatterned domed bogs lack pools.

Bog Moss Lawn
Huckleberry – Crowberry Bog
Leatherleaf Boggy Fen
Low Sedge – Buckbean Fen Lawn
Mixed Tall Sedge Fen
Mountain Holly – Alder Woodland Fen
Northern White Cedar Woodland Fen
Red Maple Wooded Fen
Sedge – Leatherleaf Fen Lawn
Sheep Laurel Dwarf Shrub Bog
Shrubby Cinquefoil – Sedge Circumneutral Fen
Spruce – Larch Wooded Bog
Sweetgale Mixed Shrub Fen

The Domed Bog Ecosystem in Maine corresponds to the Boreal – Laurentian Bog national Ecological System.

Eccentric Bog Ecosystem S3

Eccentric Bog Ecosystems are gently sloping raised bogs on the sides of shallow valleys. The slope of the bog is patterned, with ridges of dwarf shrub bog (sometimes semi-forested) alternating with troughs containing bog pools and/or moss lawns. Ridges and troughs run perpendicular to the direction of water flow. The upslope end is typically dryish for a bog and dominated by ombrotrophic vegetation; moisture and mineral availability increase downslope, and consequently the lowest areas have fen vegetation. In Maine, eccentric bogs appear to be restricted to the east central region, roughly from Lincoln north to Ashland and east to the Canadian border. See Davis and Anderson (1991) for a thorough discussion of these peatlands.

Bog Moss Lawn
Huckleberry – Crowberry Bog
Leatherleaf Boggy Fen

Low Sedge – Buckbean Fen Lawn
Mixed Tall Sedge Fen
Mountain Holly – Alder Woodland Fen
Northern White Cedar Woodland Fen
Red Maple Wooded Fen
Sedge – Leatherleaf Fen Lawn
Sheep Laurel Dwarf Shrub Bog
Shrubby Cinquefoil – Sedge Circumneutral Fen
Spruce – Larch Wooded Bog
Sweetgale Mixed Shrub Fen

The Eccentric Bog Ecosystem in Maine corresponds to the Boreal – Laurentian Bog national Ecological System.

Kettlehole Bog-Pond Ecosystem S4

Kettlehole bogs are flat peatlands in "kettles," circular or elliptical depressions, usually deeper than they are wide, formed in morainal, glaciofluvial, or coastal plain deposits by the melting of buried ice blocks. The centers of these bowl-shaped basins may be a floating peatland mat or open water ringed by peatland. Where the surface of the floating mat is sufficiently elevated by peat accumulation to be free from contact with the mineral-enriched pond water, vegetation typical of nutrient-poor conditions develops. In the southernmost part of the state, kettlehole vegetation may include species of more southern affinity such as Atlantic white cedar, sweet pepperbush, and arrow-arum.

Bog Moss Lawn
Leatherleaf Boggy Fen
Low Sedge – Buckbean Fen Lawn
Mixed Tall Sedge Fen
Mountain Holly – Alder Woodland Fen
Northern White Cedar Woodland Fen
Red Maple Wooded Fen
Sedge – Leatherleaf Fen Lawn
Spruce – Larch Wooded Bog
Sweetgale Mixed Shrub Fen

The Kettlehole Bog Ecosystem in Maine corresponds to the Boreal – Laurentian Bog national Ecological System.

Patterned Fen Ecosystem S3

Patterned fens are floristically similar to unpatterned fens, but are typically found on gentle slopes and feature low, parallel to anastomosing, peat ridges (strings) alternating with wet hollows or shallow pools (flarks). The strings and flarks are oriented across the major slope of the peatland and at right angles to the direction of water movement. Groundwater chemistry will determine whether acidic fen communities or circumneutral fen communities predominate. The ribbing may occupy only a portion of the peatland area; other portions may support dwarf shrub or forested bog vegetation. Ribbed fens are restricted to the northwestern portion of the state.

Bog Moss Lawn
Leatherleaf Boggy Fen
Low Sedge – Buckbean Fen Lawn
Mixed Tall Sedge Fen
Mountain Holly – Alder Woodland Fen
Northern White Cedar Woodland Fen
Red Maple Wooded Fen
Sedge – Leatherleaf Fen Lawn
Sheep Laurel Dwarf Shrub Bog
Shrubby Cinquefoil – Sedge Circumneutral Fen
Spruce – Larch Wooded Bog
Sweetgale Mixed Shrub Fen

The Patterned Fen Ecosystem in Maine corresponds to the following national Ecological Systems: Boreal – Laurentian – Acadian Acidic Basin Fen and Laurentian – Acadian Alkaline Fen.

Raised Level Bog Ecosystem S4
Raised Level Bog Ecosystems are flat peatlands in basins with mostly closed drainage, receiving water from precipitation and runoff from the immediate surroundings. Most parts of level bogs are somewhat raised (though not domed), in which case vegetation is almost entirely ombrotrophic (dwarf shrub heath or forested bog). Other parts of the bog are not raised; in this case, vegetation is transitional (in nutrient status) between that of ombrotrophic bogs and minerotrophic fens. In all cases, *Sphagnum* dominates the ground surface and is the main peat constituent. The surface of the bog is flat and featureless. These bogs are often at least partly treed with black spruce and larch.

Bog Moss Lawn
Huckleberry – Crowberry Bog
Leatherleaf Boggy Fen
Low Sedge – Buckbean Fen Lawn
Mixed Tall Sedge Fen
Mountain Holly – Alder Woodland Fen
Northern White Cedar Woodland Fen
Pitch Pine Bog

Red Maple Wooded Fen
Sedge – Leatherleaf Fen Lawn
Sheep Laurel Dwarf Shrub Bog
Shrubby Cinquefoil – Sedge Circumneutral Fen
Spruce – Larch Wooded Bog
Sweetgale Mixed Shrub Fen

The Raised Level Bog Ecosystem in Maine corresponds to the following national Ecological Systems: Boreal – Laurentian – Acadian Acidic Basin Fen and Laurentian – Acadian Alkaline Fen.

Unpatterned Fen Ecosystem S4

Unpatterned Fen Ecosystems are peatlands in which groundwater or water from adjacent uplands moves through the area. As a result, plants are exposed to more nutrients, and the vegetation is typically different and more diverse than that of bogs. Peat is moderately- to well-decomposed and of variable thickness. The vegetation consists predominantly of sedges, grasses, reeds, and *Sphagnum*. Bog communities, dominated by heath shrubs, may be present; but though fen and bog vegetation may co-occur, in a fen ecosystem the former is more extensive. This type is broadly defined geographically: in very few locations in southern Maine one may find an Atlantic White Cedar Bog community as a constituent, but far more common statewide would be the Northern White Cedar Woodland Fen community.

Atlantic White Cedar Bog
Bog Moss Lawn
Huckleberry – Crowberry Bog
Leatherleaf Boggy Fen
Low Sedge – Buckbean Fen Lawn
Mixed Tall Sedge Fen
Mountain Holly – Alder Woodland Fen
Northern White Cedar Woodland Fen
Red Maple Wooded Fen
Sedge – Leatherleaf Fen Lawn
Sheep Laurel Dwarf Shrub Bog
Shrubby Cinquefoil – Sedge Circumneutral Fen
Spruce – Larch Wooded Bog
Sweetgale Mixed Shrub Fen

The Unpatterned Fen Ecosystem in Maine corresponds to the following national Ecological Systems: Boreal – Laurentian – Acadian Acidic Basin Fen, Boreal – Laurentian Bog, Laurentian – Acadian Alkaline Fen, and North – Central Interior and Appalachian Acidic Peatland.

Freshwater Shoreline Ecosystems

Includes the nearshore and shoreline portions of lakes, ponds, streams, and rivers, excluding areas where peat development is sufficient (generally more than about 1 m deep) to classify as peatlands. These ecosystems include areas where vascular plants are the primary producer, but do not include open-water ecosystems. Additional work beyond the scope of this classification needs to be done to classify open-water communities in a way useful to biodiversity conservation (Gawler et al. 1996, Vaux 2001).

Appalachian – Acadian Rivershore Ecosystem S4

Appalachian – Acadian Rivershore Ecosystems are the group of communities bordering and directly influenced by the open-water portions of a river (fourth-order or higher). It is distributed statewide except for extreme southern Maine, where rivershore ecosystems are distinguished as a separate type (Coastal Plain Rivershore Ecosystem) characterized by assemblages with more southerly affinities. It includes palustrine communities such as riverside seeps or floodplain forests, as well as communities (such as the river beach community) where the soils are not saturated over most of the growing season but which are disturbed by ice-scour or flooding. Upland forests bordering rivers are included under forested upland ecosystems.

Alder Floodplain
Bluebell – Balsam Ragwort Shoreline Outcrop
Blueberry – Heath Shoreline Thicket
Bluejoint Meadow
Bulrush Bed
Circumneutral Riverside Seep
Cobble Rivershore
Dogwood – Willow Shoreline Thicket
Hardwood River Terrace Forest
Hudsonia River Beach
Mixed Graminoid – Shrub Marsh
Pickerelweed – Macrophyte Aquatic Bed
Sand Cherry – Tufted Hairgrass River Beach
Silver Maple Floodplain Forest
Water-lily – Macrophyte Aquatic Bed

The Appalachian – Acadian Rivershore Ecosystem in Maine corresponds to the following national Ecological Systems: Laurentian – Acadian Freshwater Marsh, Laurentian – Acadian Wet Meadow – Shrub Swamp.

Coastal Plain Pondshore Ecosystem S1

This group of natural communities occurs on the shoreline or the near-shore vegetated waters of a lake or pond; specifically, ponds in southern Maine where climatic influences and species affinities are allied with areas southward along the Atlantic coastal plain. It includes palustrine communities such as pondshore marshes and fens, as well as communities where the soils are not saturated over most of the growing season but which are disturbed by ice-scour, flooding, or waves. Terrestrial forests bordering lakes are included under forested upland ecosystems.

Alder Shrub Thicket
Blueberry – Heath Shoreline Thicket
Bulrush Bed
Cattail Marsh
Lakeshore Beach
Mixed Graminoid – Shrub Marsh
Mixed Tall Sedge Fen
Pickerelweed – Macrophyte Aquatic Bed
Pipewort – Water Lobelia Aquatic Bed
Sweetgale Mixed Shrub Fen
Three-way Sedge – Goldenrod Outwash Plain Pondshore
Water-lily – Macrophyte Aquatic Bed

The Coastal Plain Pondshore Ecosystem in Maine corresponds to the following national Ecological Systems: Laurentian – Acadian Freshwater Marsh, Laurentian – Acadian Wet Meadow – Shrub Swamp, and Northern Atlantic Coastal Plain Pond. The Coastal Plain Pondshore Ecosystem.

Coastal Plain Rivershore Ecosystem S2

The Coastal Plain Rivershore Ecosystem is the group of natural communities bordering and directly influenced by the open-water portion of rivers in extreme southern Maine. It is superficially similar to the Appalachian - Acadian Rivershore Ecosystem, widespread rivershore ecosystem, but its climatic influences and species affinities are allied with areas southward along the Atlantic coastal plain. It includes palustrine communities such floodplain forests, as well as terrestrial communities (such as the river beach community) where the soils are not saturated over most of the growing season but which are disturbed by ice-scour or flooding. Terrestrial forests bordering rivers are included under forested upland ecosystems.

Alder Floodplain
Blueberry – Heath Shoreline Thicket
Bulrush Bed
Cobble Rivershore

Dogwood – Willow Shoreline Thicket
Hardwood River Terrace Forest
Hudsonia River Beach
Mixed Graminoid – Shrub Marsh
Pickerelweed – Macrophyte Aqauatic Bed
Silver Maple Floodplain Forest
Water-lily – Macrophyte Aquatic Bed

The Coastal Plain Rivershore Ecosystem in Maine corresponds to the following national Ecological Systems: Laurentian – Acadian Freshwater Marsh and Laurentian – Acadian Wet Meadow – Shrub Swamp.

Lakeshore Ecosystem S5
Lakeshore ecosystems consist of the group of communities on the shoreline or the near-shore vegetated waters surrounding the open-waters of a lake or pond. It is distributed statewide except for extreme southern Maine, where lakeshore/pondshore ecosystems are distinguished as a separate type (Coastal Plain Pondshore Ecosystem) which is characterized by assemblages with more southerly affinities. It includes wetland communities such as lakeside marshes and fens, as well as communities where the soils are not saturated over most of the growing season but which are disturbed by ice-scour, flooding, or waves. Terrestrial forests bordering lakes are included under forested upland ecosystems.

Alder Shrub Thicket
Blueberry – Heath Shoreline Thicket
Bluejoint Meadow
Bulrush Bed
Cattail Marsh
Circumneutral – Alkalaine Water Macrophyte Suite
Lakeshore Beach
Mixed Graminoid – Shrub Marsh
Mixed Tall Sedge Fen
Northern White Cedar Swamp
Pickerelweed – Macrophyte Aquatic Bed
Pipewort – Water Lobelia Aquatic Bed
Red Maple – Sensitive Fern Swamp
Sweetgale Mixed Shrub Fen
Tussock Sedge Meadow
Water-lily – Macrophyte Aquatic Bed

The Lakeshore Ecosystem in Maine corresponds to the following national Ecological Systems: Laurentian – Acadian Freshwater Marsh, Laurential – Acadian

Lakeshore Beach, Laurentian – Acadian Wet Meadow – Shrub Swamp, and Northern Atlantic Coastal Plain Pond.

Streamshore Ecosystem S4

The Streamshore Ecosystem is the group of communities bordering and directly influenced by the open-water portion of a stream (first-order through third or fourth-order). It includes vegetated aquatic communities as well as the emergent and bordering communities. Most communities are palustrine; streams are generally too small to exert many disturbance effects on adjacent terrestrial areas. Terrestrial forests bordering streams are included under forested upland ecosystems.

Alder Floodplain
Alder Shrub Thicket
Black Ash Swamp
Black Willow – Alder Floodplain
Bluejoint Meadow
Bulrush Bed
Cattail Marsh
Mixed Graminoid – Shrub Marsh
Mixed Tall sedge Fen
Northern White Cedar Swamp
Pickerelweed – Macrophyte Aquatic Bed
Red Maple – Sensitive Fern Swamp
Silver Maple Floodplain Forest
Sweetgale Mixed Shrub Fen
Tussock Sedge Meadow
Water-lily – Macrophyte Aquatic Bed

The Streamshore Ecosystem in Maine corresponds to the following national Ecological Systems: Boreal Ice – Scour Rivershore, Eastern Boreal Floodplain, Laurentian – Acadian Floodplain Forest, Laurentian – Acadian Freshwater Marsh, and Laurentian – Acadian Wet Meadow – Shrub Swamp.

Tidal Ecosystems

Includes all areas immediately associated with or affected by tidal waters, i.e., estuaries and coastal shores.

Coastal Dune – Marsh Ecosystem S3

Coastal Dune – Marsh Ecosystems are low-lying coastal areas with sand beaches, dunes, and salt marshes behind the dunes. This ecosystem is bounded landward by terrestrial ecosystems, either forests or cultural ecosystems, and seaward by the high tide line of the open shore. Coastal dune – marsh ecosystems are most common along the southwestern coast; they become uncommon east of Penobscot Bay.

Beach Strand
Dune Grassland
Pitch Pine Dune Woodland
Rose – Bayberry Maritime Shrubland
Spartina Saltmarsh

The Coastal Dune – Marsh Ecosystem in Maine corresponds to the following national Ecological Systems: Acadian Coastal Salt Marsh, Northern Atlantic Coastal Plain Dune and Swale, Northern Atlantic Coastal Plain Sandy Beach, Northern Atlantic Coastal Plain Tidal Salt Marsh.

Coastal Headland Ecosystem S4

Coastal Headland Ecosystems are moderate- to high-relief coastlines of bedrock or consolidated marine sediments. This ecosystem is bounded landward by upland forests or cultural ecosystems and seaward by the high tide line. It includes communities which are directly affected by the proximity of the ocean through salt spray and storm tides; these are characterized by low vegetation that can withstand wind, sun, and salt. Constituent communities and species vary geographically.

Crowberry – Bayberry Headland
Heath – Crowberry Maritime Slope Bog
Mixed Graminoid – Forb Saltmarsh
Rose – Bayberry Maritime Shrubland
Seaside Goldenrod – Goosetongue Open Headland

The Coastal Headland Ecosystem in Maine corresponds to the following national Ecological Systems: Acadian – North Atlantic Rocky Coast, North Atlantic Cobble Shore, and North Atlantic Rocky Intertidal.

Tidal Marsh Estuary Ecosystem S3

Tidal Marsh Estuary Ecosystems are intertidal portions of bays and rivers, from the head of tide to the coastline. Salinity can vary greatly depending on season, weather, and other factors, but generally increases downriver, with portions near the head of tide almost completely freshwater. Saltmarshes are restricted to the higher salinity areas.

Brackish Tidal Marsh
Freshwater Tidal Marsh
Mixed Graminoid – Forb Saltmarsh
Spartina Saltmarsh

The Tidal Marsh Estuary Ecosystem in Maine corresponds to the following national Ecological Systems: Acadian Coastal Salt Marsh, Acadian Estuary Marsh, North Atlantic Intertidal Mudflat, North Atlantic Tidal Sand Flat, Northern Atlantic Coastal Plain Fresh and Oligohaline Tidal Marsh, and North Atlantic Coastal Plain Seagrass Bed.

Keys to Natural Community Types

The key is a standard dichotomous key that includes characteristics to progressively separate the natural community types. Refer to the Glossary (Appendix F) for unfamiliar terms, and bear in mind that percentages used in the key are guidelines and not absolute cutoffs.

For readers not familiar with using keys, this is a series of paired statements (couplets) representing mutually exclusive choices. Start with the first paired statement, read both choices, and decide which best describes your sample area. Each statement will direct you to another couplet number. Keep working through the couplets until your choice brings you to a particular community type. Because of ecological variation and the fact that this key is based on a limited number of samples, you may reach a point where neither choice seems right. Make a note of where you are, try one direction, and then come back and try the other direction to see which of the community types you arrive at seems more appropriate to your sample area. You will have better results in identifying natural communities if you avoid transition zones.

First, a Key to the Keys:

1. Trees over 10 feet tall form ≥ 30% cover: forests and woodlands 2
1. Trees over 10 feet tall form <30% cover; some small islands of trees may be present: open vegetation 3

2. **UPLAND**: soils are not saturated or seasonally flooded. Exception: in summit or ledgy types the "soil" may be thin pockets of peat in bedrock depressions and may be saturated; these are treated as uplands .... **WOODED UPLANDS KEY** (p. 42)
2. **WETLAND**: soils are saturated or seasonally flooded; usually basins or streamsides ....**WOODED WETLANDS KEY** (p. 48)

3. **UPLAND**: soils are not saturated or seasonally flooded. Exception: sometimes in open summit or other ledgy types the "soil" is thin peat pockets over bedrock, and often saturated; these are treated as uplands... **OPEN UPLANDS KEY** (p. 51)
3. **WETLAND**: soils are saturated or seasonally flooded; usually basins or streamsides ....**OPEN WETLANDS KEY** (p. 55)

WOODED UPLANDS KEY

1. Conifer forests and woodlands: broad-leaved trees < 25% cover 2
1. Broad-leaved or mixed forests and woodlands: broad-leaved trees > 25% cover 24

2. Conifer forests: tree cover usually > 65%, lower layers generally more sparse than canopy (except for dense patches of tree regeneration in canopy openings); area lacks a well-developed low shrub layer of lowbush blueberry, velvet-leaf blueberry, black huckleberry, sheep laurel, or black chokeberry; substrate various, but often mineral soil, usually not a thin organic layer over bedrock 3
2. Conifer woodlands: tree cover usually < 65%, as low as 20% (occasionally to ~80%), trees more-or-less open grown; low shrub layer of lowbush blueberry, huckleberry, sheep laurel, and/or chokeberry prominent (>15% cover), or sometimes bracken fern prominent (>10% cover) instead; ground cover may feature fruticose lichens; substrate bedrock, with a thin layer of mostly organic soil typically < 15 cm deep 14

Conifer forests

3. Hemlock and/or pines dominant (near the coast, northern white cedar occasionally co-dominant) 4
3. Spruce and/or fir dominant 8

4. Hemlock dominant, red spruce, white pine, and/or cedar often present, sometimes co-dominant, stands may be either coniferous or mixed, strongly coniferous stands with very sparse herb and bryoid layers. .... **Hemlock Forest** (p. 76)
4. Hemlock not dominant 5

5. Northern white cedar dominant, rocky hillslopes (some very closed canopy examples of the type) .... **White Cedar Woodland** (p. 122)
5. Pines dominant 6

6. Jack pine dominant, or co-dominant with black spruce; known only from western Maine .... **Jack Pine Forest** (p. 78)
6. Red pine or white pine dominant; more widespread 7

7. White pine dominant (highest basal area); red spruce, hemlock, and/or northern white cedar often present and may be nearly co-dominant
.... **White Pine – Mixed Conifer Forest** (p. 126)

7. Red pine dominant (highest basal area); red spruce, white pine, (or coastally, northern white cedar) often present, may be nearly co-dominant
.... **Red Pine – White Pine Forest** (p. 108)

8. Balsam fir dominant, heart-leaved paper birch and/or mountain ash often common, subalpine or coastal 9

8. Red and/or black spruce dominant.. 10

9. Coastal; white spruce often present..........**Maritime Spruce – Fir Forest** (p. 84)

9. Montane; white spruce absent
.... **Fir - Heart-leaved Birch Subalpine Forest** (p. 118)

10. Variable canopy closure; black spruce or red-black hybrids often present; red pine occasionally present; heath shrubs prominent in herb layer (often >25% cover, may be very dense); usually on level, well-drained to xeric glacial deposits, sometimes with patches of saturated-soil and peat moss pockets in hummocky topography...**Spruce – Heath Barren** (p. 66)

10. More or less closed canopy with mixtures of red spruce, black spruce, or other conifers; heath shrubs and herbs sparse (less than 20% cover), or if more, then with northern wood-sorrel, bluebead lily, tree regeneration may be dense in gaps... 11

11. Red spruce, black spruce, or hybrids on level, poorly drained soils; wetland plants and peat mosses may be frequent, bryoids abundant (>50% cover)
....**Spruce – Fir – Cinnamon Fern Forest** (p. 158)

11. Upland sites on till, bedrock, or talus, with red spruce or white spruce dominant; yellow birch and balsam fir may be occasional; wetland plants absent except in small seeps and ravines 12

12. Yellow birch often present and may form >25% cover. Herb layer with moderate to extensive cover of herbaceous species (often in patches), including northern wood-sorrel, bluebead lily, rose twisted-stalk, or mountain wood-fern; bryoid layer >20%, with feather-mosses prominent; northern/montane areas
....**Spruce – Fir – Wood-sorrel – Feather-moss Forest** (p. 86)

12. Yellow birch, if present, forms less than 25% cover. Herbs and dwarf shrubs sparse (<10% cover each); herb layer may include dense regeneration patches
.... 13

13. Coastal (within ¼ mi. of the coast); white spruce often present, especially near coastline; bryoid layer >15% cover and not dominated by broom-mosses .... **Maritime Spruce – Fir Forest** (p. 84)

13. Widespread inland, may be near-coastal; both herb and bryoid layers very sparse; what bryoids are present include a large proportion of broom-mosses and pincushion moss.... ....**Spruce – Fir – Broom-moss Forest** (p. 82)

Conifer woodlands

14. Northern white cedar the dominant tree species, usually twice as abundant as any other tree species .... **White Cedar Woodland** (p. 122)

14. Other species dominate tree layer.. 15

15. On talus, spruce dominant.............................**Spruce Talus Woodland** (p. 116)

15. On bedrock (or, less commonly, sand or glacial till), various dominants........ 16

16. Mixture of conifer species, each < 50% cover; or woodlands dominated by white pine or red pine.. 17

16. Pitch pine, jack pine, <u>or</u> spruce each ≥ 60% cover 18

17. Canopy dominated by red spruce or white pine (> 60% cover), or by a mixture of conifers with none dominant, red spruce usually present ....**Red Spruce – Mixed Conifer Woodland** (p. 114)

17. Red pine dominant in canopy..............................**Red Pine Woodland** (p. 110)

18. Jack pine dominates canopy...................................**Jack Pine Woodland** (p. 80)

18. Other conifers dominate canopy; jack pine usually absent altogether............ 19

19. Black spruce the most abundant canopy tree..20

19. Pitch pine the most abundant canopy tree...21

20. Woodlands on bedrock, often coastal or on upper summits inland ....**Black Spruce Woodland** (p. 68)

20. Woodlands or forests on level or gently sloping till, occasionally with a thick organic layer ....**Spruce – Heath Barren** (p. 66)

21. Open woodlands on ledge; trees often stunted.. **Pitch Pine Woodland** (p. 106)

21. Open woodlands or sometimes almost closed forest, on sandy soil 22

22. Dunes or backdune areas along the immediate coast; bayberry almost always present...**Pitch Pine Dune Woodland** (p. 100)

22. On glacial deposits, not dunes; inland or at least not at the immediate shoreline...23

23. Scrub oak extensive and locally dominant, making foot-travel difficult .... **Pitch Pine – Scrub Oak Barren** (p. 104)
23. Scrub oak absent or sparse, foot-travel easier ....**Pitch Pine – Heath Barren** (p. 102)

Broad-leaved or mixed forests and woodlands

24. Forest seasonally saturated and/or discharging groundwater for significant periods during the growing season... 25
24. Forest not seasonally saturated or discharging groundwater............................28

25. Forest on gentle to moderate slope or flat bench within upland setting; discharging groundwater for significant periods during the growing season; dominant trees include ashes, sugar maple, red maple, yellow birch, or hemlock; jewel-weed often present..... ....**Hardwood Seepage Forest** (p. 142)
25. Forest may be saturated and flooded in the spring and adjacent to larger streams and rivers in a floodplain setting, or may be a somewhat elevated riverside terrace; ferns frequent... 26

26. Floodplain forests with silver maple dominant and conifers absent or very sparse; usually along medium to large rivers. Herb layer includes ostrich fern and other ferns.. .... **Silver Maple Floodplain Forest** (p. 156)
26. Other hardwoods dominant (not silver maple)..27

27. High floodplain or terrace forests along medium to larger rivers in southern and central Maine, most often with sugar maple or red oak dominant, sometimes with yellow birch or ash dominant; on larger rivers this type is often found adjacent to but at slightly higher elevation than silver maple floodplain forests .... **Hardwood River Terrace Forest** (p. 160)
27. Floodplain with balsam poplar dominant or co-dominant with American elm and ash; northern Maine.................. **Balsam Poplar Floodplain Forest** (p.134)

28. Open broad-leaved woodlands (tree cover <65%) with birches, red oak, red maple, or ironwood dominant (Ironwood – Oak - Ash Woodlands may occasionally have a forest canopy >65% but they will key here: check description in couplet 31 below). 29
28. Broad-leaved or mixed closed-canopy forests (tree cover >65%) or mixed woodlands 32

29. Hillside or low summit woodlands dominated by red oak, without much ironwood or sugar maple, and with a heath shrub and bracken fern understory
.... **Oak – Pine Woodland** (p. 98)

29. Not as above 30

30. Open to closed canopy woodlands on glacial till or outwash; early successional broad-leaved trees dominant (aspen, white/gray birches, red maple), red oak <40% cover (rarely up to 50%), conifers <30% cover
.... **Aspen – Birch Woodland/Forest Complex** (p. 72)

30. Open woodlands on talus or thin-soiled hilltops rather than till or outwash 31

31. Woodlands on talus or thin-soiled hilltops and upper slopes; ironwood and oaks dominant (the former as either canopy or subcanopy), sugar maple occasionally common, and basswood often present but not abundant; with enriched site indicators e.g. round-leaved dogwood, herb Robert, hepatica, columbine, blue-stem goldenrod, and wide-leaved sedge
.... **Ironwood – Oak – Ash Woodland** (p. 90)

31. Woodlands on talus; birches the most abundant tree; oaks and ironwood may be present but less abundant; enriched site indicators absent or very incidental
.... **Birch – Oak Talus Woodland** (p. 64)

32. Broad-leaved component dominated by northern hardwood species (beech, yellow birch, sugar maple, red maple); red oak < 30% cover 33

32. Broad-leaved component dominated by oak, white or gray birch, red maple, shagbark hickory, and/or aspen, rather than northern hardwood species 39

33. Mixed forests with conifers >25% cover 34

33. Broad-leaved forests, conifers <25% cover 37

34. Red spruce the dominant conifer
.... **Spruce – Northern Hardwoods Forest** (p. 112)

34. Hemlock or white pine the dominant conifer 35

35. Hemlock the dominant conifer .... **Hemlock Forest** (p. 76)

35. White pine the dominant conifer 36

36. One or more northern hardwood species (beech, yellow birch, sugar maple) mixed with red oak and conifers; northern hardwood cover > 15%
.... **Red Oak – Northern Hardwoods – White Pine Forest** (p. 94)

36. Canopy almost entirely red oak and white pine (occasionally red spruce), sometimes mixed with red maple; northern hardwoods sparse
.... **Oak – Pine Forest** (p. 96)

37. Some combination of beech, sugar maple, red maple, and yellow birch dominates the canopy, with sugar maple always <50% cover; ash and basswood absent or very minor; herb layer rarely >40% cover except for tree seedlings; enriched site herbs absent. .... **Beech – Birch – Maple Forest** (p. 88)
37. Sugar maple dominant or co-dominant (≥50% cover); in southern or central Maine, white ash and/or basswood are usually present; herbs >40% cover. ... 38

38. Several moderate rich site indicators are present, including doll's eyes, jack in the pulpit, grape fern, alternate-leaved dogwood, spikenard, and false Solomon's seal.. **Semi-rich Northern Hardwood Forest** (p. 120)
38. In addition to the species noted above, numerous uncommon rich site herbs such as leatherwood, Dutchman's breeches, maidenhair fern, and blue cohosh, and flat sedge are frequent. Sites are typically small acreages in sheltered sites (coves) or where calcium-enriched soils have accumulated .... **Maple – Basswood – Ash Forest** (p. 74)

39. Early successional broad-leaved trees dominant (aspen, white/gray birches, red maple), red oak <40% cover (rarely up to 50%), canopy closure variable from very open to almost closed; conifers <30% cover .... **Aspen – Birch Woodland/Forest Complex** (p. 72)
39. Oaks >50% cover, or if less, then canopy cover >35% coniferous.................. 40

40. White oak, chestnut oak, and/or hickories present; southern Maine............ 41
40. White oak, chestnut oak, hickories absent or virtually so; statewide.............. 43

41. Chestnut oak dominant or at least common; open-canopy woodlands (<65% canopy)........... .... **Chestnut Oak Woodland** (p. 70)
41. Other oaks dominant... 42

42. White oak and red oak dominate canopy; hickory essentially absent .... **White Oak – Red Oak Forest** (p. 124)
42. White oak and red oak present and may be dominant, but shagbark hickory (and sometimes bitternut hickory) also present .... **Oak – Hickory Forest** (p. 92)

43. Closed-canopy forests with northern hardwood species (usually beech or sugar maple) >20% cover, may be co-dominant with red oak .... **Red Oak – Northern Hardwoods – White Pine Forest** (p. 94)
43. Beech, sugar maple, and white ash <10% cover each...................................... 44

44. Woodland structure: open canopy (<60% closure, occasionally more), trees well-spaced and often stunted; well-developed herb layer (>15% cover of dwarf shrubs and >5% cover of herbs), with graminoids often abundant and typical closed-forest species essentially absent .... **Oak – Pine Woodland** (p. 98)
44. Forest structure: canopy closure >70% and understory somewhat sparse, with dwarf shrubs much less than 15% cover; herbs include typical forest species such as Indian cucumber-root and wild-oats. .... **Oak – Pine Forest** (p. 96)

WOODED WETLANDS KEY

1. Peat-substrate wetlands dominated by conifers or, less commonly, red maple. The herb layer is dominated by typical bog plants such as leatherleaf, sheep laurel, or other heath plants, cotton-grasses, white beak-rush, or sedges. Peat moss usually occupies more than 25% of ground surface. Sites occur on peat substrates in basins (peat >30 cm deep, usually more), or rarely with thinner peat over mineral soil; not on slopes..2
1. Mineral-soil wetlands or floodplains dominated by hardwoods or, less commonly Atlantic White Cedar or red spruce. The herb layer is dominated by plants other than the bog species listed above. Peat moss may be present on ground surface but generally occupies < 20% cover and rarely forms a peat deposit of more than 15 cm; in drainages or on gentle slopes.. 10

2. Red maple dominant in canopy, or co-dominant with larch (less commonly, spruce); occasionally, larch may be dominant and red maple far less abundant; canopy closure usually <50%, occasionally somewhat higher (to 65%) ....**Red Maple Wooded Fen** (p. 152)
2. Conifers other than larch dominant in canopy...3

3. Northern white cedar dominant in canopy...4
3. Other conifers dominant..5

4. Closed-canopy or nearly closed-canopy forests (almost always >60% canopy); heath shrubs <15% cover; usually in poorly drained basins where they may occupy most of the basin, rather than occurring as part of a larger peatland vegetation complex.............................. **Northern White Cedar Swamp** (p. 144)
4. Canopy more open (usually <50%); partially wooded fens that are part of a larger peatland vegetation complex; heath shrubs or other dwarf shrubs usually >15% cover............................. **Northern White Cedar Woodland Fen** (p. 146)

5. Atlantic white cedar dominant or at least common..6
5. Atlantic white cedar absent (or very incidental)..7

6. Closed-canopy (or nearly so) forests, >60% tree cover; shrubs only patchily abundant in openings .... **Atlantic White Cedar Swamp** (p. 132)

6. More like an open bog, with trees patchy to sparse (<50% cover) and mostly stunted; abundant heath shrubs .... **Atlantic White Cedar Bog** (p. 130)

7. Pitch pine is the dominant tree (though black spruce may be present); southern and coastal Maine; uncommon .... **Pitch Pine Bog** (p. 148)

7. Black spruce is the dominant tree (may be stunted) 8

8. Organic soil (peat) is usually > 50 cm deep; larch may be co-dominant; statewide and common .... **Spruce – Larch Wooded Bog** (p. 138)
NOTE: **Spruce – Fir – Cinnamon Fern Forests** with an unusually deep peat layer may key here (p. 158)

8. Mineral-soil flats in colder regions; substrate usually includes at least patches of poorly drained mineral soil, sometimes with a shallow peat layer (< 30 cm deep)... 9

9. Sites are primarily well-drained/xeric but often undulate with both wetland and upland patches; tree cover is typically less than 60% .... **Spruce – Heath Barren** (p. 66)

9. Sites are primarily poorly drained flats, tree cover is typically greater than 60% .... **Spruce – Fir – Cinnamon Fern Swamp** (P. 158)

10. Conifers dominant (>50% of canopy) 11

10. Broad-leaved or mixed forests or shrublands with conifers <50% of the canopy 13

11. Atlantic white cedar dominant .... **Atlantic White Cedar Swamp** (p. 132)

11. Other conifers dominant 12

12. Red spruce or black spruce the most abundant conifer; red maple may be up to 40% cover .... **Spruce – Fir – Cinnamon Fern Forest** (p. 158)

12. Northern white cedar the most abundant conifer; red maple may be common here as well .... **Cedar – Spruce Seepage Forest** (p. 140)

13. Alder is the dominant cover; the shrub layer is usually well developed .... **Alder Floodplain** (p. 204)

13. Alder is absent or very minor, sometimes forming a fringe on the channel edge of some Silver Maple Floodplain Forests 14

14. Silver maple, red maple, or balsam poplar are the most abundant hardwoods, black gum absent; wetlands on alluvial flats bordering rivers or permanent streams..15

14. Other hardwoods (yellow birch, sugar maple, black gum, black or green ash, red oak) are more abundant than silver or red maple, or at least co-dominant; wetland setting various.. 18

15. Floodplain forests with silver maple dominant, and conifers absent or very sparse; usually along third order or larger streams/rivers
.... **Silver Maple Floodplain Forest** (p. 156)

15. Red maple, American elm, or balsam poplar dominant in floodplain or basin adjacent to smaller stream.. 16

16. Balsam poplar dominant or co-dominant with American elm and ash. Floodplain species such as ostrich fern and virgin's bower present; northern Maine
....**Balsam Poplar Floodplain Forest** (p. 134)

16. Red maple, yellow birch, or hemlock dominant. 17

17. Red maple dominant, silver maple and musclewood more or less absent; conifers may be >25% of canopy; in various settings, such as low basins and along small to medium-sized rivers and streams
....**Red Maple – Sensitive Fern Swamp** (p. 154)

17. High floodplain or terrace forests along medium to large rivers, most often with sugar maple or red oak dominant, sometimes with yellow birch or ash dominant; herb layer includes ostrich fern and a mixture of wetland and upland species; often found adjacent to but at slightly higher elevation than silver maple floodplain forests ....**Hardwood River Terrace Forest** (p. 160)

18. Hardwood or hemlock-hardwood swamps in small (typically less than 3 acre), often isolated basins; shrub layer of highbush blueberry or winterberry; red maple common, black gum often present. Sites are in southern and coastal Maine ....**Hemlock – Hardwood Pocket Swamp** (p. 150)

18. Swamps or seepage forests in broader basins; highbush blueberry, winterberry, and similar shrubs absent or very sparse; sites are statewide........................... 19

19. Hardwood or hemlock-hardwood seepage forests on gentle slopes or discharge areas at level breaks in a slope; yellow birch, green ash, brown ash and/or white ash may also occur in the canopy and may be co-dominant with red maple; statewide..**Hardwood Seepage Forest** (p. 142)

19. Basin wetlands on mineral soils or shallow peat; black ash is dominant or co-dominant with northern white cedar and red maple; more common in northern and eastern Maine.....................................**Black Ash Swamp** (p. 136)

OPEN UPLANDS KEY

1. Vegetation along rivershores, lakeshores, or the immediate Atlantic coast. 2
1. Vegetation not associated with shores, most often on mountains, lower summits and rocky slopes....... 8

2. Vegetation along the immediate Atlantic coast, subject to salt spray or occasional storm tides... 3
2. Inland vegetation. 7

3. Sandy or gravelly beach vegetation, either sparse or grass-dominated. 4
3. Vegetation on shoreline bedrock, or with >25% shrub cover. 5

4. Sparse forb-dominated vegetation on loose sand, gravel or cobble near and above the high tide line; sea-kale and beach-pea characteristic. .... **Beach Strand** (p. 220)
4. Grass-dominated dune and tidal-edge vegetation on sand, dominated by beach grass or (Downeast) Virginia wild rye. .... **Dune Grassland** (p. 178)

5. Sparse herb-dominated vegetation in bedrock crevices and depressions; seaside goldenrod and goosetongue typical ....**Seaside Goldenrod – Goosetongue Open Headland** (p. 186)
5. Cover more extensive, with shrubs or dwarf shrubs >25% cover....................... 6

6. Mats of dwarf shrubs, such as crowberry, lowbush blueberry, velvet-leaf blueberry and cranberries punctuated by sparse taller shrubs and forbs; more common Downeast............. .... **Crowberry – Bayberry Headland** (p. 176)
6. Mat-forming shrubs limited, most shrubs 1 m or so tall (or taller); bayberry and roses most characteristic; coastwide ....**Rose – Bayberry Maritime Shrubland** (p. 192)

7. Lakeshore with broad bands of vegetation zones, sometimes associated with sandy or gravelly berms. .... **Lakeshore Sand/Cobble Beach** (p. 232)
7. Shoreline rock outcrops, mostly along rivers, with sparse forb-dominated vegetation; bluebell, three-toothed cinquefoil, tufted hairgrass characteristic; substrate typically vertically fissured slaty or shaly rock (Note: for steep cliff faces, along streams and gorges, see couplet #22, these are usually on granitic rock as compared to vertically fissured slate) ....**Bluebell – Balsam Ragwort Shoreline Outcrop** (p. 188)

8. Open talus slopes (talus slopes with tree cover will key here, as well as in Wooded Uplands key)..9
8. Substrate not talus. 13

9. Vegetation dominated by carpets of heath shrubs, such as Labrador tea, sheep laurel, pale laurel, over *Sphagnum* mosses and/or reindeer lichens, with scattered conifers.. 10
9. Vegetation sparse, conditions more xeric... 11

10. Vegetation cover nearly continuous over the talus, with *Sphagnum* dominating the bryoid layer; wetland vegetation in an upland subalpine setting, near treeline.. **Heath – Lichen Subalpine Slope Bog** (p. 270)
10. Vegetation very patchy, much exposed talus; bryoid layer has reindeer lichens more extensive than *Sphagnum* cover; usually at lower elevations, though in cool microsites (type may be partially forested, primarily with spruce, or open) .... **Labrador Tea Talus Dwarf-shrubland** (p. 174)

11. Sparsely vegetated patches in partially forested talus, trees primarily coniferous .... **Spruce Talus Woodland** (p. 116)
11. Sparsely vegetated patches in partially forested talus, trees primarily broad-leaved.. 12

12. Ironwood, sugar maple, and/or basswood present in the tree layer; open spots with enriched site indicators e.g. round-leaved dogwood, herb Robert, hepatica, columbine, blue-stem goldenrod, wide-leaved sedge .... **Ironwood – Oak – Ash Woodland** (p. 90)
12. Paper birch, yellow birch, and/or red oak are the major trees; enriched site indicators absent or very incidental......... **Birch – Oak Talus Woodland** (p. 64)

13. Substrate sandy outwash; flat or rolling plains with lowbush blueberry and/or velvet-leaf blueberry among the dominant plants.. 14
13. Substrate rocky; mountains or hills 15

14. Blueberry (mostly lowbush blueberry) growing with little bluestem, other grasses, forbs, and occasional pitch pine; southern Maine ....**Little Bluestem – Blueberry Sandplain Grassland** (p. 194)
14. Blueberry (mostly lowbush blueberry) growing among extensive carpets of reindeer lichens; pitch pine absent; spruces, red pine, or white pine may be scattered; Downeast and extreme northwestern Maine ....**Blueberry – Lichen Barren** (p. 170)

15. Alpine areas: rocky summits and slopes near or above treeline, usually >3500' elevation; if tree species present, then less than 2 m tall; vegetation may be patchy or more continuous.......... 16

15. Open areas with patchy vegetation on hills or low mountains, below treeline; scattered trees often present.......... 22

16. Vegetation dense masses or patches of stunted conifers, mostly 1 – 2 meters tall, forming a band between treeline and more open summit areas .... **Spruce - Fir - Birch Krummholz** (p. 196)

16. Vegetation not krummholz. 17

17. Vegetation dominated by deciduous non-heath shrubs > 0.5 m tall; shrublands on subalpine tablelands or somewhat protected slopes between treeline and the dwarf – heath alpine zone, sometimes around subalpine pondshores; dominants include mountain alder, bush-honeysuckle and meadowsweet .... **Mountain Alder – Bush-honeysuckle Subalpine Meadow** (p. 198)

17. Vegetation mixed dwarf shrubs and herbs; shrubs mostly < 0.5 m tall, often < 0.2 m tall, forming low mats, and primarily heath shrub species subalpine to alpine elevations.......... 18

18. Alpine bilberry and crowberry common, highland rush often present, strictly alpine species (see list in next half of couplet) absent; patches of vegetation over bare rock, usually below or near treeline, but may extend higher .... **Crowberry – Bilberry Summit Bald** (p. 184)

18. One or more strictly alpine species present (not necessarily dominant): diapensia, boreal bentgrass, Bigelow's sedge, alpine bearberry, Lapland rosebay, bearberry willow, tundra dwarf birch, moss plant, mountain heath, star saxifrage or alpine bistort.......... 19

19. Herb-dominated cliffs and narrow benches on steep slopes (typically cirque walls), with constant seepage; rare alpine/boreal plants such as star saxifrage, alpine bistort, hairy arnica, and/or northern painted cup typically present though not abundant.......... **Alpine Cliff** (p. 166)

19. Vegetation primarily dwarf shrubs; not on seepy cirque walls. 20

20. Protected areas where snow lingers, such as at summit edge, upper slopes, or base of cliffs; alpine bilberry typically dominant and mountain-heath, tundra dwarf birch, or moss-plant characteristic .... **Bilberry – Mountain-heath Alpine Snowbank** (p. 168)

20. More exposed areas; alpine bilberry may be prominent but mountain-heath, tundra dwarf birch, and moss-plant absent 21

21. Cushions of diapensia dominate vegetation (>30% cover, or >40% relative cover), with alpine bilberry; herbs <10% cover .... **Diapensia Alpine Ridge** (p. 200)

21. Diapensia if present is not dominant; vegetation a variable mixture of dwarf evergreen shrubs and herbs, with graminoid cover (especially Bigelow's sedge) locally extensive .... **Dwarf Heath – Graminoid Alpine Ridge** (p. 180)

22. Sparsely vegetated, nearly vertical rock faces. 23

22. Summits or upper slopes, more extensive and not cliffs. 24

23. Sparsely vegetated circumneutral rock faces, usually small in area; composition variable but includes some species indicative of higher pH, such as shrubby cinquefoil, ebony sedge, rock whitlow-cress, lance-leaved draba, bird's-eye primrose, smooth woodsia, etc. .... **Boreal Circumneutral Open Outcrop** (p. 172)

23. Cliff faces with sparse vegetation and without circumneutral indicator species; granitic or other acidic bedrock. .... **Acidic Cliff – Gorge** (p. 164)

24. Vegetation a mosaic of tree or tall shrub islands within the predominant heath shrub cover; tree cover 5-25%; tall shrub and stunted tree patches form 10% - 70% cover and typically include mountain holly; at moderate elevations (usually > 2000') inland, or at lower elevations near the coast .... **Rocky Summit Heath** (p. 190)

24. Patches of heath shrubs and herbs over bare rock, almost all under 0.5 m tall, with shrub and tree cover (>1 m tall) <10%. 25

25. Bald summits or upper slopes of inland hills or mountains, > 1800', with patchy vegetation dominated by crowberry and/or alpine bilberry; herbs, other than three-toothed cinquefoil, are subalpine species such as highland rush, mountain sandwort or dwarf rattlesnake root .... **Crowberry – Bilberry Summit Bald** (p. 184)

25. Bald summits, mostly near the coast, without crowberry and bilberry (broom-crowberry may be present); a mixture of dwarf shrubs and herbs, including lower-elevation species such as common hairgrass, Rand's goldenrod, pinweed, orange-grass, or smooth sandwort .... **Three-toothed Cinquefoil – Blueberry Low Summit Bald** (p. 182)

OPEN WETLANDS KEY

1. Tidal wetlands. 2

1. Non-tidal wetlands. 6

2. Saltmarshes: vegetation varies, saltmeadow cordgrass and/or smooth cordgrass present and often a major component; dominants also may include black-grass, sedges, etc. 3
2. Brackish to freshwater marshes; saltmeadow cordgrass and/or smooth cordgrass not prominent. 4

3. Saltmarshes with saltmeadow cordgrass and/or smooth cordgrass totaling >35% cover, or with black-grass >35% cover; most other species clearly less abundant (not including low-growing species like goosetongue that may be extensive beneath the graminoids) .... **Spartina Saltmarsh** (p. 260)
3. Saltmarsh cordgrasses and/or black-grass are not strongly dominant, "canopy" vegetation more a mixture of graminoids and forbs; chair-maker's rush typically present, and may be dominant .... **Mixed Graminoid – Forb Saltmarsh** (p. 244)

4. Shrubs dominant .... **Alder Floodplain** (p. 204)
4. Herbaceous plants dominant. 5

5. Brackish tidal setting; vegetation a mix of tall graminoids and rosette-forming forbs; freshwater cordgrass and/or wire rush usually present; obligate freshwater species such as cardinal flower, sweet flag, and pickerelweed absent .... **Brackish Tidal Marsh** (p. 210)
5. Freshwater, near the upstream end of the tidal reach; vegetation graminoid-dominated, with wild rice and/or softstem bulrush typical; some obligate freshwater plants such as pickerelweed, cardinal flower, and/or sweet flag present .... **Freshwater Tidal Marsh** (p. 228)

6. Submerged or floating-leaved aquatic vegetation; emergent plants, if present, are mostly those that die back below the water in autumn. 7
6. Not as above; lakeshores, rivershores, bogs, fens, marshes that are not permanently underwater or, if so, are vegetated with emergent vegetation that remains through the winter. 12

7. Plants with emergent leaves dominant. 8
7. Plants with floating or submerged leaves dominant. 10

8. Pickerelweed dominant among the emergent species .... **Pickerelweed – Macrophyte Aquatic Bed** (p. 252)
8. Graminoids dominant among emergent species. 9

9. Bulrushes and/or bayonet rush dominant. .... **Bulrush Bed** (p. 212)
9. Cattails dominant. .... **Cattail Marsh** (p. 214)

10. Some plants indicative of higher pH waters present, e.g. tapegrass, common waterweed, water stargrass, white water crowfoot, Robbins' pondweed, alpine pondweed, Vasey's pondweed, and straight-leaved pondweed; water-shield, pipewort, water lobelia, and pickerelweed absent or virtually so .... **Circumneutral – Alkaline Water Macrophyte Suite** (p. 218)

10. Alkaline indicators absent; vegetation usually includes pipewort, water lobelia, pondweed species other than those listed above, and/or water-lilies.............. 11

11. Vegetation mostly floating-leaved plants and/or submerged plants with aquatic stems floating in the water column; water-lilies and pondweeds typically dominant; depth varies. .... **Water-lily – Macrophyte Aquatic Bed** (p. 248)

11. Vegetation dominated by submerged plants (flower stalks may protrude) with leaves mostly on the substrate, often in rosettes; pipewort and water lobelia characteristic; water depth usually < 1 m .... **Pipewort – Water Lobelia Aquatic Bed** (p. 262)

12. Gently-sloping sandy or gravelly pondshores (usually of small ponds in outwash basins) of Southern Maine where natural water levels usually drop by late summer, exposing progressive bands of vegetation. From the upland edge, a band of shrubs is followed by three-way sedge, bayonet rush, and narrow-leaved goldenrod in the upper shore zone, then pipewort or other submerged aquatic species in the most flooded zone. Golden pert and meadow beauty are usually present as indicators (see couplet 22 for herbaceous lakeshore vegetation that is not in distinct bands .... **Three-way Sedge – Goldenrod Outwash Plain Pondshore** (p. 250)

12. Vegetation not obviously banded around a central pond whose water level drops through the season... 13

13. *Sphagnum* and dwarf shrub dominated vegetation forming a thin layer over sloping bedrock (sometimes talus); only in subalpine or extreme maritime zones; peat may not remain saturated through the summer (may not be true wetlands)... 14

13. Basin wetlands, or vegetation along lakeshores or rivershores; widespread..... 15

14. Bog-like vegetation forming a "blanket" on slightly sloping bedrock at the immediate coast, from Washington County east .... **Heath – Crowberry Maritime Slope Bog** (p. 242)

14. Bog-like vegetation in subalpine setting, on steep slopes over bedrock or talus of mountain slopes (see couplet #24 for basin wetlands on summit plateaus) .... **Heath – Lichen Subalpine Slope Bog** (p. 270)

15. Vegetation of rivershores or lakeshores, with the substrate primarily mineral rather than organic; mostly in linear bands following the shoreline; may be flooded seasonally, but out of the water for most of growing season. 16

15. Wetlands in basins or along broad drainages, with organic soils or with an organic layer over mineral substrate; vegetation often covering a large part of the substrate; saturated through all or most of the year. 24

16. On rivershores. 17

16. On lakeshores or pondshores. 23

17. Shrubs, mostly >1 m tall, predominate. 18

17. Shrubs < 1 m tall, or mixture of shrubs and herbaceous plants, predominate 19

18. Alder is dominant or co-dominant; virgin's bower is present .... **Alder Floodplain** (p. 204)

18. Dogwood and willows predominate .... **Dogwood – Willow Shoreline Thicket** (p. 254)

19. Mixed shrub-herb vegetation on sloping, eroding river shores where substrate is constantly saturated by groundwater seepage; calciphilic fen species present, e.g., grass-of-Parnassus, Kalm's lobelia, sticky false asphodel; bryophyte layer at least locally well developed, with species other than *Sphagnum* dominant .... **Circumneutral Riverside Seep** (p. 256)

19. Vegetation of almost-flat rivershores or Lakeshore Sand/Cobble Beaches (see couplet 23), substrate gravelly to sandy and not constantly saturated to the surface by seepage; bryophytes sparse or absent. 20

20. Tall graminoids and forbs dominant, forming a dense meadow; shrubs, if present, are rarely taller than the herbs; bluejoint, spotted joe-pye weed, and flat-topped white aster characteristic. .... **Bluejoint Meadow** (p. 272)

20. Vegetation more sparse and not dominated by bluejoint. 21

21. Beach heather a locally prominent dwarf shrub; islands of gray birch usually present; little bluestem one of the more common graminoids; documented in Maine only from Saco River drainage. .... **Hudsonia River Beach** (p. 258)

21. Beach heather and little bluestem are absent or incidental. 22

22. Sand cherry and roses the dominant dwarf shrubs, Laurentian/Cordilleran plants such as Huron tansy, alpine sweet broom, and alpine milk-vetch often present; in Maine documented only on far northern rivers ....**Sand Cherry – Tufted Hairgrass River Beach** (p. 234)

22. Tufted hairgrass, twisted sedge, or other forbs and graminoids are dominant; statewide ....**Twisted Sedge Cobble Rivershore** (p. 224)

23. Sparse vegetation on lakeshore cobble or sand beaches; silverweed typical, beach heather, or golden heather may be present .... **Lakeshore Sand/Cobble Beach** (p. 232)

23. Patchy or sparse vegetation on subalpine gravelly pondshores .... **Mountain Alder – Bush-honeysuckle Subalpine Meadow** (p. 198)

24. Bogs and fens: substrate is accumulated peat (undecayed to partially decayed), usually > 0.5 m deep and with extensive Sphagnum on the surface, sometimes floating over water; constantly saturated. 25

24. Marshes: substrate is mineral soil, often with a surface layer of well-decomposed organic matter (peat, typically sedge-derived, may be > 0.5 m thick but is generally less); *Sphagnum* may be present but does not form an extensive deposit; some remain saturated, but many dry out for at least part of the growing season. 37

25. Alpine or subalpine basin peatlands, near or above treeline .... **Cotton-grass – Heath Alpine Bog** (p. 208)

25. Lower elevation wetlands. 26

26. Bryophyte-dominated, often in raised peatlands; substrate wet and unstable; vascular plants usually <25% cover and limited to low-growing species e.g. cranberries, horned bladderwort, and white beak-rush; very dwarfed leatherleaf or other ericads may be present. .... **Bog Moss Lawn** (p. 246)

26. Shrub and/or herb cover more extensive; vegetation not primarily bryophytes 27

27. Shrub cover (including dwarf shrubs) exceeds graminoid cover. 28

27. Graminoid cover equals or exceeds shrub cover, including dwarf shrubs. 33

28. Vegetation dominated by tall shrubs (mostly > 1.5 m); mountain holly, alder, and wild calla characteristic; standing water usually present among hummocks of *Sphagnum*; in peatlands often at the upland/peatland interface .... **Mountain Holly – Alder Woodland Fen** (p. 276)

28. Vegetation dominated by herbs or shrubs mostly under 1m tall, typically including abundant ericads; mountain holly and alder may be present but are not dominant. 29

29. Shrubby cinquefoil a prominent shrub, and other circumneutral indicators present, e.g. livid sedge, grass-of-Parnassus, and/or Kalm's lobelia .... **Shrubby Cinquefoil – Sedge Circumneutral Fen** (p. 216)
29. Shrubby cinquefoil absent (or scarce) and other circumneutral indicators lacking. 30

30. Shrubs relatively tall (often >1 m) with sweetgale, hardhack, and/or meadowsweet prominent; leatherleaf often present but other ericads not abundant; typically in standing shallow water without a continuous *Sphagnum* carpet. .... **Sweetgale Mixed Shrub Fen** (p. 268)
30. Vegetation dominated by ericads (though sweetgale often present), usually well under 1 m tall; bogs and fens on well developed *Sphagnum* carpets. 31

31. Low bog vegetation with dwarf huckleberry locally dominant (often patchy), >20% cover; black crowberry and/or deer-hair sedge also present though not necessarily at high cover; graminoid cover may be relatively high (often > 25%); coastal (up to 15 miles or so inland)..... **Huckleberry – Crowberry Bog** (p. 240)
31. Other shrubs dominant, or if dwarf huckleberry dominant then neither black crowberry nor deer-hair sedge present; both coastal and inland...................... 32

32. Weakly minerotrophic fen conditions (vegetation mostly in contact with the water table), with leatherleaf (or a combination of leatherleaf with bog rosemary and/or sweetgale) the dominant shrub; other shrubs and sedges mixed in; graminoid cover variable.................... **Leatherleaf Boggy Fen** (p. 236)
32. Ombrotrophic bog conditions (vegetation mostly raised above the water table), with sheep laurel, Labrador tea, and/or rhodora together more abundant than leatherleaf and sweetgale; graminoid cover rarely exceeds 15% .... **Sheep Laurel Dwarf Shrub Bog** (p. 226)

33. Sedge-dominated lawns, deer-hair sedge and/or coast sedge characteristic; centrally located on raised (ombrotrophic) plateau bogs within about ten miles of the coast, Hancock County east. .... **Deer-hair Sedge Bog Lawn** (p. 222)
33. Fen (minerotrophic) setting, areas sometimes transitional to bogs, but not the central lawn of a coastal plateau bog. 34

34. Fens with circumneutral indicators e.g. livid sedge, yellow sedge, grass-of-Parnassus, Kalm's lobelia, and/or shrubby cinquefoil, usually with more than one indicator species .... **Shrubby Cinquefoil – Sedge Circumneutral Fen** (p. 216)
34. Circumneutral indicator species absent or virtually so. 35

35. Vegetation dominated by tall sedges: dominant species are slender sedge, beaked sedge, and/or lake bank sedge; or rarely tussock sedge or inflated sedge .... **Mixed Tall Sedge Fen** (p. 274)

35. Slender sedge, beaked sedge, lake bank sedge, and inflated sedge not the dominant sedges, vegetation mostly lower. 36

36. Carpets of low sedges over very wet *Sphagnum* substrate, with mud sedge, podgrass and white beak-rush characteristic; dwarf or creeping shrubs may be mixed with the sedges but are sparse (<20% cover). .... **Low Sedge – Buckbean Fen Lawn** (p. 238)

36. Sedges mixed with dwarf shrubs, usually >20% shrub cover; white beak-rush often present, but mud sedge and podgrass not; dominant sedges include few-seeded sedge, coast sedge, Michaux's sedge white beak-rush, and/or narrow-leaved cotton-grass .... **Sedge – Leatherleaf Fen Lawn** (p. 264)

37. Tussock sedge is the dominant herbaceous species (>30% cover, and usually >50%); shrub cover usually <30%; standing water between hummocks for much of season .... **Tussock Sedge Meadow** (p. 266)

37. Herb component dominated by species other than tussock sedge. 38

38. Shrub cover exceeds herb cover. 39

38. Herb cover exceeds shrub cover. 41

39. Mixture of herbs and shrubs; dominant shrubs are not alder, sweetgale, or ericads, but are more often buttonbush or bog willow .... **Mixed Graminoid – Shrub Marsh** (p. 230)

39. Alder, sweetgale, and/or ericads are the dominant shrubs. 40

40. Alders strongly dominant, >20% cover; ericads absent or virtually so .... **Alder Shrub Thicket** (p. 206)

40. Sweetgale and/or heath shrubs dominant, alders clearly subordinate .... **Sweetgale Mixed Shrub Fen** (p. 268)

41. Bluejoint dominates herb component, >50% cover; shrubs <25% cover .... **Bluejoint Meadow** (p. 272)

41. Mixture of herbs and shrubs (herbs 25-95%, shrubs 0-70% cover), without dominance of tussock sedge, bluejoint, or alder .... **Mixed Graminoid – Shrub Marsh** (p. 230)

Wooded Uplands

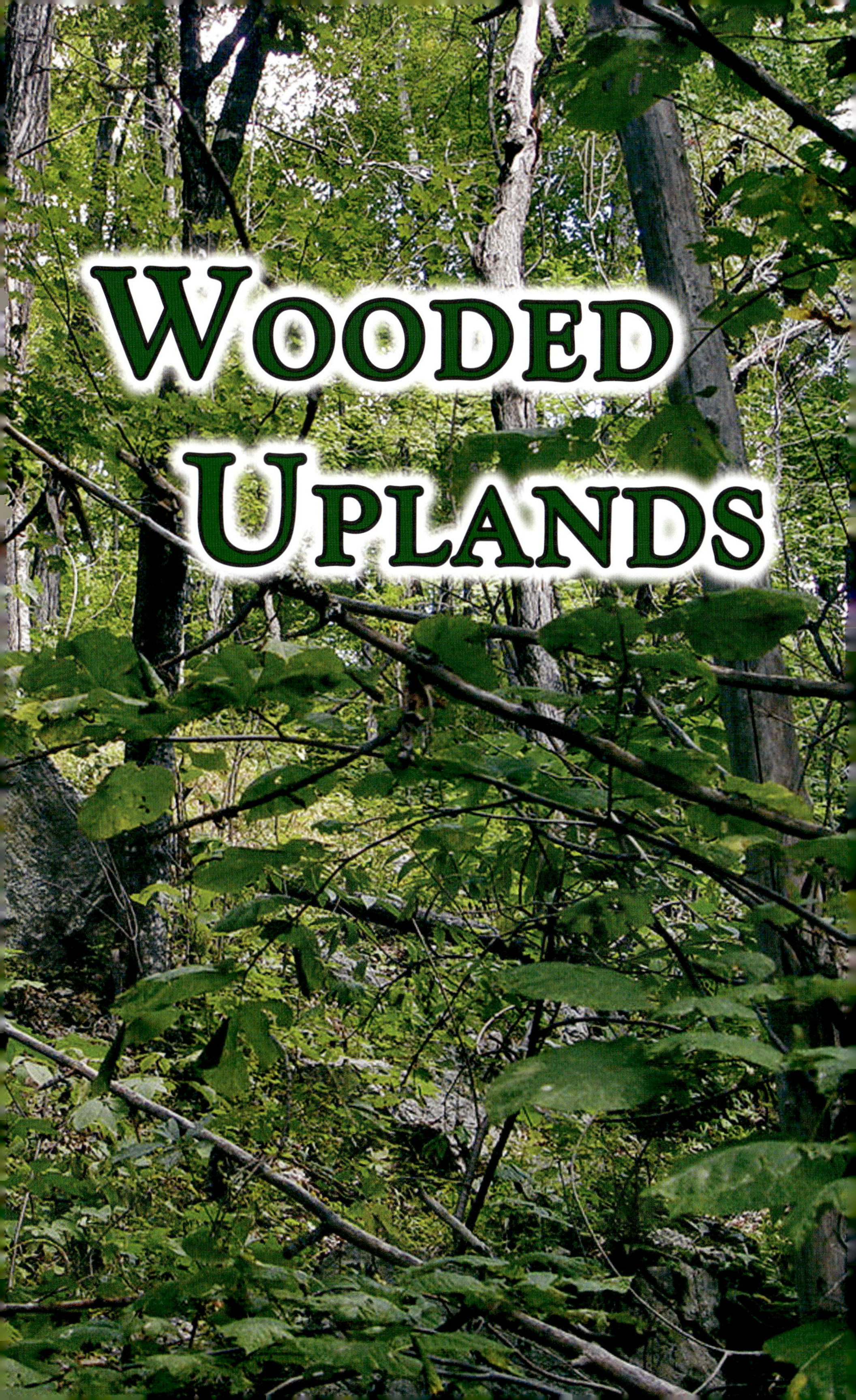

Birch - Oak Rocky Woodland

State Rank S3

Community Description

These community types are partial canopy deciduous woodlands or patches of woodland among talus areas. Overall canopy closure may be <25% when the open areas are included. Paper birch, red oak, and/or yellow birch are dominant. Sugar maple, if present, is not abundant. Marginal wood-fern, rock polypody, and poison ivy are characteristic of the herb layer, which is best developed in open patches. Vegetation is generally very patchy, developing in pockets among the rocks.

Soil and Site Characteristics

Sites typically occur on dry, acidic talus substrates with various aspects. Documented elevations are from nearly sea level to about 1500'.

Diagnostics

Sites occur on talus substrate; conifer cover can be one-fourth of the canopy, ironwood is absent and birches are common.

Striped Maple

Similar Types

Ironwood - Oak - Ash Woodlands are similar. They may have a high cover of red oak, but also include ironwood and sometimes sugar maple. Birches are less common, and the herbaceous flora includes rich site indicators that are absent from this type. Other community types dominated by red oak occur on thin soil over bedrock or on deeper soils. Oak - Pine Woodlands, the most similar, typically include bracken fern and lowbush blueberry, which are rarely (or only sparsely) found in talus woodlands.

Location Map

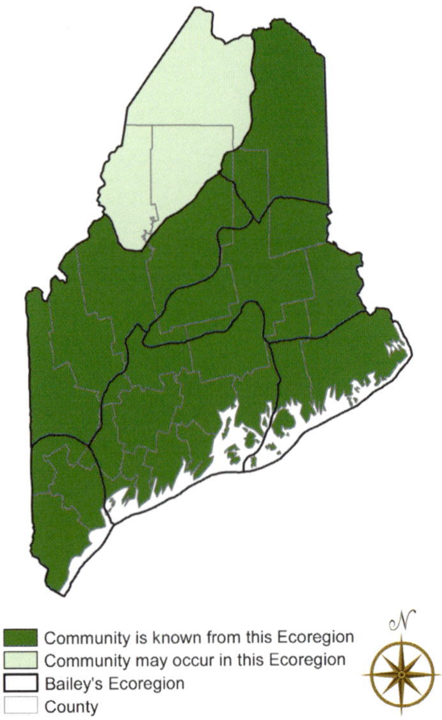

Paper Birch

Conservation, Wildlife, and Management Considerations

Talus woodlands receive little human use because of their inaccessibility and low timber value; however, areas at the base of talus slopes that receive water and nutrients from above sometimes have enough large trees to make logging economical. Conservation of these sites should include the range of talus forest cover, from the base of the slope on up, with a buffer of adjacent forest cover.

South facing occurrences of this type in the southern part of the state may have provided historical habitat for the timber rattlesnake, which is believed to have been extirpated from Maine.

Distribution

Not well documented. Extends westward from Maine, and possibly in other directions.

Landscape Pattern: Small Patch

Characteristic Plants

These plants are frequently found in this community type. Those with an asterisk are often diagnostic of this community.

Canopy

Big-toothed aspen
Blue birch*
Paper birch*
Red maple
Red oak*
Striped maple
Yellow birch*

Sapling/shrub

Mountain maple*
Striped maple*
Yellow birch*

Dwarf Shrub

Lowbush blueberry
Poison-ivy

Herb

Big-leaved aster
Common hairgrass
Marginal woodfern
Rock polypody
Wild sarsaparilla

Bryoid

Dicranum moss
Large hair-cap moss
Pincushion moss

Associated Rare Plants

Purple clematis

Examples on Conservation Lands You Can Visit

- Acadia National Park – Hancock Co.
- Caribou Mountain, Donnell Pond Public Lands – Hancock Co.
- Horse Mountain, Baxter State Park – Penobscot Co.
- Little Kineo Mountain Public Lands –Piscataquis Co.

Black Spruce Barren

State Rank S2

Community Description

Black spruce, sometimes mixed with red spruce (and hybrids between the two), forms a variable canopy over heath shrubs, mosses, and lichens. Canopy closure is usually 25-60%, occasionally greater. Associates include balsam fir, larch, and white spruce. The shrub/sapling layer is usually well developed (>25%) and may be very dense; mountain holly and wild-raisin are characteristic shrubs. Dwarf shrubs, herbs, and regenerating trees cover the ground layer, with heath shrubs prominent. Bryoid cover is close to 100%; mosses dominate many areas, but in drier sites reindeer lichens may be abundant. Openings with blueberry and lichens may occur within the barrens. Wetter sites may approach concepts of boreal 'muskeg', though they are not true peatlands.

Soil and Site Characteristics

Sites occur on flat to rolling terrain in cold lowlands (usually <1200' elevation) characteristic of nutrient-poor or highly acidic sites. The sandy to clayey soils over till can vary (even within the same site) from well drained to very poorly drained, reflecting the microtopography. Wet areas may have an organic layer of up to 25 cm over the mineral soil. Sites often contain evidence of fire.

Diagnostics

A somewhat open spruce canopy, usually including black spruce, occurs with prominent shrub and herb layers (typically > 25% cover each). Heath shrubs may form dense thickets. Sites are underlain by mineral soil sometimes overlain by a thin organic layer.

Similar Types

Spruce - Fir - Broom-moss Forests and Maritime Spruce - Fir Forests have far lower shrub and herb cover and generally lack black spruce. Black Spruce Woodlands and Red Spruce - Mixed Conifer Woodlands share many species but are on bedrock rather than mineral soil. Black Spruce Woodlands

Location Map

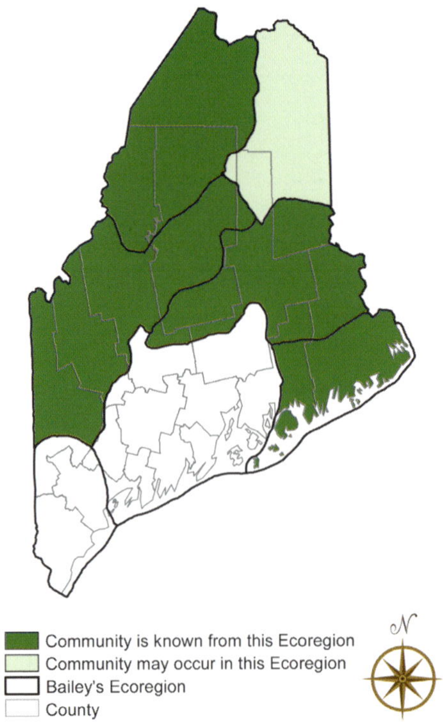

Black Spruce Branch

in particular may co-occur with this type, and further work is needed to distinguish the two types. Spruce - Fir - Cinnamon Fern Forests have more closed canopies with red spruce dominant and lack the abundant heath shrubs. Spruce - Larch Wooded Bogs are on deeper (>1 m) peat deposits.

Conservation, Wildlife, and Management Considerations

The extent of this type in Maine has not been well documented, nor have site and successional relationships. This type appears mostly in areas that have burned within the last century or so and that also have a cool microclimate. Several known sites are in conservation ownership.

Birds that may nest in this habitat include the Lincoln's sparrow, fox sparrow, palm warbler, and blackpoll warbler.

Distribution

Eastern and far northern Maine, extending into Canada; part of the New England - Adirondack Province.

Landscape Pattern: Large Patch

Characteristic Plants

These plants are frequently found in this community type. Those with an asterisk are often diagnostic of this community.

Canopy
Balsam fir
Big-toothed aspen*
Black spruce*
Red spruce*
White spruce
White pine

Sapling/shrub
Balsam fir*
Gray birch*
Mountain holly
Red spruce*
White pine*

Dwarf Shrub
Labrador tea*
Lowbush blueberry*
Rhodora*
Sheep laurel*

Herb
Bracken fern
Bunchberry
Canada mayflower
Starflower
Wild sarsaparilla*

Bryoid
Reindeer lichen
Sphagnum mosses*
Wavy broom-moss

Examples on Conservation Lands You Can Visit

- St. John River Preserve – Aroostook Co.

Black Spruce Woodland

State Rank S2

Community Description

These are boreal, open canopy woodlands in which black spruce is strongly dominant, though it is sometimes mixed with red spruce or white pine. Trees are stunted and canopy closure is usually less than half, although occasionally a site may have a more closed canopy (~85%). The shrub layer often has smaller black spruce and a scattering of evergreen and deciduous shrubs. The herb layer is usually extensive (>50% cover) and strongly dominated by heath shrubs. Herbs comprise <5% of the herb layer cover; bracken fern and bunchberry are typical. The bryoid layer is fairly well developed (>15% cover) and may be prominent, with abundant reindeer lichens. Peat mosses may be present in low pockets but are not abundant.

Soil and Site Characteristics

Sites are upland or transitional wetland-upland and are on flat or slightly sloping ground. Substrate is bedrock with only a thin patchy layer (<12 cm) of sandy soil or poorly decomposed organic duff. At some sites, the sandy substrate is up to 30 cm over till. Soils are acidic (pH ~5.0) and well drained; moisture may accumulate in pockets of the substrate. Sites occur on cool and moist microclimates, mostly at the immediate coast or in northwest Maine at up to ~1200'.

Bunchberry in Fruit

Diagnostics

Black spruce is strongly dominant, with canopy closure 25-60%; the dwarf shrub component of the herb layer is well developed and dominated by heaths.

Location Map

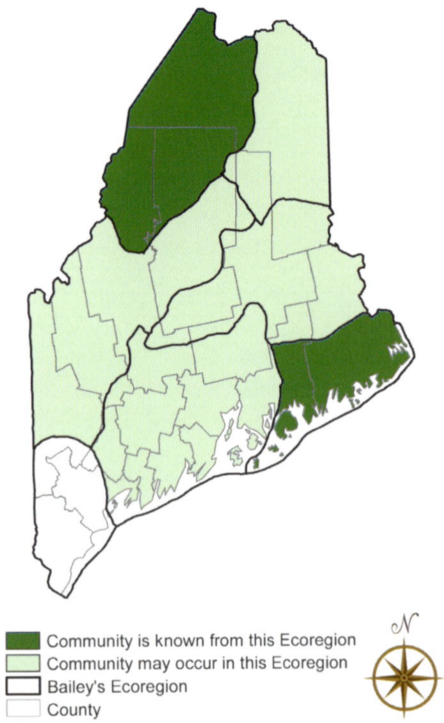

Black Huckleberry and Black Crowberry

Characteristic Plants

These plants are frequently found in this community type. Those with an asterisk are often diagnostic of this community.

Canopy

Black spruce*
Red spruce

Sapling/shrub

Black huckleberry*
Black spruce*
Gray birch*
Wild-raisin*

Dwarf Shrub

Black huckleberry*
Lowbush blueberry*
Sheep laurel*

Bryoid

Dicranum moss
Reindeer lichen

Associated Rare Animals

Purple lesser fritillary

Similar Types

Spruce - Heath Barrens have higher shrub cover and generally occur on deeper, poorly drained mineral soil rather than on bedrock, but the two types may co-occur, and further work is needed to distinguish the two. Spruce - Larch Wooded Bogs can have very similar species composition but grow on saturated peat moss, as part of a peatland. Labrador Tea Talus Dwarf-shrublands can have similar species but have <25% tree cover.

Conservation, Wildlife, and Management Considerations

Sites appear to receive little human impact other than some light recreational use. Some sites may be of fire origin.

A variety of conifer-nesting birds may use this habitat, including blackpoll warbler, bay-breasted warbler, and black-backed woodpecker. The only modern occurrence in Maine of the rare purple lesser fritillary butterfly is found within the northernmost example of this community type.

Distribution

Known sites occur on either Downeast islands or extreme northwest Maine. More broadly, this type occurs in the Laurentian Mixed Forest Province, extending northward and eastward.

Landscape Pattern: Small Patch

Chestnut Oak Woodland

State Rank S1

Community Description

This partial canopy woodland type is dominated by chestnut oak in association with other oak species, white pine, and rarely shagbark hickory. The sapling layer typically includes small oaks as well as various shrub species. Beneath the trees, a mixture of low heath shrubs and herbs covers much of the ground surface, with some bare rock patches. Bracken fern and lowbush blueberry are most common. Bryoids are virtually absent.

Soil and Site Characteristics

These woodlands occupy dry ridges and south facing slopes on thin, excessively well drained and stony soils. Known sites are in extreme southern Maine only, on granite-syenite bedrock.

Red Oak Twig with Acorns

Diagnostics

These woodlands (25 to 65% canopy cover) are dominated by chestnut oak.

Similar Types

Other hardwood forest types are dominated by trees other than chestnut oak.

Conservation, Wildlife, and Management Considerations

The single documented site in Maine is in public ownership, with attention given to conserving this rare type. In other parts of the range of this type, fire appears to have played a role in preventing the invasion of fire sensitive

Location Map

Chestnut Oak Woodland

hardwood trees and shrubs. The suppression of fire could therefore result in the gradual conversion of these woodlands to a more mesic oak-pine type. In Maine, chestnut oak does not seem to regenerate well under its own canopy.

Mature occurrences of this type offer excellent potential sites for cavity dwellers such as the southern flying squirrel.

Distribution

Characteristic of the Eastern Broadleaf Forest Province, extending southward and southwestward from Maine and only barely reaching into the state.

Landscape Pattern: Small to Large Patch in Maine; matrix-forming further south.

Examples on Conservation Lands You Can Visit

- Mt Agamenticus – York Co.

Characteristic Plants

These plants are frequently found in this community type. Those with an asterisk are often diagnostic of this community.

Canopy
Black cherry
Chestnut oak*
Red maple
Red oak
White oak
White pine

Sapling/shrub
Black huckleberry
Maple-leaved viburnum
Meadowsweet
Shadbush
Witch-hazel

Dwarf Shrub
Lowbush blueberry*
Velvet-leaf blueberry

Herb
Bracken fern*
Common hairgrass
Partridgeberry
Rough-leaved ricegrass
Wild-oats
Wintergreen
Woodland sedge

Bryoid
Hair cap moss

Associated Rare Plants

Chestnut oak
Flowering dogwood

Associated Rare Animals

Red-winged sallow
Whip-poor-will

Aspen - Birch Woodland/Forest Complex

State Rank S5

Community Description

This complex of post fire associations of aspen, birch, and other species can occur as open canopy woodlands, as closed forest, or, in very exposed areas, as stunted, dense shrublands. Paper birch, big-toothed aspen, quaking aspen, and red maple are the most common trees; other trees may be common at some sites. The shrub layer, usually <50% cover except in the shrubland variant, is variable; shadbush (at low cover) and gray birch (locally abundant) are the most consistent species. Herb cover is higher under more open canopies, where more light reaches the ground. Many sites have patches of lowbush blueberry or black huckleberry; bracken fern is the most characteristic herbaceous species. The bryoid layer is sparse.

Soil and Site Characteristics

Occurring in various settings, nearly all sites are post-fire and/or or post-harvest and typically occur on nutrient-poor soils. The mineral soils are usually <25 cm deep, and some stands occur on thin glacial till or bare granite. Documented examples all have well drained to excessively drained soils with pH 5.0-5.4. In time, most sites will transition to one of several matrix-forming forest types such as northern hardwood forest or spruce – northern hardwood forest. The shrubland form occurs on exposed low elevation summits (900-1200') and is currently documented from Acadia National Park and a few other sites Downeast.

Yellow Warbler

Diagnostics

The canopy is dominated by early successional deciduous trees (poplars, birches, red maple). Conifers and red oak may be present but are not dominant. Lowbush blueberry and bracken fern are usually present below. Sites are on thin mineral soil over till or bedrock.

Similar Types

Other deciduous forest/woodland types have greater amounts of red oak (Oak - Birch Talus Woodland) or have northern hardwood species dominant (Beech - Birch - Maple Forest). Some Oak - Pine

Location Map

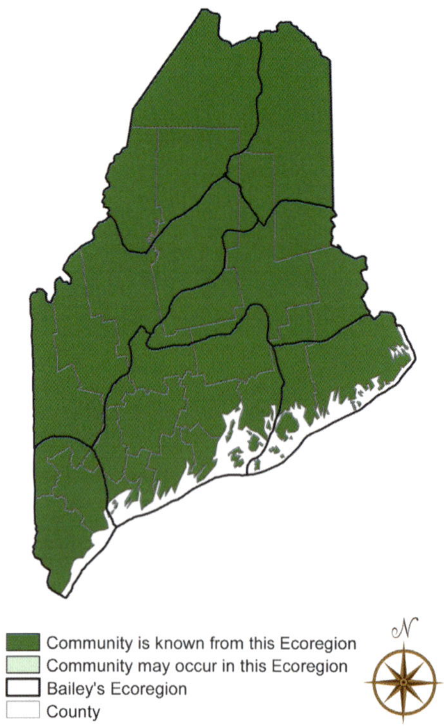

Woodlands can be strongly deciduous and can resemble this type, but red oak cover exceeds that of birches, aspen, and red maple combined.

Early Successional Forest

Conservation, Wildlife, and Management Considerations

Aspen - birch stands originating after heavy harvesting are frequent in Maine. Exemplary sites, however, are those that originated after fire or other natural disturbance, since those sites best reflect the composition and condition of natural processes. Natural succession dictates that the aspen - birch dominance will not be long lived, and in time, most sites will transition to one of several matrix-forming forest types such as northern hardwood forest or spruce – northern hardwood forest. Conservation efforts should not attempt to maintain the aspen - birch dominance, but rather to let natural succession proceed. Good examples are known from public lands and private conservation lands.

Ruffed grouse commonly use young stands with dense sapling cover and little herbaceous cover. Snags remaining after wildfire in northern or eastern occurrences of this community type provide foraging strata for three-toed woodpeckers and perches for the rare olive-sided flycatcher.

Distribution

Statewide, less common in southern Maine. Extends eastward, westward, and northward from Maine.

Landscape Pattern: Large Patch, or temporary matrix in successional areas.

Characteristic Plants

These plants are frequently found in this community type. Those with an asterisk are often diagnostic of this community.

Canopy
Big-toothed aspen*
Paper birch*
Quaking aspen*
Red maple*
Red spruce*

Sapling/shrub
Balsam fir*
Red maple*
Striped maple*
Wild-raisin*

Dwarf Shrub
Black huckleberry*
Lowbush blueberry*

Herb
Bracken fern*
Canada mayflower
Sheep fescue*

Bryoid
Dicranum moss
Large hair-cap moss
Reindeer lichen

Associated Rare Animals

Olive-sided flycatcher

Examples on Conservation Lands You Can Visit

- Acadia National Park – Hancock Co.
- Black Mountain, Donnell Pond Public Lands – Hancock Co.
- Deadwater Brook, Appalachian Trail – Piscataquis Co.

Enriched Northern Hardwoods Forest

State Rank S3

Community Description

Sometimes referred to as "cove forests," these closed canopy forests are dominated by sugar maple, with beech and/or yellow birch subordinate. Basswood and white ash are typical indicators but are not necessarily abundant, and they are often absent in northwest Maine. The shrub layer is usually sparse and dominated by saplings of the canopy species. The lush herb layer may contain species that are strong indicators of this forest type, such as maidenhair fern, blue cohosh, Dutchman's breeches, grape fern, spring beauty, and silvery spleenwort. These and many rare species are characteristic of forests with relatively nutrient rich soils. Bryoids are virtually absent.

Soil and Site Characteristics

Sites occur on sheltered hillsides, ravines, stream drainages, or slope bases where nutrients accumulate, often over calcium-bearing bedrock. Slopes often grade from moderate to flat as these forests straddle the base of a hillslope. Small drainage channels may occur in the lower portions, maintaining saturated soils over at least part of the site. Forests upslope often grade to typical northern hardwood forest (Beech - Birch - Maple Forest).

Squirrel Corn

Diagnostics

Sugar maple is dominant or co-dominant; white ash and basswood are present in central and southern Maine sites (basswood is absent in northwest Maine); conifers and oaks sparse or absent. Silvery spleenwort, maidenhair fern, blue cohosh, grape fern, Christmas fern, Braun's holly fern, and Dutchman's breeches are good herb indicators.

Similar Types

The more typical and extensive Beech - Birch - Maple Forest is dominated by beech, birches, and sugar maple, occurs on less nutrient rich sites, and lacks substantial ash. Sugar Maple Forests contain some moderate rich site indicators, e.g., zig-zag goldenrod and

Location Map

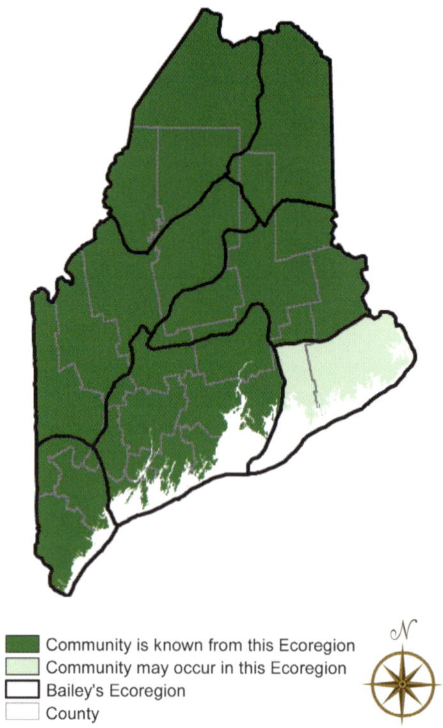

Enriched Northern Hardwoods Forest

Solomon's seal, but lack the abundance and variety of true rich site indicators present in this type.

Conservation, Wildlife, and Management Considerations

There are many known mature occurrences of Maple - Basswood - Ash Forests in the state, most with a history of harvesting. However, the market pressures for hardwoods have recently led to heavy cutting of several sites. Typical sites where this community occurs are naturally small and should be buffered from surrounding forest uses. Since this natural community type is most often an inclusion within larger northern hardwood forests, many of the species using northern hardwood forests will also use this type.

Distribution

New England - Adirondack Province and Laurentian Mixed Forest Province, with many of the known sites concentrated in the western mountain region of the state.

Landscape Pattern: Small Patch, typically occur as 2-20 acres within a larger matrix of northern hardwood forests.

Characteristic Plants

These plants are frequently found in this community type. Those with an asterisk are often diagnostic of this community.

Canopy

American beech
Basswood*
Ironwood
Sugar maple*
White ash

Sapling/shrub

Alternate-leaved dogwood*
American beech
Striped maple
Sugar maple*
White ash
Yellow birch

Herb

Blue cohosh
Christmas fern
Doll's eyes
Grape fern
Maidenhair fern
Round-leaved violet
Silvery spleenwort*
Wild sarsaparilla

Associated Rare Plants

American ginseng
Broad beech fern
Goldie's wood-fern
Male fern
Pale jewel-weed
Squirrel-corn
Wild ginger
Wild leek

Examples on Conservation Lands You Can Visit

- Albany Notch, White Mountain National Forest - Oxford Co.
- Hastings Mountain, White Mountain National Forest - Oxford Co.
- Miles Notch, White Mountain National Forest - Oxford Co.
- Peter Mountain, White Mountain National Forest - Oxford Co.
- Square Dock Mountain, White Mountain National - Oxford Co.

Hemlock Forest

State Rank S4

Community Description

This closed canopy forest type is dominated by hemlock (>50% cover) or, less often, hemlock is co-dominant with red spruce, red oak, yellow birch, red maple, or sugar maple (very rarely with northern white cedar, near the coast). White pine may be co-dominant in stands that are transitional, with pine giving way to hemlock in time. The conifer canopy allows little light to reach below, and the shrub, herb, and bryoid layers are sparse (each usually <25%, and sometimes absent altogether). Small conifers are present in the herb layer, as well as scattered individuals of typical upland conifer forest plants such as Canada mayflower, starflower, Indian cucumber-root, partridgeberry, wild sarsaparilla, and wintergreen. Graminoids are rarely very apparent. The ground layer is mostly conifer litter, with spotty bryophyte cover.

Soil and Site Characteristics

Hemlock forests are usually on slopes (typically 5-50%) and ravines, with well drained loamy soil. On lower slopes and flats, soils may grade to imperfectly drained. Soils tend to be shallow (<50 cm) and acidic (pH 4.8-5.6). Sites are from sea level to 1200' and often in cool microsites, although aspect varies.

Pin Cushion Moss

Diagnostics

Hemlock is the dominant tree, occasionally co-dominant with red spruce, white pine, or red maple; soils are typically not saturated; sparse herbaceous and bryoid layers.

Similar Types

White Pine - Mixed Conifer Forests have more white pine than hemlock. Mixed examples of this type can be transitional to Beech - Birch - Maple Forests, but hemlock will have >33% cover. Hemlock - Hardwood Pocket Swamps have wetland soils and vegetation, red maple and/or black gum co-dominant with the hemlock, and more well developed shrub, herb, and bryoid layers.

Conservation, Wildlife, and Management Considerations

Demand in the 1700s -1800s for hemlock

Location Map

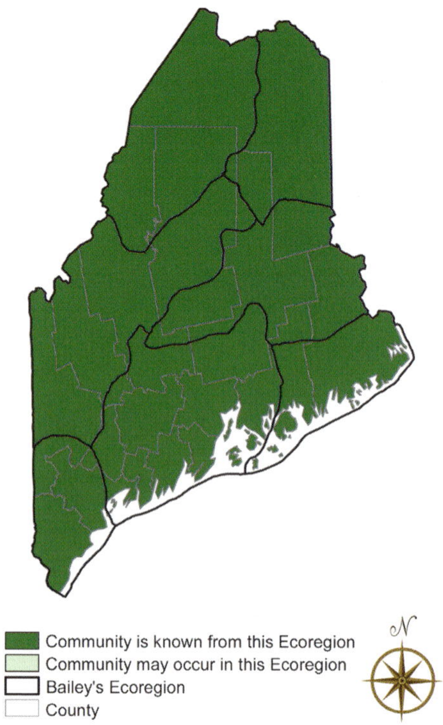

Hemlock Forest

considerably reduced mature, undisturbed examples of this type, yet poor market conditions more recently have caused hemlock to be left in partial harvests; many of these legacy trees are quite old. Some evidence suggests that hemlock is less successful at maintaining itself in the face of human-caused disturbance than are northern hardwoods. Most sites known to be of high ecological quality are in southern and central Maine and lack formal protection. Maintaining the surrounding lands as forest is important in conserving particular stands of this type, particularly given that many known examples are small (<50 acres).

South of Maine, the hemlock woolly adelgid (an introduced insect) has decimated hemlock stands. It has recently been documented in York County, and efforts are underway to limit its impact.

This community type may be used as nesting habitat by a number of coniferous forest

Hemlock Branch with Cone

Characteristic Plants

These plants are frequently found in this community type. Those with an asterisk are often diagnostic of this community.

Canopy
Eastern hemlock*
Paper birch
Red maple
Red oak*
Red spruce*
Sugar maple*
White pine*
Yellow birch*

Sapling/shrub
American beech*

Herb
Canada mayflower

Bryoid
Dicranum moss
Three-lobed bazzania

specialist bird species, such as the yellow-bellied flycatcher, black-throated green warbler, blackburnian warbler, red crossbill, and northern parula.

Distribution

Statewide, less common northward; extends in all directions from Maine.

Landscape Pattern: Small to Large Patch

Examples on Conservation Lands You Can Visit

- Cooper Brook, Appalachian Trail – Piscataquis Co.
- Fourth Machias Lake, Duck Lake Public Lands – Washington Co.
- Little Concord Pond Public Lands – Oxford Co.
- Magoon Pond Public Lands – Penobscot Co.
- North Of Carlton Notch, White Mountain National – Oxford Co.
- Scraggly Lake Public Lands – Penobscot Co
- Squa Pan Mountain, Squa Pan Public Lands – Aroostook Co.

Jack Pine Forest

State Rank S1

Community Description

This closed canopy forest is dominated by jack pine. Black or red spruce and balsam fir are common associates (up to 20% cover and may be more common in the understory), and red pine may be mixed with jack pine in some areas. With the dense canopy, shrubs and herbs are limited in extent. Lowbush blueberry and boreal herbs such as bunchberry and Canada mayflower are typically present. The bryoid layer is very well developed, with an almost continuous carpet of feather-mosses in places. In these forests, jack pine requires natural or human-caused disturbance to stimulate seed germination, and hence to regenerate. In Maine, Jack Pine Forests would eventually succeed to spruce and fir without clearcuts or fire.

Soil and Site Characteristics

These forests are found on flat or rolling terrain at moderate elevations north of 45 degrees latitude. The coarse textured soils are acidic, nutrient poor, and well drained to excessively well drained. In the one area sampled, charcoal was found between the shallow organic layer and the mineral soil horizons.

Canada Mayflower

Diagnostics

These are closed canopy forests with jack pine dominant in canopy; black spruce and/or red pine are common and locally abundant associates.

Similar Types

Jack Pine Woodlands occur on bedrock, have a partial canopy with a well developed shrub or dwarf shrub layer, and are found either near the Downeast coast or on the shores of certain inland

Location Map

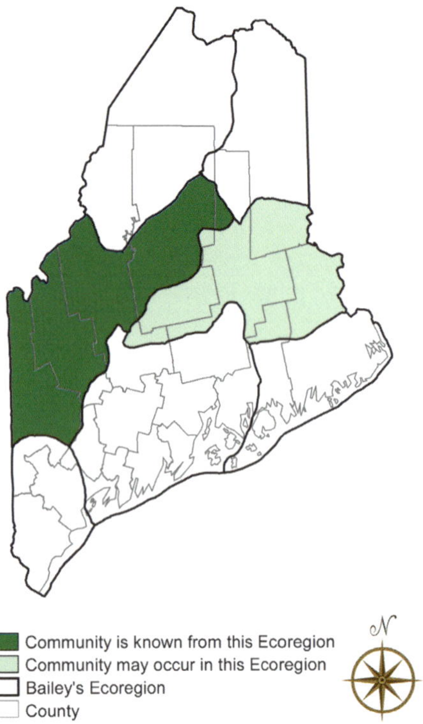

Red-stemmed Moss

lakes. Spruce - Heath Barrens are ecologically similar, with many of the same species of boreal affinities. They may be either open or closed canopy and lack jack pine.

Conservation, Wildlife, and Management Considerations

This is likely a disturbance dependant community, with fire the primary natural disturbance agent. In the limited area from which this type is known in Maine, most of the jack pine forest has been clearcut within the last 25 years and is now growing back as young jack pine forest. In areas where mature jack pine has not been cut, regeneration is primarily red or black spruce. Most of the known jack pine forest is on commercial forest land, with a small proportion on conservation land.

Jack pine forests provide nesting habitat for coniferous forest specialists. They may also be inhabited by uncommon moths such as the western pine elfin, which often uses jack pine as a larval host plant in the Midwest and black spruce in the east.

Characteristic Plants

These plants are frequently found in this community type. Those with an asterisk are often diagnostic of this community.

Canopy

Balsam fir
Black spruce
Heart-leaved paper birch
Jack pine*
Red pine

Sapling/shrub

Black spruce
Dwarf Shrub
Lowbush blueberry
Sheep laurel

Herb

Bracken fern
Bunchberry
Canada mayflower
Creeping snowberry

Bryoid

Red-stemmed moss*

Distribution

Limited to a small portion of northwestern Maine (New England-Adirondack Province); more widespread in the uppermost midwest and adjacent Canada.

Landscape Pattern: Large Patch

Examples on Conservation Lands You Can Visit

- Holeb - Attean Pond Public Lands – Somerset Co.

Jack Pine Woodland

State Rank S3

Community Description

These are open canopy woodlands (<60% closure) in which the dominant tree is always jack pine. Red spruce, black spruce, or white pine are common associates. The canopy trees are generally stunted and have poor growth form. Below the canopy, smaller jack pines are common, with scattered shrubs. The extensive herb layer is mostly heath shrubs that may form a thick tangle in canopy openings. At some maritime sites, black crowberry or mountain cranberry reflect the coastal influence. Herbs are very sparse. The bryoid layer varies from extensive to quite sparse, and is dominated by reindeer lichens.

Soil and Site Characteristics

Sites are dry and occur in cooler climate regions, on gentle slopes with very well drained flats, or on low ridges, usually at <900' elevation. The substrate is a thin layer (<20 cm) of sandy soil or poorly decomposed organic duff over bedrock; some occurrences grow on deeper sands. Substrates are acidic (pH ~5.0) and nutrient poor. Most sites contain evidence of past fire, which in most cases is required to open the serotinous cones of jack pine.

Jack Pine Shoot

Diagnostics

Jack pine is dominant; canopy is generally less than 60% cover.

Similar Types

Black Spruce Woodlands are floristically and structurally similar, with the exception of the dominant canopy species. Both types tend to occur in extremely nutrient poor and often coastal environments. Pitch Pine Woodlands occur in somewhat more temperate settings and are dominated by pitch pine. Jack Pine Forests have a closed canopy and much lower cover of heath shrubs and are known from only one area of western Maine.

Location Map

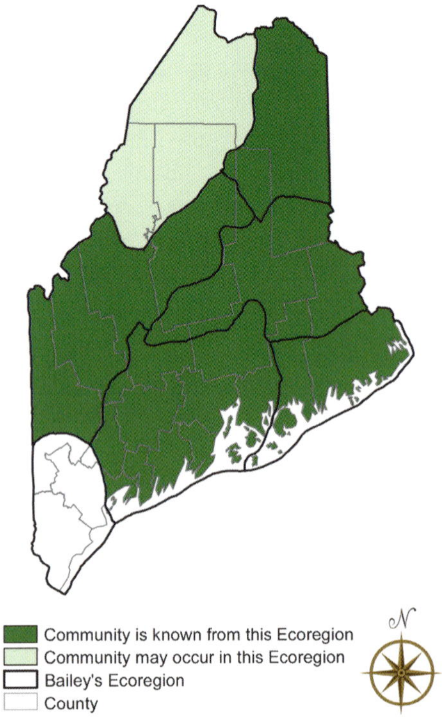

Jack Pine Woodland

Characteristic Plants

These plants are frequently found in this community type. Those with an asterisk are often diagnostic of this community.

Canopy
Black spruce*
Jack pine*
Red spruce*

Sapling/shrub
Black chokeberry
Black spruce*
Jack pine
Mountain holly*

Dwarf Shrub
Black huckleberry*
Lowbush blueberry
Mountain cranberry
Sheep laurel*

Herb
Bunchberry
Starflower

Bryoid
Reindeer lichen*

Conservation, Wildlife, and Management Considerations

The coastal occurrences of this woodland type appear to be self maintaining and more or less stable in their extent. Perpetuation of the inland sites is probably dependent upon occasional fire or other disturbance. Natural fires in these woodlands may increase the ability of jack pine to maintain its dominance over potentially invasive or fire sensitive tree species. Several sites are on public lands or private conservation lands.

Jack pine woodlands may be inhabited by uncommon moths such as the western pine elfin, which uses jack pine as a larval host plant in the Midwest and more often black spruce in the east.

Distribution

Immediate coastline of eastern Maine and lakeshores of north-central Maine (Laurentian Mixed Forest Province and New England - Adirondack Province), extending north, east, and west into Canada.

Landscape Pattern: Small Patch, usually less than 40 acres, occasionally 100-200 acres.

Examples on Conservation Lands You Can Visit

- Cadillac Mountain, Acadia National Park – Hancock Co.
- Great Wass Island Preserve – Washington Co.
- Lake Umbagog National Wildlife Refuge – Oxford Co.
- Petit Manan National Wildlife Refuge – Washington Co.
- Schoodic Point, Acadia National Park – Hancock Co.

Lower-elevation Spruce - Fir Forest

State Rank S5

Community Description

These closed canopy (>75% closure) forests are dominated by red spruce (>60% cover), typically with few other tree species in any of the layers. Fir is often a minor canopy component (up to 20% cover), particularly in open gaps or in younger stands. Hemlock is occasionally mixed with the spruce in southern or central Maine. The lower layers are sparse or patchy, consisting mostly of tree regeneration. In the sparse herb layer, dwarf shrubs are virtually absent except for spotty lowbush blueberry; herbaceous species cover well under 10% of the ground surface, and usually consist of scattered plants of Canada mayflower, starflower, and bunchberry. Most of the ground surface is bare conifer litter, although at some sites (particularly Downeast Maine), bryophytes may form patchy to full cover. Broom-mosses are the most frequent and abundant bryoids.

Soil and Site Characteristics

Sites are typically on hill slopes (lower, middle, or upper) at elevations up to 2200'. Slopes are gentle to moderately steep; aspect is various. The podzolic soils are quite rocky and/or shallow (<40 cm to obstruction) and may be very acidic (pH 4.1-5.2), creating low nutrient conditions. The somewhat xeric to mesic soils range from well drained to imperfectly drained; mottling is often present. Many sites have charcoal in the soil.

Diagnostics

Red spruce is dominant, and fir is usually present but much less abundant. Regeneration is dense in patches, and herbaceous species are almost absent (< 2% cover, up to 10%). Bryoids are more abundant than herbs and are dominated by broom-moss species.

Similar Types

Maritime Spruce - Fir Forests may be very similar, but generally feature more balsam fir, white spruce, and mountain ash and bryoids other than broom-mosses (typically three-lobed bazzania or red-stemmed moss). Spruce - Fir - Wood-sorrel - Feather-moss Forests have more herbs and a bryoid layer dominated by feather-mosses. Spruce - Heath Barrens have patchier canopies and more extensive shrub and herb layers. At higher elevations, this type can grade into Fir - Heart-leaved Birch Subalpine Forest, where spruce is far less abundant than fir.

Location Map

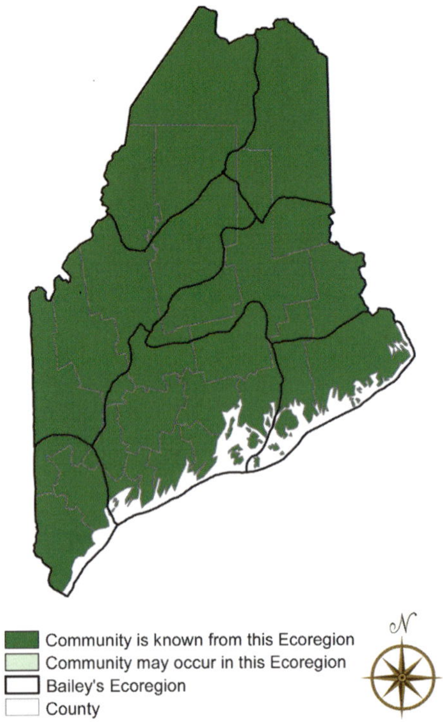

Lower Elevation Spruce – Fir Forest

Characteristic Plants

These plants are frequently found in this community type. Those with an asterisk are often diagnostic of this community.

Canopy
Balsam fir
Eastern hemlock
Red spruce*
White pine

Sapling/shrub
Balsam fir*
Red spruce*

Herb
Balsam fir*
Red spruce*

Bryoid
Dicranum moss
Red-stemmed moss
Reindeer lichen
Three-lobed bazzania

Conservation, Wildlife, and Management Considerations

This is the dominant spruce - fir type in Maine and is therefore extensively harvested and managed. In addition, spruce-budworm and past harvesting have played significant roles in the age dynamics of this type; some studies suggest that many current stands are more even-aged than they would be in the absence of past harvesting. Large (>1000 acres) examples free from human disturbance are scarce. Some areas of high ecological quality, in the hundreds of acres, are known but not necessarily designated as areas reserved from cutting. Almost all are within a landscape of managed forest rather than surrounded by land that has been permanently cleared and converted to other uses.

This community type may be utilized as nesting habitat by a number of coniferous forest specialist bird species such as the sharp-shinned hawk, yellow-bellied flycatcher, bay-breasted warbler, Cape May warbler, blackpoll warbler, northern parula, blackburnian warbler, boreal chickadee, Swainson's thrush, red crossbill, and white-winged crossbill.

Distribution

Statewide, characteristic of the Laurentian Mixed Forest Province and New England - Adirondack Province. Extends eastward, westward, and northward from Maine.

Landscape Pattern: Matrix

Examples on Conservation Lands You Can Visit

- Big & Little Squaw Mountain Public Lands – Piscataquis Co.
- Borestone Mountain Sanctuary – Piscataquis Co.
- East Nubble, Bigelow Preserve – Somerset Co
- Spruce Hill, White Mountain National Forest – Oxford Co.
- Squa Pan Mountain, Squa Pan Public Lands – Aroostook Co
- Tunk Mountain & Wizard Pond, Donnell Public Lands – Hancock Co.

Maritime Spruce - Fir Forest

State Rank S4

Community Description

Red spruce, white spruce, balsam fir, and/or larch are dominant in this Downeast coastal type. Composition is variable from the mid-coast to the Downeast coast. Red and white spruce are the most typical dominants; northern white cedar or hemlock are rarely co-dominant. The canopy may contain gaps with regenerating red maple, paper birch, mountain-ash, heart-leaved paper birch, and fir. Herbs and dwarf shrubs are typically <10% cover each, though in the canopy openings species such as raspberries, rough-stemmed goldenrod, whorled aster, and hay-scented fern may be locally abundant. The bryoid layer is >15% cover, dominated by mosses and liverworts rather than lichens.

Soil and Site Characteristics

Sites are along the immediate coast, often foggy and cool, on flats or lower to mid slopes (0-15%, may be steeper). Soils are shallow (<40 cm) over bedrock or till, with a well developed organic layer, acidic (pH 4.8-5.2) and mesic. Texture is sandy to loamy.

Usnea Lichen

Diagnostics

White spruce, bayberry, hay-scented fern, and mountain cranberry are indicators, though not always present. Sites contain relatively little or no bluebead lily, wood-ferns, or painted trillium. Broom-mosses do not dominate the bryoid layer, though they are often present.

Similar Types

Spruce - Fir - Broom-moss Forests are the most similar. They occur in more inland settings and, like this type, often have only sparse herbs, but unlike this type they are dominated by red spruce rather than white spruce and balsam fir, and their bryoid layer is dominated by broom-mosses. In poorly drained areas, Maritime Spruce – Fir Forests may grade into the Spruce - Fir - Cinnamon Fern Forest, which is distinguished by seasonally flooded or saturated soils and a more prominent cover of herbs and bryoids; along the coast, it usually occurs in small bedrock basins.

Location Map

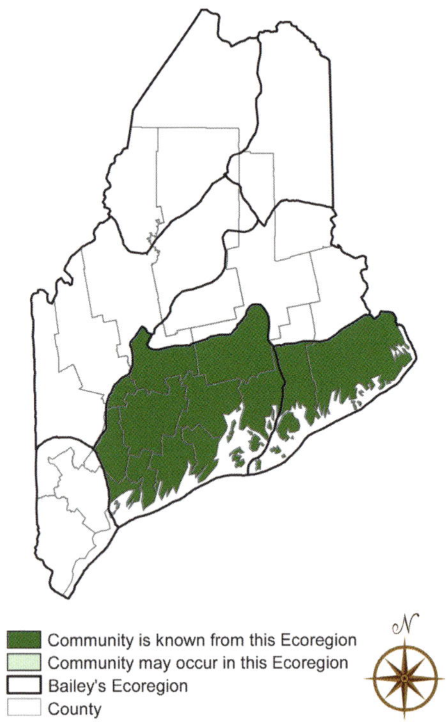

White Spruce

Conservation, Wildlife, and Management Considerations

After centuries of intensive use, almost no original coastal forest remains. Many now mature forests are on old pastureland. Many good (albeit secondary-growth) sites are in conservation ownership. Acadia National Park contains a variety of successional stages of this type, including stands that burned in 1947 and stands that did not. Maritime forests are subject to higher wind and weather stress than inland sites, and as a result the disturbances tend to be higher intensity and more frequent, and the trees do not grow as old.

This community type may be utilized as nesting habitat by a number of coniferous forest specialist bird species such as the sharp-shinned hawk, yellow-bellied flycatcher, Cape May warbler, blackpoll warbler, bay-breasted warbler, northern parula, boreal chickadee, Swainson's thrush, red crossbill, and white-winged crossbill.

Distribution

Coastal, primarily from mid-coast Maine eastward into the Canadian Maritimes (Laurentian Mixed Forest Province).

Landscape Pattern: Large Patch

Characteristic Plants

These plants are frequently found in this community type. Those with an asterisk are often diagnostic of this community.

Canopy

Balsam fir*
Red spruce*
White spruce*
Eastern hemlock
Mountain ash*
Northern white cedar
Paper birch*

Sapling/shrub

Balsam fir*
Red spruce
White spruce*
Mountain ash*

Herb

Balsam fir
Bayberry*
Raspberries
Mountain cranberry*
Red spruce
Rough-stemmed goldenrod
Hay-scented fern

Bryoid

Dicranum moss
Pincushion moss
Three-lobed bazzania

Associated Rare Plants

Swarthy sedge
White adder's-mouth

Examples on Conservation Lands You Can Visit

- Black Point Brook, Cutler Public Lands – Washington Co.
- Great Wass Island Preserve – Washington Co.
- North Cutler Coast, Cutler Public Lands – Washington Co.
- West Quoddy Head State Park – Washington Co.

Montane Spruce - Fir Forest

Spruce - Fir - Wood-sorrel - Feather-moss Forest

State Rank S4

Community Description

These closed canopy or sometimes patchy canopy forests are dominated by red spruce (50-95% cover); fir is a common associate (up to 35% cover) in younger stands and in canopy gaps, and yellow birch is the most common hardwood. Other conifers (northern white cedar, hemlock, or white pine) occasionally reduce the spruce dominance to as low as 40% cover. Striped maple is typical in the shrub layer, along with tree saplings. The herb layer is well developed (>15% cover, and often >30%), with tree regeneration and an assortment of herbs. Dwarf shrubs are conspicuously absent, except for a bit of velvet-leaf blueberry. Most of the ground surface is a lush mosaic of feather-mosses and leafy liverworts.

Soil and Site Characteristics

These forests occur on cool and moist microsites at moderate elevations (600'-2500', perhaps slightly higher), and north of 45 degrees latitude. Slopes are moderate to steep (5-50%), and usually north, west, or east facing. Soils are mostly well drained (some imperfectly drained), sandy to loamy, of moderate depth (25-50 cm), with pH 5.0-5.5.

Diagnostics

Red spruce is dominant, and yellow birch is the most abundant hardwood. Herbaceous species exceed 15% cover, with montane/boreal herbs such as bluebead lily, northern wood-sorrel, creeping snowberry, mountain wood fern, and/or rose twisted stalk locally common. Byoids exceed 40% cover, with a large proportion of feather-mosses.

Similar Types

Fir - Heart-leaved Birch Subalpine Forests can share many species and often grade into this type as elevation decreases, but will have fir more abundant than spruce in the canopy, shorter trees, and canopy gaps more frequent. Spruce - Fir - Broom-moss Forests have similar canopies but much more depauperate herb and bryoid layers. They usually occur on somewhat drier sites and lack the assortment of montane/boreal herbs and the most common mosses will be broom-mosses rather than feather-mosses. Some Maritime Spruce - Fir Forests have a similar herb layer, but if so they have more canopy fir and occur along the immediate coast.

Location Map

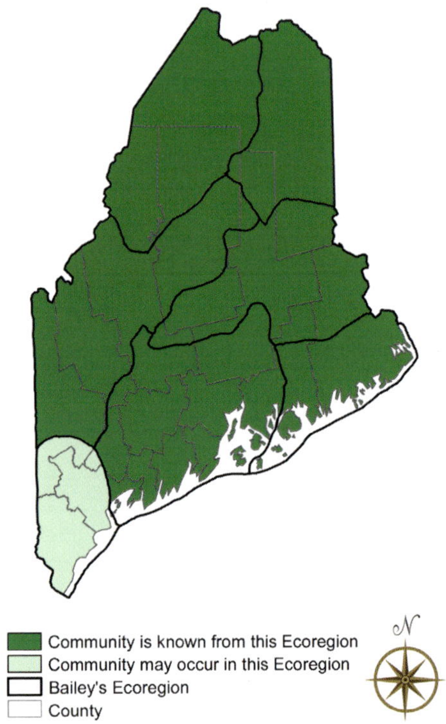

Montane Spruce – Fir Forest

Characteristic Plants

These plants are frequently found in this community type. Those with an asterisk are often diagnostic of this community.

Canopy
Balsam fir*
Red spruce*
Yellow birch*

Sapling/shrub
Balsam fir*
Red maple
Striped maple

Dwarf Shrub
Velvet-leaf blueberry

Herb
Bluebead lily*
Bunchberry
Canada mayflower
Creeping snowberry*
Goldthread
Northern wood-sorrel*
Painted trillium
Starflower

Bryoid
Common broom-moss*
Mountain fern moss
Red-stemmed moss
Three-lobed bazzania

Associated Rare Plants

Boreal bedstraw
Lesser wintergreen

Associated Rare Animals

Bicknell's thrush

Conservation, Wildlife, and Management Considerations

This is the characteristic spruce - fir type of mountain slopes just below the subalpine zone, and it is extensively harvested and managed. Spruce budworm has impacted many sites as well, creating patchy forest structure. Some areas of high ecological quality, in the hundreds of acres, are known but not necessarily designated as areas reserved from harvesting. Almost all are within a landscape of managed forest rather than surrounded by land that has been permanently cleared and converted to other uses.

This community type may be utilized as nesting habitat by a number of coniferous forest specialist bird species, such as the sharp-shinned hawk, yellow-bellied flycatcher, bay-breasted warbler, Cape May warbler, blackpoll warbler, northern parula, blackburnian warbler, boreal chickadee, Swainson's thrush, red crossbill, white-winged crossbill, gray jay, and spruce grouse.

Distribution

Western Maine westward (New England - Adirondack Province).

Landscape Pattern: Large Patch, mostly as hundreds of acres.

Examples on Conservation Lands You Can Visit

- Deboullie Ponds Public Lands – Aroostook Co.
- Elephant Mountain, Appalachian Trail – Franklin Co.
- Lower Horns Pond Trail, Bigelow Preserve – Franklin Co.
- Traveler Mountain, Baxter State Park – Piscataquis Co.
- Whitecap Mountain, Appalachian Trail – Piscataquis Co.

Northern Hardwoods Forest

Beech - Birch - Maple Forest

State Rank S5

Community Description

These closed canopy forests are dominated by a combination of beech, yellow birch, and sugar maple. Paper birch, red maple, conifers, and red oak may be present at lower cover. Conifers and red oak can each have <25% cover. Striped maple is a common subcanopy tree. The variable shrub layer is dominated by tree regeneration. Cover, richness and composition vary with site conditions.

Soil and Site Characteristics

Sites are typically found on the lower to middle portion of hillslopes (slopes generally 10-50%). Soils are generally mesic and well drained, though not deep (typically 15-50 cm) silt loams to sandy loams to loamy sands formed over glacial till, with pH 5.0-5.6; some occur on stabilized talus. Elevations range up to 2000'.

Diagnostics

A combination of beech, sugar maple, and yellow birch distinguishes this type. Though red oak is often entirely absent, conifers and red oak can be present and have up to 25% cover each. The herb layer lacks rich site indicators such as Dutchman's breeches, maidenhair fern, and blue cohosh.

Diseased Beech Bark

Similar Types

Maple – Basswood – Ash Forests and Sugar Maple Forests are similar to, and often contiguous with, this type. In them, beech is far less abundant, white ash is usually well represented in the canopy, and the herb layer contains species indicative of rich-soil areas. Red Oak – Northern Hardwoods – White Pine Forests have a higher proportion of red oak and can have a higher proportion of conifers (>25%). Spruce – Northern Hardwoods Forests also have >25% conifers in the canopy.

Conservation, Wildlife, and Management Considerations

This is the dominant hardwood type in Maine, and therefore it is extensively harvested and managed. Most management techniques diverge from the natural gap

Location Map

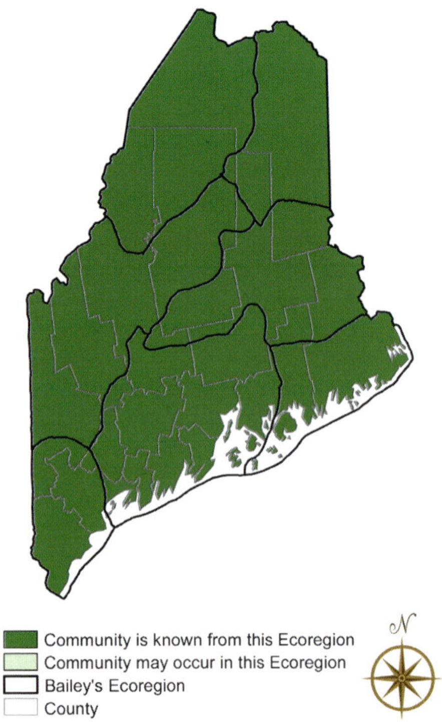

Northern Hardwoods Forest

pattern, which is at the scale of single trees or small groups of trees. Large (>1000 acres) examples reflecting only natural disturbance are scarce statewide, and intact examples in central and southern Maine tend to be smaller and more isolated.

Beech scale disease (Nectria) has devastated beech in many stands in eastern Maine. Although beech regenerates vigorously from sprouts after the trees have died, most sprouts succumb to the disease by the time they reach maturity. There are indications that some trees may express a genetic resistance to this disease.

Distribution

One of the predominant forest types in the New England - Adirondack Province and Laurentian Mixed Forest Province. Extends east, west, and north from Maine; occurs only as scattered areas southward.

Landscape Pattern: Matrix, typically hundreds of acres; high-quality patches usually now smaller.

American Beech with Beech Nuts

Characteristic Plants

These plants are frequently found in this community type. Those with an asterisk are often diagnostic of this community.

Canopy

American beech*
Eastern hemlock*
Paper birch*
Sugar maple*
Yellow birch*

Sapling/shrub

American beech*
Hobblebush*
Striped maple*
Sugar maple*
Yellow birch*

Herb

Bluebead lily*
Canada mayflower
Shining clubmoss*
Starflower
Striped maple*
Sugar maple*

Associated Rare Plants

Autumn coral-root
Boreal bedstraw
Broad beech fern
Cut-leaved toothwort
Nodding pogonia
Tall white violet

Associated Rare Animals

Early hairstreak

Examples on Conservation Lands You Can Visit

- Baxter State Park – Piscataquis Co.
- Big & Little Squaw Mountain Public Lands – Piscataquis Co.
- Bigelow Preserve Public Lands – Franklin/Somerset Co.
- Deboullie Ponds Public Lands – Aroostook Co.
- Grafton Notch State Park & Mahoosuc Public Lands – Oxford Co.
- White Mountain National Forest – Oxford Co.

Oak - Ash Woodland

State Rank S3

Community Description

These partial canopy deciduous woodlands are dominated by red oak and ironwood (the latter often as subcanopy). Basswood is an indicator species. Sugar maple may be codominant at some sites. The herb layer features species typical of somewhat enriched sites, such as Venus' looking-glass, herb Robert, round-lobed hepatica, plantain-leaved pussytoes, and wild-licorice, among an often dense cover of graminoids. Marginal wood-fern is characteristic of the herb layer. Vegetation may be patchy, developing in pockets among the rocks, or more continuous along upper slopes and ridges.

Soil and Site Characteristics

Sites occur on upper hill slopes and ridges or on talus slopes, often with some southerly exposure. Soils are thin and very well drained. The exposure and thin soils create dry conditions. These woodlands usually develop over bedrock that is not strongly acidic.

Ebony Spleenwort

Diagnostics

Ironwood is well represented and may be codominant with red oak. Sugar maple and/or basswood are present if not abundant. Some rich site indicators are present in herb layer.

Similar Types

Birch - Oak Talus Woodlands are similar but have far less ironwood, ash, and sugar maple, generally lack basswood, and have less extensive graminoid cover. Intermediates can be difficult to classify. Maple – Basswood – Ash Forests share some rich woods species but typically contain less oak and ironwood and occur on deeper, more mesic soils.

Location Map

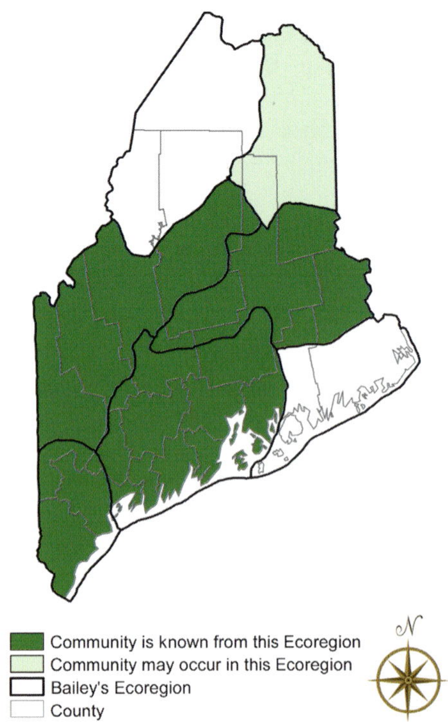

Oak – Ash Woodland

Characteristic Plants

These plants are frequently found in this community type. Those with an asterisk are often diagnostic of this community.

Canopy
Basswood*
Ironwood*
Red oak
Sugar maple
White Ash

Sapling/shrub
Beaked hazelnut
Ironwood*

Herb
Columbine
False spikenard
Fibrous-rooted sedge
Marginal woodfern
Rough-leaved ricegrass
Round-lobed hepatica
Woodland sedge*

Bryoid
Dicranum moss
Large hair-cap moss
Pincushion moss

Conservation, Wildlife, and Management Considerations

These types are subject to fragmentation from development pressure in southern Maine. The hilltop setting of many makes them attractive as house sites, although the shallow soils generally will not support intensive residential use. Some sites were probably pastured at some time in the 1700s - 1800s but have regrown. Almost all known sites are on private land with individual ownership.

Sites that have eastern red cedar present may host the rare juniper hairstreak butterfly, which uses cedar as its larval host plant. Two other insects, the columbine dusky wing butterfly and the aureolaria seed borer moth, may have historically inhabited this community type where they used uncommon larval host plants, columbine and false foxglove, respectively.

Associated Rare Plants

Blunt-lobed woodsia
Bottlebrush grass
Douglas' knotweed
Dry land sedge
Early crowfoot
Ebony spleenwort
Fern-leaved false foxglove
Hairy wood brome-grass
Missouri rockcress
Summer grape

Associated Rare Animals

Juniper hairstreak

Distribution

Eastern Broadleaf Forest Province, extending southwestward from Maine.

Landscape Pattern: Small Patch

Examples on Conservation Lands You Can Visit

- Derry Mountain, Camden Hills State Park – Knox Co.

Oak – Hickory Forest

State Rank S1

Community Description

This dry forest type, characteristic of the Central Appalachian Mountains, occurs in small patches or as inclusions within broader expanses of oak-pine forest. It is dominated by a mixture of shagbark hickory and oaks (white, black, red, or chestnut) over park-like sedge lawn. Sugar maple, white pine, or white ash may be canopy associates, and hop-hornbeam is a characteristic sub-canopy species. Additional species in the subcanopy or tall-shrub layer may include witch hazel, shadbushes, striped maple, and maple-leaved viburnum. Low shrubs can include blueberries, and the herb layer is primarily a lawn of woodland sedge with some other grass and sedge species. Moderately enriched sites may support tick-trefoils, hepatica, and the rare bottlebrush grass.

Soil and Site Characteristics

Sites occur on low-elevation, south- or west-facing sideslopes with well-drained loams or sandy loams. Known sites are within 10 miles of the coast.

Shagbark Hickory Bud

Diagnostics

Moderately open to closed canopy forests are dominated by a mixture of shagbark hickory (at least 30% cover) and oak species.

Similar Types

White Oak - Red Oak Forests and Oak - Pine Forests lack shagbark hickory.

Conservation, Wildlife, and Management Considerations

The few mature sites known in Maine were probably cleared in the past. Sites are small and subject to further fragmentation from development. Community dynamics are not well known, but there are some indications that shagbark hickory and white oak are

Location Map

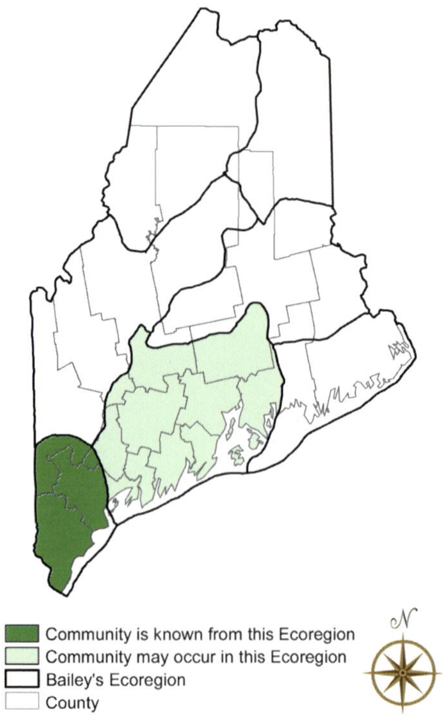

adapted to disturbance – likely fire – though there is no research on this topic in Maine. Most occurrences of this type are on private lands.

Bitternut Hickory Bark

This type offers habitat for a variety of birds, including scarlet tanager and ovenbird. Mature occurrences of this community type offer excellent potential sites for cavity dwellers such as the southern flying squirrel. The rare red-winged sallow moth uses red oak as one of its host plants and may be found in this community.

Distribution

Restricted to southern and coastal Maine, characteristic of the Eastern Broadleaf Forest Province. Extends south and west from Maine.

Landscape Pattern: Small patch (in Maine), generally 20 acres or less.

Bitternut Hickory Leaves

Characteristic Plants

These plants are frequently found in this community type. Those with an asterisk are often diagnostic of this community.

Canopy

Black oak
Red oak*
Shagbark hickory*
Sugar maple
White oak

Sapling/Shrub

Low-bush blueberry
Maple-leaved viburnum*
Shadbushes
Witch hazel*

Herb

Asters
Canada mayflower
Carex (Laxiflorae group)
Panic grasses
Sarsaparilla
Silverrod
Whorled loosestrife*
Wild oats
Woodland sedge*

Associated Rare Plants

Bitternut hickory
Bottlebrush grass
Chestnut oak
Scarlet oak

Associated Rare Animals

Red-winged sallow
Whip-poor-will

Oak - Northern Hardwoods Forest

State Rank S4

Community Description

This is a mixed upland forest type with red oak and northern hardwoods in the canopy. Some stands are almost entirely deciduous (typically oak - beech), while others are mixed with white pine, red spruce, hemlock, or (especially along the coast) northern white cedar. Red oak comprises up to 25-85% cover; beech is less than half that of red oak. Large red oak trees are prominent. Red maple is frequent. The shrub/sapling layer is usually sparse (<25%, but occasionally up to 50% cover). The herb layer is likewise spotty (usually <10% cover, sometimes 20-50% cover), with very few dwarf shrubs aside from lowbush blueberry, and with typical forest herbs and tree regeneration (red maple, red oak, white pine, beech). Few bryoids are found on the leaf litter covered forest floor.

Soil and Site Characteristics

Sites occur on gently to somewhat steeply sloping (15-35%) mid- and lower slopes, occasionally upper slopes, but usually not highly exposed sites. Known sites are at low elevations (<1200') on moderately well drained mineral soils, often rocky but not extremely shallow (typically 25-50 cm) and loamy, with pH 5.0-5.4.

Diagnostics

These are closed canopy forests in which red oak and at least one northern hardwood species (beech, sugar maple, or, infrequently, yellow birch) dominate the canopy. In southern Maine, maple-leaved viburnum is an indicator shrub.

Maple-leaved Viburnum

Similar Types

Oak - Pine Forests, the most similar type, generally lack sugar maple, white ash, and ironwood. Beech - Birch - Maple Forests have only minor amounts of red oak, and generally are strongly deciduous (>75%) rather than mixed. Spruce - Northern Hardwoods Forests lack red oak.

Conservation, Wildlife, and Management Considerations

Most sites in Maine are on lands with a long settlement history, and have apparently been timbered, pastured, or burned in the past. Several sites occur on public lands but are not necessarily designated as areas to be set aside from timber harvest. Small and

Location Map

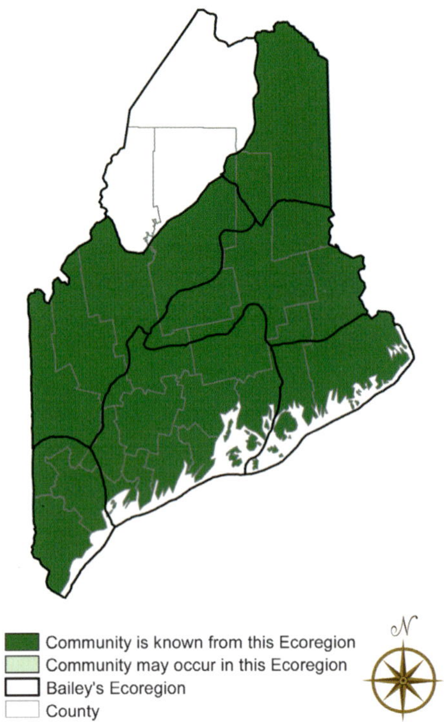

Oak – Northern Hardwoods Forest

isolated protected stands (on the order of 25 acres or less) may not be viable in the long run; though larger stands, or naturally small stands protected within a managed forest matrix, could be viable.

The community provides nesting habitat for a large number of passerine bird species, such as black-throated blue warbler, black-throated green warbler, scarlet tanager, and ovenbird. Mature stands offer excellent potential sites for cavity nesters. The rare red-winged sallow moth uses red oak as one of its host plants and may be found in this community. The globally uncommon early hairstreak butterfly uses beech as its larval host plant.

Distribution

Primarily the Laurentian Mixed Forest Province (except for the far northern portion) and southern portion of the New England - Adirondack Province, extending eastward and westward from Maine.

Landscape Pattern: Matrix-forming in southern Maine, though currently many occurrences exist as large patches due to fragmentation.

Examples on Conservation Lands You Can Visit

- Albany Mountain, White Mountain National Forest – Oxford Co.
- Alonzo Garcelon Wildlife Management Area – Kennebec Co.
- Center Hill, Mount Blue State Park – Franklin Co.
- Mt Megunticook, Camden Hills State Park – Knox Co.
- Patte Hill, White Mountain National Forest – Oxford Co.
- Sebago Lake State Park – Cumberland Co.

Characteristic Plants

These plants are frequently found in this community type. Those with an asterisk are often diagnostic of this community.

Canopy

American beech
Paper birch
Red maple
Red oak*
Striped maple
Sugar maple*
White pine

Sapling/shrub

American Beech
Maple-leaved viburnum
Striped maple
Sugar maple
White pine

Herb

Bracken fern
Canada mayflower
Starflower
Wild sarsaparilla
Wild-oats

Associated Rare Plants

American chestnut
Mountain-laurel
Nantucket shadbush
Ram's-head lady's-slipper

Associated Rare Animals

Early hairstreak
Red-winged sallow
Whip-poor-will

Oak - Pine Forest

State Rank S4

Community Description

This type is a closed canopy forest (>75% closure) in which red oak or a mixture of oak and white pine (rarely red spruce or hemlock) dominate. Red maple (up to 30% cover) and paper birch (up to 15% cover) can be common in younger stands. Striped maple is a common subcanopy associate; several other shrubs may be frequent. The herb layer is usually somewhat sparse (<30% cover), and features bracken fern, lowbush blueberry, and various herbaceous species; dwarf shrubs contribute 0-15% cover. The herb layer often includes forest species such as wild-oats and Indian cucumber-root that are seldom found in more open Oak - Pine Woodlands. Bryoids are sparse and are almost exclusively mosses rather than liverworts or lichens.

Soil and Site Characteristics

Sites occur on lower to mid-slopes or occasionally upper slopes on low hills. Slopes are typically 10-25% and aspect varies. Sites are characterized by well drained mineral soils that are somewhat shallow (10-50 cm to obstruction), usually sandy loams or loamy sands, and acidic (pH ~5.0).

Diagnostics

These are more or less closed canopy forests with dominance of red oak or red oak - white pine mixture (occasionally red spruce or hemlock replace white pine); there is an absence or at least low cover of northern hardwood species and other oaks.

Similar Types

Oak - Pine Woodlands are similar and sometimes contiguous with this type. Their canopy is more open and the dwarf shrub layer much more well developed (usually >15% cover of dwarf shrubs). Red Oak - Northern Hardwoods - White Pine Forests occur on more mesic sites and feature at least 10% cover of other tolerant hardwoods (beech, sugar maple, white ash, or ironwood). White Oak - Red Oak Forests contain white oak in the canopy.

Conservation, Wildlife, and Management Considerations

Most Oak - Pine Forests in Maine are on land that was once cleared or pastured. The known sites are subject to fragmentation by timber harvesting, clearing for agriculture, and residential development. Fire or other soil disturbance may be important in maintaining this type.

Location Map

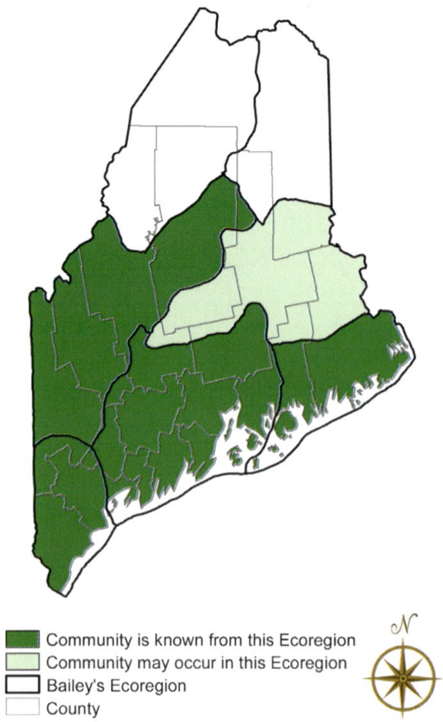

Oak – Pine Forest

The community provides nesting habitat for a large number of passerine bird species such as the wood thrush, scarlet tanager, ovenbird, and pine warbler.

Mature stands with a high proportion of oaks offer excellent potential sites for cavity nesters. The state rare red-winged sallow moth uses red oak as one of its host plants and may be found in this community. Historical butterfly species like the persius duskywing and the frosted elfin were known from oak-pine forests and woodlands in southern Maine where their host plants, wild indigo and wild lupine, were formerly present. Similarly, this community type historically included chestnut as a canopy constituent and may include the chestnut clearwing moth, which uses chestnut as a host plant.

Distribution

Extending southward and southwestward from southern and central Maine; this is common, matrix-forming forest type in the Eastern Broadleaf Forest Province.

Small Whorled Pogonia

Landscape Pattern: Large Patch; formerly a matrix type in southern Maine and lower New England.

Characteristic Plants

These plants are frequently found in this community type. Those with an asterisk are often diagnostic of this community.

Canopy
American beech*
Balsam fir*
Eastern hemlock
Paper birch
Red maple*
Red oak*
Red spruce
White pine*

Sapling/shrub
Beaked hazelnut*
Black huckleberry*
Witch-hazel*

Dwarf Shrub
Black huckleberry*
Lowbush blueberry*

Herb
Big-leaved aster*
Bracken fern*
Canada mayflower
Starflower

Bryoid
Dicranum moss
Large hair-cap moss

Associated Rare Plants

American chestnut
Mountain laurel
Small whorled pogonia
Variable sedge
Wild indigo

Associated Rare Animals

Red-winged sallow
Whip-poor-will

Examples on Conservation Lands You Can Visit

- Bald Mountain, Little Concord Pond Public Lands – Oxford Co.
- Sebago Lake State Park – Cumberland Co.

Oak - Pine Woodland

State Rank S4

Community Description

These woodlands support a partial canopy (20-70%) dominated by red oak, or red oak with white pine or red spruce (rarely with red pine). The trees are widely spaced and often stunted, with an open understory. Gray birch is a common small tree. The herb layer is well developed (>25% cover), with one-third to nearly all of the layer consisting of dwarf shrubs. Lowbush blueberry is the most abundant dwarf shrub; sheep laurel is also common. Herbs form up to 20% cover among the dwarf shrubs. The bryoid layer is patchy, sparse, and variable in composition. While this is categorized as a "mixed" type, individual examples may be either mixed (>25% conifer and >25% deciduous) or deciduous (<25% conifer); they are otherwise very similar.

Soil and Site Characteristics

Sites occupy upper hillsides and low ridgelines, with slopes up to 30% and elevations up to 1500'. South aspect is more likely than north. The substrate is typically thin sandy to loamy soil (<25 cm) over bedrock or coarse till, occasionally with a layer of poorly decomposed duff over the mineral horizons. Soils are acidic (pH 5.0-5.2). Many sites were former pasture and/or have evidence of past fire.

Diagnostics

Canopy closure is less than 65% and red oak is dominant. Trees are short and spreading, or pines are dominant with lesser amounts of red oak, beech, sugar maple, or ash. The herb layer exceeds 20%, with lowbush blueberry, sheep laurel, or sweetfern prominent; graminoids are often abundant.

Similar Types

Oak - Pine Forests are similar and may be contiguous with this type. They are distinguished by their higher canopy cover (>75%), taller and straighter trees, and more sparse herb layer (generally <30%) with little cover of heath shrubs.

Conservation, Wildlife, and Management Considerations

This community appears to be relatively stable in Maine, with little habitat conversion. Fire has apparently played a role at some sites by preventing the invasion of fire sensitive hardwood

Location Map

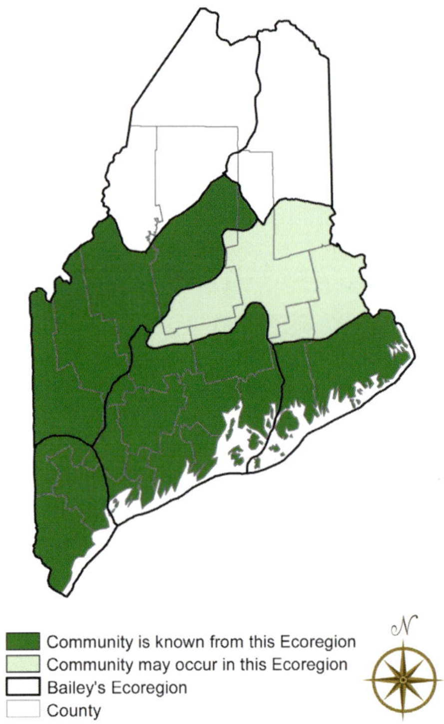

Oak – Pine Woodland

trees and shrubs. Many sites receive recreational use. In a few locations that use is heavy enough to have degraded the community, but most recreational foot traffic is compatible. Communications or wind power towers have degraded some sites on mid-elevation summits.

This community type hosts several rare Lepidopteran species that feed on oaks, including red-winged sallow, barrens chaetaglaea, barrens xylotype, similar underwing, and oblique zale.

Distribution

Eastern Broadleaf Forest Province and southern portions of the New England - Adirondack Province and Laurentian Mixed Forest Province, extending southward and westward from Maine.

Landscape Pattern: Small Patch

Examples on Conservation Lands You Can Visit

- Blueberry Mountain, White Mountain National Forest - Oxford Co.
- Round Mountain, Donnell Pond Public Lands - Hancock Co.
- Spring River Mountain, Donnell Pond Public Lands - Hancock Co.
- Styles Mountain, White Mountain National Forest - Oxford Co.

Characteristic Plants

These plants are frequently found in this community type. Those with an asterisk are often diagnostic of this community.

Canopy

Red maple
Red oak*
Red pine
Red spruce
Shadbush*
White pine*

Sapling/shrub

Black huckleberry*
Lowbush blueberry*
Shadbush

Dwarf Shrub

Lowbush blueberry*
Sheep laurel*

Herb

Bracken fern*
Canada mayflower
Sheep fescue*
Starflower
Wintergreen

Bryoid

Dicranum moss
Large hair-cap moss
Reindeer lichen

Associated Rare Plants

New Jersey tea
Scarlet oak

Associated Rare Animals

Barrens chaetaglaea
Barrens xylotype
Edward's hairstreak
Oblique zale
Red-winged sallow
Similar underwing
Sleepy duskywing
The buckmoth
Whip-poor-will

Pitch Pine Dune Woodland

State Rank S1

Community Description

These partial canopy woodlands (typically ~35% closure) support stunted pitch pine trees on sand dunes, where the density of the pines and the composition of the understory vary within and between sites. Bayberry, wild raisin, and/or alder form a patchy shrub layer. The herb layer is typically well developed (>35%), often with extensive patches of grasses or sedges. These are frequently disturbed forests on more dynamic dunes and may have patches of lichens and beach heather beneath the spotty canopy. Older examples on more stable dunes can be nearly closed canopy forests in which pitch pine is mixed with other species such as red maple, red oak, and paper birch, with less understory vegetation.

Soil and Site Characteristics

Sites occur on open coastal dunes. The sandy substrate of this community is very dry, acidic, and nutrient poor, resulting in stunted tree growth. Only limited herbaceous species are able to tolerate the oceanside conditions.

Beach Plum in Bud

Diagnostics

These woodlands are distinguished by stunted pitch pine growing on stabilized dunes along the immediate coast.

Similar Types

Pitch Pine - Scrub Oak Barrens and Pitch Pine - Heath Barrens are also partially forested with pitch pine, but occur more inland and have a well-developed shrub layer of scrub oak and/or ericaceous shrubs. Pitch Pine Woodlands occur on a bedrock substrate.

Location Map

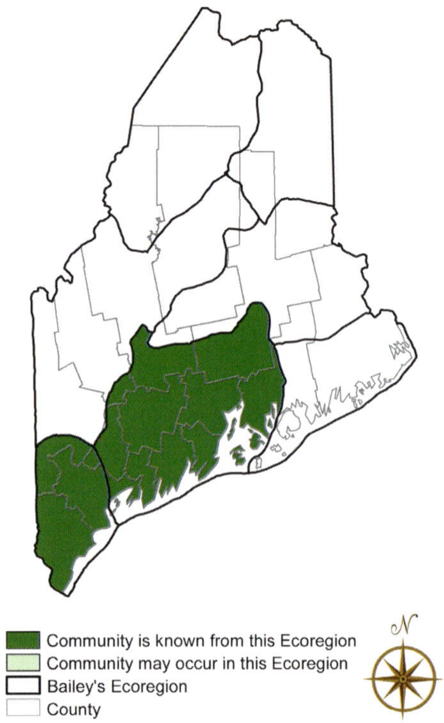

Conservation, Wildlife, and Management Considerations

This community appears to have been reduced in extent due to extensive beachfront development. Some of the remaining sites are ecologically intact, and while dynamic, are moderately well buffered from human impacts. They may, however, be vulnerable to ATV traffic, intensive recreational use, or over the longer term, sea level rise and severe storms. Most known sites are on public land or private conservation land.

Birds such as the pine warbler and prairie warbler may prefer this open habitat. This community type may include rare moths that utilize hard pines as larval host plants such as the pine pinion, oblique zale, and southern pine sphinx, which use pitch pine as a host plant. The pine-devil moth, a historical species for Maine, probably also inhabited this community type, where its larvae fed on pitch pine.

Distribution

South-coastal Maine (Eastern Broadleaf Forest Province), extending south along the Atlantic coastal plain. This type achieves its greatest extent on Cape Cod.

Landscape Pattern: Small Patch, typically less than 50 acres.

Characteristic Plants

These plants are frequently found in this community type. Those with an asterisk are often diagnostic of this community.

Canopy
Pitch pine*

Sapling/shrub
Bayberry
Speckled alder
Wild-raisin

Dwarf Shrub
Beach heather*
Golden heather*
Lowbush blueberry
Sheep laurel*

Herb
Canada mayflower
Common hairgrass*
Indian pipe
Starflower*
Wild sarsaparilla
Woodland sedge

Bryoid
Reindeer lichen*

Associated Rare Plants

Beach plum

Associated Rare Animals

Oblique zale
Pine pinion
Pine-devil moth
Southern pine sphinx

Examples on Conservation Lands You Can Visit

- Crescent Beach State Park – Cumberland Co.
- Popham Beach State Park – Sagadahoc Co.
- Seawall Beach, Morse Mountain Preserve – Sagadahoc Co.

Pitch Pine - Heath Barren

Pitch Pine - Heath Barren

State Rank S1

Community Description

This is an open canopy type in which pitch pine dominates the tree layer, with an understory of dwarf shrubs and herbs, and without an extensive tall shrub layer. Canopy cover is usually <50%, and openings with blueberry and lichens may occur within the barrens. Scrub oak may be present but at only low cover. Sites typically have an herb layer of lowbush blueberry and woodland sedge, with scattered bracken fern and forbs. Sharp-pointed ricegrass is characteristic, although it is rarely abundant. The lack of a shrub layer gives these barrens a park like appearance. Occasionally these may occur as almost closed canopy forests, but the pitch pines with the heath understory and little scrub oak retain the character of the type.

Soil and Site Characteristics

Sites occur on well drained to very well drained sandy soils on outwash plains. Soils have little organic matter and are acidic. Topography is flat or very gently sloping. Sites generally have a history of periodic fire.

Edwards' Hairstreak

Diagnostics

These are pitch pine dominated, partially forested areas which develop on sands or glacial outwash deposits, not on stabilized coastal dunes. Scrub oak is lacking or sparse.

Similar Types

Pitch - Pine Dune Woodlands are also pitch pine dominated, but they occur on stabilized sand dunes along the coast and lack a well developed heath shrub layer. Pitch Pine - Scrub Oak Barrens are similar but have areas dominated by scrub oak or at least have scrub oak as a common component in the shrub layer.

Location Map

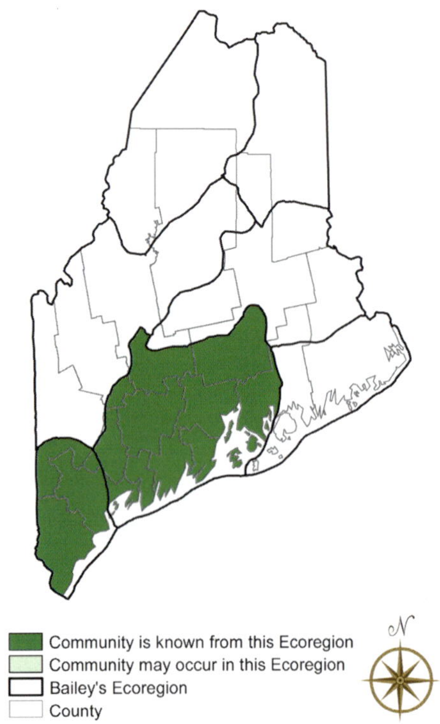

Pitch Pine

Conservation, Wildlife, and Management Considerations

This community type is dependent upon periodic fires to eliminate competing tree species and prevent succession to an oak-pine forest. Because of habitat loss and fragmentation and a history of fire suppression, this forest type has become very rare. Most of the sites in the state have been degraded by permanent conversion to residential areas or sand and gravel mines.

Birds such as the eastern towhee, whip-poor-will, pine warbler, and prairie warbler may prefer this open habitat. Rare moths such as the pine pinion, oblique zale, and pine sphinx, which utilize pitch pine as a larval host plant, may be found in this community type. The pine-devil Moth, a historical species for Maine, probably also inhabited this community type, where its larvae fed on pitch pine.

Characteristic Plants

These plants are frequently found in this community type. Those with an asterisk are often diagnostic of this community.

Canopy
Pitch pine*

Sapling/shrub
Scrub oak
Sweetfern

Dwarf Shrub
Lowbush blueberry*

Herb
Sharp-pointed ricegrass
Whorled loosestrife
Woodland sedge*

Associated Rare Plants

Branching needle-grass
Fern-leaved false foxglove
Unicorn root
Wild chess

Associated Rare Animals

Edward's hairstreak
Oblique zale
Pine barrens itame
Pine barrens zale
Pine barrens zanclognatha
Pine pinion
Pine-devil moth
Pink sallow
Similar underwing
Sleepy duskywing
Southern pine sphinx
The buckmoth
Twilight moth
Whip-poor-will

Distribution

Southern Maine (Eastern Broadleaf Forest Province), with a few sites in central Maine. Extends southward and southwestward from the state.

Landscape Pattern: Large Patch

Examples on Conservation Lands You Can Visit

- Brunswick Town Commons – Cumberland Co.
- Jugtown Plains – Cumberland Co.

Pitch Pine - Scrub Oak Barren

State Rank S2

Community Description

This woodland type ranges from very open to nearly closed canopy (25-75% closure) in which pitch pine is dominant (up to 50% cover). Red maple is frequent but rarely abundant in the canopy. In openings among the trees, a dense shrub/sapling layer of scrub oak is typical. Gray birch may be a prominent feature of the shrub layer, and shrubs are locally dense. A low layer of heath shrubs dominated by lowbush or velvet-leaf blueberry is usually present. Bracken fern and woodland sedge are characteristic herbs. Bryoids are virtually absent. Vegetation is typically very patchy, with some areas clearly pitch pine dominated and others areas extensive thickets of scrub oak. Nonforested openings with blueberry and lichens may occur within the barrens.

Soil and Site Characteristics

Sites occur on nutrient poor soils of glacial outwash plains or moraines south of 44 degrees latitude. Topography is flat to undulating. The xeric to dry-mesic, sandy soils are acidic (pH usually <5.0) and have little organic matter. Fire is an important factor in maintaining this community.

Northern Blazing Star

Diagnostics

These are pitch pine dominated partially forested areas which develop on sands or glacial outwash deposits, not on stabilized coastal dunes. Scrub oak is common and locally dominant in the shrub layer.

Similar Types

Pitch Pine Woodlands can be floristically similar but occur on bedrock, not on deep sandy soils. Pitch Pine Dune Woodlands occur on stabilized sand dunes along the coast. They also lack a well developed heath shrub layer. Pitch Pine - Heath Barrens share many species but lack the scrub oak layer (scrub oak may be present but only at low cover). Pitch Pine Bogs are wetlands, with at least a shallow peat substrate.

Conservation, Wildlife, and Management Considerations

This community type is dependent upon periodic fires to eliminate competing tree species and prevent succession to an Oak

Location Map

Pitch Pine – Scrub Oak Barren

- Pine Forest. Because of fire suppression in the last century, this community type has become very rare. Relatively large areas are required to maintain this dynamic community and its associated rare animal species. Most of the large sites in the state have been fragmented by permanent conversion to residential areas or to sand and gravel pits.

Birds such as the whip-poor-will, eastern towhee, pine warbler, and prairie warbler may prefer this open habitat. This community type includes a rich array of rare butterflies and moths that use pitch pine or scrub oak as their larval host plant, including the southern pine sphinx, pine pinion, oblique zale, the buckmoth, Edward's hairstreak, pine barrens zale, pine barrens itame, and sleepy dusky wing.

Distribution

Primarily southern Maine (Eastern Broadleaf Forest Province). Extends southward and southwestward from the state along the Atlantic coastal plain.

Landscape Pattern: Large Patch

Examples on Conservation Lands You Can Visit

- Brownfield Bog Wildlife Management Area – Oxford Co.
- Kennebunk Plains Preserve – York Co.
- Killick Pond Wildlife Management Area – York Co.
- Waterboro Barrens Preserve – York Co.

Characteristic Plants

These plants are frequently found in this community type. Those with an asterisk are often diagnostic of this community.

Canopy

Gray birch
Pitch pine*
Red maple

Sapling/shrub

Gray birch
Pitch pine
Scrub oak*
Shadbush
Sweetfern
Wild-raisin

Dwarf Shrub

Lowbush blueberry*
Sheep laurel
Velvet-leaf blueberry

Herb

Bracken fern*
Canada mayflower
Mayflower
Sharp-pointed ricegrass
Wintergreen*
Woodland sedge

Bryoid

Large hair-cap moss

Associated Rare Plants

Butterfly weed
Fern-leaved false foxglove
Northern blazing star
Wild chess
Wild indigo
Wild lupine

Associated Rare Animals

Edward's hairstreak
Oblique zale
Pine barrens itame
Pine barrens zale
Pine barrens zanclognatha
Pine pinion
Pine-devil moth
Pink sallow
Similar underwing
Sleepy duskywing
Southern pine sphinx
The buckmoth
Twilight moth
Whip-poor-will

Pitch Pine Woodland

State Rank S3

Community Description

These very open to semi open woodlands (25-65% canopy, occasionally to 75%) are dominated by pitch pine, often with a much smaller component of red oak, red or white pine, or black or red spruce. The well spaced pines allow a substantial amount of light to reach the understory. The sapling/shrub layer is usually <40% cover, with smaller pitch pines, mountain holly, or black huckleberry. The herb layer is well developed (>30% cover) and strongly dominated by dwarf, mostly heath, shrubs. At some sites, broom-crowberry is a prominent species. Herbs contribute <10% cover, and the composition varies. The bryoid layer may be 0-50% cover (rarely more) and is typically dominated by reindeer lichens.

Soil and Site Characteristics

Typical sites are ledges or rock outcrops in coastal areas. They may be flat to gently sloping, at elevations up to 1500'. Soils are usually very thin, consisting of a coarse mineral fraction or a layer of poorly decomposed duff over bedrock, with pH 4.6-5.4. Many sites have evidence of past fire.

Pitch Pine Cones

Diagnostics

These pitch pine dominated woodlands (25-65% canopy cover) grow on bedrock with very little soil.

Similar Types

Pitch Pine - Scrub Oak Barrens, Pitch Pine - Heath Barrens, and Pitch Pine Dune Woodlands differ in that they develop on sandy outwash or dunes, rather than on thin soil over bedrock. Pitch Pine Bogs are wetlands, with wetland plants, including peat mosses.

Conservation, Wildlife, and Management Considerations

This community appears to be relatively stable in Maine, with little habitat conversion. Fire has apparently played

Location Map

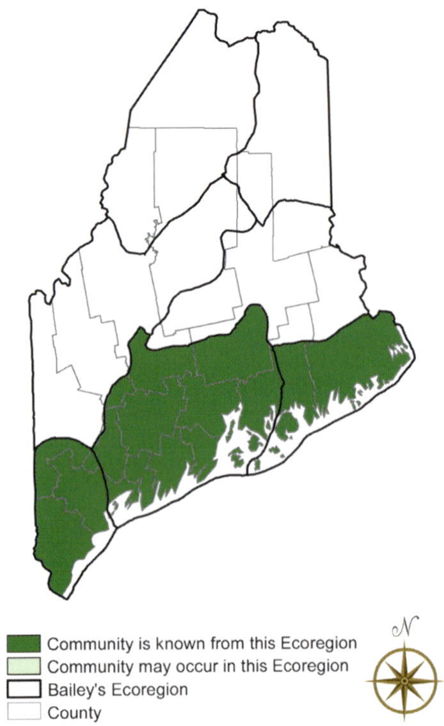

Pitch Pine Woodland

a role in maintaining this woodland type by preventing the invasion of fire sensitive hardwood trees and shrubs. The suppression of fire may result in the conversion of these woodlands to a different type. Many sites receive recreational use. In a few locations use is heavy enough to have degraded the community, but most foot traffic recreational use is compatible. Communications towers could impact some sites on mid-elevation summits.

Birds such as the pine warbler and prairie warbler may prefer this open habitat. This community type may include rare moths that utilize pitch pines as a larval host plant such as the oblique zale, southern pine sphinx, and pine-devil moth, a historical species for Maine.

Distribution

Coastal Maine, east to Mount Desert Island; extending southward along the Atlantic coastal plain and Appalachian foothills.

Landscape Pattern: Small Patch; size range variable from a few acres to nearly 100 acres.

Characteristic Plants

These plants are frequently found in this community type. Those with an asterisk are often diagnostic of this community.

Canopy

Red spruce
Pitch pine*
Red oak*
Red pine*
White pine*

Sapling/shrub

Black huckleberry*
Gray birch*
Mountain holly*
Pitch pine*
Red spruce

Dwarf Shrub

Black huckleberry*
Broom-crowberry*
Lowbush blueberry*
Rhodora*
Sheep laurel*

Herb

Bracken fern

Bryoid

Reindeer lichen

Associated Rare Plants

Mountain sandwort
Smooth sandwort

Associated Rare Animals

Pine-devil moth
Southern pine sphinx

Examples on Conservation Lands You Can Visit

- Bald Head Preserve – Sagadahoc Co.
- Champlain Mountain, Acadia National Park – Hancock Co.
- Dorr Mountain, Acadia National Park – Hancock Co.
- Reid State Park – Sagadahoc Co.

Red and White Pine Forest

State Rank S3

Community Description

These are upland forests with red pine as the dominant tree; white pine, red spruce, or, near the coast, northern white cedar may be co-dominant. The canopy may be somewhat open but is more typically >70%. Especially in post-fire sites, the canopy may include deciduous trees such as paper birch, red maple, or big-toothed aspen. Lower layers are generally sparse (<25% cover) and contain few species; some sites may have scattered heath shrubs such as huckleberry, lowbush blueberry, or sheep laurel. Bracken fern and wintergreen are almost always present in the herb layer, but at low cover. Graminoids are virtually absent. The ground is typically covered with conifer litter and patches of bryophytes, or less commonly, lichens.

Soil and Site Characteristics

Sites are usually on flats, slopes of <25% or low ridges (<1000'), on dry-mesic to xeric soils that are somewhat to very shallow (10-50 cm to obstruction, usually bedrock). Soils are coarse (sandy loams to sands) and acidic (pH 4.8-5.2). Many sites have evidence of past fires.

Red Pine Needles and Cone

Diagnostics

The forest canopy typically exceeds 65%, with red pine dominant or co-dominant (at least 33% cover) with other conifers; the shrub layer is usually sparse (<15%).

Similar Types

Red Pine Woodlands have a more open canopy (usually <50%), a more well developed heath shrub layer (>25%), and are usually on sites with only a very thin soil layer, or merely an organic layer, over bedrock. White Pine - Mixed Conifer Forests may contain red pine, but white pine is more dominant.

Location Map

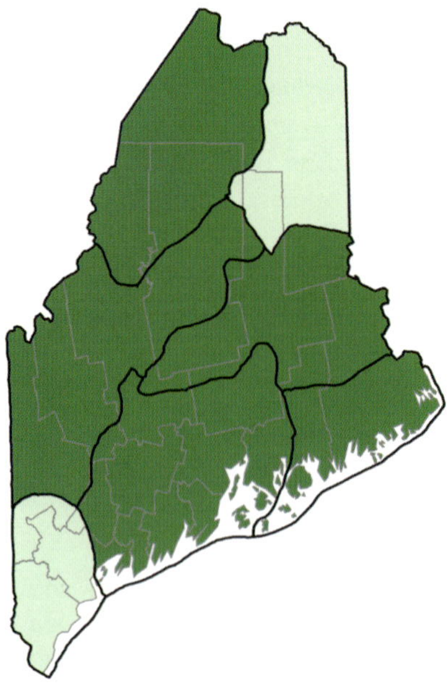

Community is known from this Ecoregion
Community may occur in this Ecoregion
Bailey's Ecoregion
County

Red and White Pine Forest

Characteristic Plants

These plants are frequently found in this community type. Those with an asterisk are often diagnostic of this community.

Canopy
Big-toothed aspen*
Eastern hemlock
Red pine*
Red spruce*
White pine

Sapling/shrub
Balsam fir
Red spruce

Dwarf Shrub
Black huckleberry*

Herb
Bracken fern
Canada mayflower
Round-leaved pyrola
Starflower
Wintergreen

Bryoid
Dicranum moss
Red-stemmed moss

Associated Rare Animals

Oblique zale
Southern pine sphinx
Whip-poor-will

Conservation, Wildlife, and Management Considerations

Red pine has been widely planted in the past, but natural occurrences of this type are fairly rare outside of eastern Maine. Under natural conditions, these forests apparently require fire for persistence or regeneration, but community dynamics are not well documented, and at some known sites clearcut harvesting has perpetuated the type. Most known sites are small, lack formal protection, and could be maintained within a forested matrix.

This community type may be used as nesting habitat by a number of coniferous forest specialist bird species, such as the pine warbler and red crossbill. It may also include rare moths such as the oblique zale and the southern pine sphinx, whose larvae feed on red pine.

Distribution

Eastern Broadleaf Forest Province and New England - Adirondack Province, extending both east and west from Maine.

Landscape Pattern: Small to Large Patch

Examples on Conservation Lands You Can Visit

- Attean Pond at Moose River – Somerset Co.
- East Machias River, Rocky Lake Public Lands – Washington Co.
- Fifth Lake Stream, Duck Lake Public Lands – Hancock Co.
- Gassabias Lake, Duck Lake Public Lands – Hancock Co.

Red Pine Woodland

State Rank S3

Community Description

These open canopy woodlands (30-75% closure) are dominated by red pine. Associated canopy species vary among sites and include white pine, red spruce, or paper birch. The shrub layer includes scattered red spruce, red maple, paper birch, or gray birch. The herb layer varies in extent, but usually features heath shrubs and scattered forbs or bracken fern; graminoids are virtually absent. Bryoids are patchy and usually consist of types associated with somewhat dry conditions such as reindeer lichens.

Soil and Site Characteristics

Sites are flat to moderately sloping and occupy low ridges or upper slopes. Most sites are inland. The substrate is usually thin soil or organic duff (<20 cm) over bedrock. Evidence of past fire is often found.

Diagnostics

Sites are distinguished by open canopy (<65%) woodlands, usually on bedrock; with red pine dominant and oaks and northern hardwoods lacking; the herb layer is dominated by dwarf shrubs and few graminoids.

Wintergreen (aka Teaberry)

Similar Types

Red Pine - White Pine Forests have more complete canopy (>75%), usually lack a well developed dwarf shrub layer, and occur on deeper sandy soils. Oak - Pine Woodlands may have red pine as a component species but will have red oak or northern hardwoods as dominant trees and usually possess a more extensive graminoid cover.

Conservation, Wildlife, and Management Considerations

These occur as small patches, usually on

Location Map

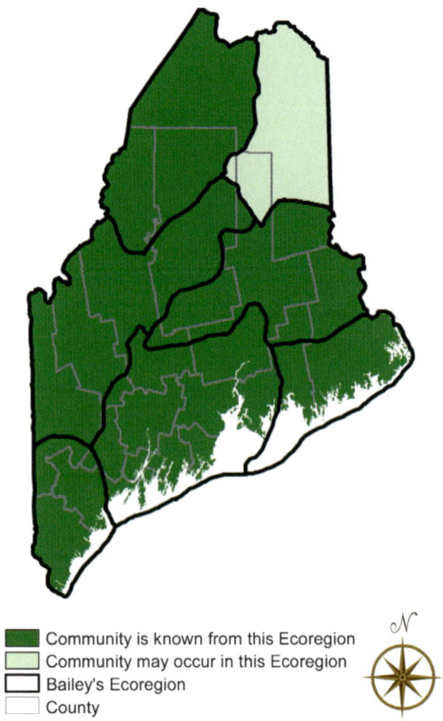

Red Pine Woodland

upper slopes or hilltops. Maintaining representative examples is best accomplished by retaining adjacent forest cover as buffer. Most sites have fire evidence, and fire may be required for regeneration or persistence of this type.

Common nighthawks and whip-poor-wills may nest in open patches within red pine woodlands. This community type may include rare moths that utilize hard pines as larval host plants such as the oblique zale, southern pine sphinx, and pine pinion.

Distribution

New England - Adirondack Province and Laurentian Mixed Forest Province, extending westward, northward, and presumably eastward from Maine.

Landscape Pattern: Small Patch

Examples on Conservation Lands You Can Visit

- Albany Notch, White Mountain National Forest – Oxford Co.
- Norumbega Mountain, Acadia National Park – Hancock Co.
- Tunk Lake Area, Donnell Pond Public Lands – Hancock Co.

Characteristic Plants

These plants are frequently found in this community type. Those with an asterisk are often diagnostic of this community.

Canopy

Paper birch
Red maple
Red pine*
Red spruce
White pine

Sapling/shrub

Black chokeberry
Gray birch*
Mountain holly
Paper birch*
Red maple
Red spruce
White pine

Dwarf Shrub

Black huckleberry
Lowbush blueberry*
Sheep laurel*

Herb

Bracken fern
Bunchberry
Canada mayflower
Wintergreen

Bryoid

Grey reindeer-lichen
Grimmia rock-moss

Associated Rare Plants

Canada mountain-ricegrass
Swarthy sedge

Associated Rare Animals

Oblique zale
Pine pinion
Southern pine sphinx

Spruce - Northern Hardwoods Forest

State Rank S5

Community Description

This mixed forest type is characterized by red spruce and yellow birch, or less often another hardwood (sugar maple, red maple, or beech). Scattered large supercanopy white pine trees are occasional. Balsam fir and paper birch are common, typically as smaller trees, and hemlock may be an associate at some sites. The sapling/shrub layer may be fairly well developed (20-40% cover), with striped maple and saplings of canopy species; shrub species vary among sites. The herb layer ranges from sparse to dense but is usually >15% cover, divided between forbs, ferns, and regenerating trees, with dwarf shrubs virtually absent. The bryoid layer is patchy and locally well developed, with bryophytes far more abundant than lichens. As is typical in mesic forests in Maine, three-lobed bazzania is a frequent bryophyte.

Soil and Site Characteristics

These forests occur on cooler microsites from near sea level to 2200'. They are usually on hillslopes, ranging from lower to upper slopes and from gentle to steep (up to 50%). The soils are typically well drained, sometimes somewhat excessively drained, sandy to loamy in texture, with pH 5.0-5.4.

Balsam Fir

Diagnostics

Sites are distinguished by a mixture of red spruce and northern hardwoods (most often yellow birch) in the canopy; conifer and deciduous components exceed 25% cover each.

Similar Types

Beech - Birch - Maple Forests are more strongly deciduous. Spruce - fir forest types can be similar but have <25% cover of northern hardwood species. Both of these types can be contiguous with this type and may intergrade with it.

Conservation, Wildlife, and Management Considerations

Nearly all forests of this type have been harvested in the past, and at many sites the

Location Map

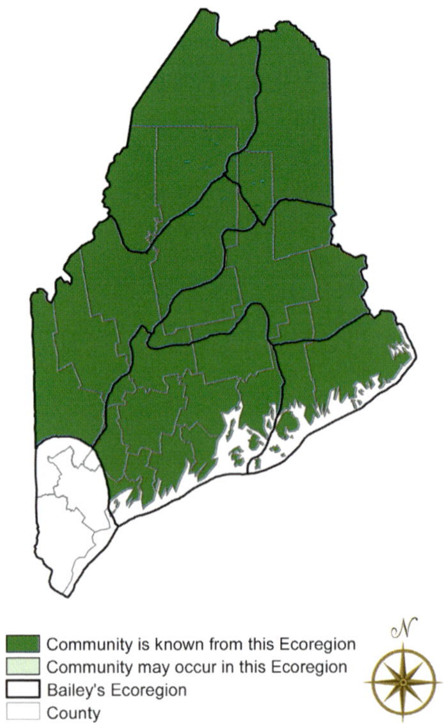

Three-lobed Bazzania

spruce has been selectively removed. As a result, the canopies of such sites are more often indicative of Beech - Birch - Maple Forests, with spruce and fir more common in the understory than in the canopy. Sites with relatively little human disturbance are rare but are moderately well represented on conservation lands.

This type provides nesting habitat for a large number of passerine bird species, including sharp-shinned hawk, Cape May warbler, black-throated blue warbler, black-throated green warbler, Blackburnian warbler, scarlet tanager, spruce grouse, Swainson's thrush, northern parula, and ovenbird. The globally uncommon early hairstreak butterfly uses beech as its larval host plant.

Distribution

Most characteristic of the New England - Adirondack Province, and extending westward from Maine; found to a lesser extent in the Laurentian Mixed Forest Province.

Landscape Pattern: Matrix. This type is intended to represent forests that are truly 'mixed' at a stand scale, rather than large blocks containing a mosaic of distinct conifer and hardwood stands. For the latter example, the 'Spruce - Northern Hardwood Forest Ecosysystem' is a more appropriate mapping unit.

Characteristic Plants

These plants are frequently found in this community type. Those with an asterisk are often diagnostic of this community.

Canopy

Balsam fir
Red spruce*
White pine*
Yellow birch*

Sapling/shrub

American beech
Balsam fir*
Hobblebush*
Mountain maple*
Red maple
Red spruce*
Striped maple*
Yellow birch

Herb

Northern wood-sorrel
Spinulose wood fern*
Starflower

Bryoid

Dicranum moss
Flat-tufted feather-moss
Pincushion moss
Three-lobed bazzania

Associated Rare Plants

Giant rattlesnake-plantain

Associated Rare Animals

Early hairstreak

Examples on Conservation Lands You Can Visit

- Big Reed Pond Preserve – Piscataquis Co.
- Black Mountain, Mahoosuc Public Lands – Oxford Co.
- Chamberlain Lake Public Lands – Piscataquis Co.
- Cranberry Brook, Moosehorn National Wildlife – Washington Co.
- Western Mountain, Acadia National Park – Hancock Co.

Spruce - Pine Woodland

State Rank S4

Community Description

This type is a mixed canopy woodland (25-70% closure) in which red spruce and/or white pine is always present and associated species vary. Red spruce or white pine is strongly dominant at some sites; at others, the canopy is mixed, with no one tree species strongly dominant. White spruce may rarely replace red spruce at coastal sites. The shrub layer is typically very sparse (and variable in composition), and the herb layer has mostly 15-50% cover. Heath shrubs are the dominant feature of the herb layer; herb species rarely exceed 8% cover. The bryoid layer is sparse at some sites (<25%) and well developed at others (35-70%). Fruticose lichens typically make up half or more of the bryoid cover.

Soil and Site Characteristics

Sites occur on mid to upper slopes (usually 10-20% slope) and low summits at elevations up to 2000'. Soils are thin (<25 cm), consisting of coarse mineral soil or poorly decomposed duff, and form patches over the bedrock substrate. The very well drained soils are acidic (pH 4.6-5.2) and nutrient poor. Some sites show evidence of past fire, but many do not.

Wild Raisin

Diagnostics

Sites are woodlands on bedrock, with conifer cover exceeding deciduous cover. Red spruce is typically dominant, or occasionally co-dominant with white pine or red spruce.

Similar Types

Other upland coniferous woodlands may include red spruce but will have other tree species (northern white cedar, pitch pine, red pine, jack pine, or black spruce) in greater abundance. Oak - Pine Woodlands may have considerable red

Location Map

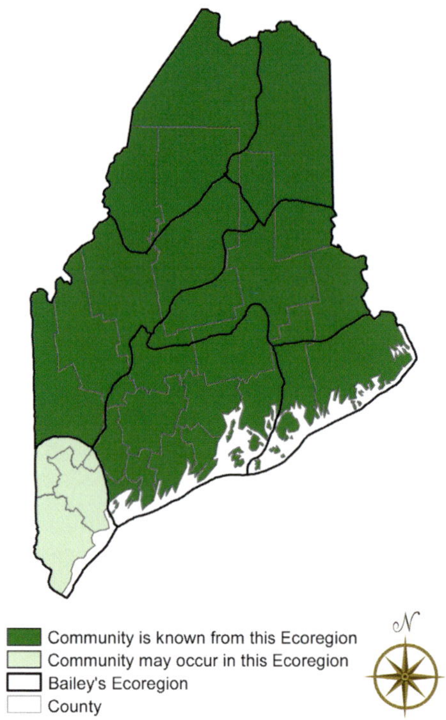

Spruce – Pine Woodland

spruce (an oak - spruce mix), but have more deciduous than coniferous tree cover. Moving downslope, or into areas of greater soil development, these woodlands can grade into spruce or pine forests, but those have more continuous canopy and less shrub and herb cover.

Conservation, Wildlife, and Management Considerations

Most sites have little pressure from development or timbering; the primary impacts are from recreational use. Communications towers or wind turbines could have an impact on some of these woodlands on mid-elevation summits. Several sites are in public or private conservation ownership.

Birds that may nest in this habitat include the sharp-shinned hawk, gray jay, yellow-bellied flycatcher, boreal chickadee, Blackburnian warbler, red crossbill, and northern parula.

Distribution

New England - Adirondack Province and Laurentian Mixed Forest Province, extending eastward, westward, and northward from Maine.

Landscape Pattern: Small Patch

Characteristic Plants

These plants are frequently found in this community type. Those with an asterisk are often diagnostic of this community.

Canopy
Balsam fir*
Black spruce*
Northern white cedar*
Paper birch*
Red spruce*
White pine*
White spruce*

Sapling/shrub
Bayberry*
Shadbush
Wild-raisin*

Dwarf Shrub
Black huckleberry*
Lowbush blueberry*
Sheep laurel*

Herb
Bracken fern

Bryoid
Dicranum moss
Red-stemmed moss
Reindeer lichen*

Examples on Conservation Lands You Can Visit

- Holbrook Island Sanctuary State Park – Hancock Co.
- Mahoosuc Mountain, Mahoosuc Public Lands – Oxford Co.
- Mansell Mountain, Acadia National Park – Hancock Co.
- Nahmakanta Public Lands – Piscataquis Co.
- Petit Manan Point, Petit Manan National Wildlife – Washington Co.

Spruce Rocky Woodland

State Rank S4

Community Description

These partial canopy woodlands (usually less than 50% cover) support red spruce, mixed with lesser amounts of other conifers, birches, red oak, or beech. Vegetation tends to be very patchy due to the substrate and may include large expanses with little to no vascular vegetation. Beneath the scattered trees and smaller, sapling sized trees, heath shrubs or herbs are found in pockets (15-45% cover overall). Typical species include lowbush blueberry, common polypody, rusty cliff fern, and crinkled hairgrass. Bryoids may include typical forest species in areas where the tree canopy is fairly well developed; in more open areas, rock-tripe lichens and patches of reindeer lichens are characteristic.

Soil and Site Characteristics

Sites occur on talus slopes, with vegetation developing in patches among the rocks. Sites are usually steep (>20% slope), have very limited soil, and are typically on acidic rocks such as granite.

Rock Tripe Lichen

Diagnostics

These are open canopy woodlands or barren talus slopes in which red spruce is dominant.

Similar Types

Red Spruce - Mixed Conifer Woodlands have similar canopies but occur on bedrock versus talus and have more heaths and fewer ferns and rock tripe lichens. Other talus woodlands (Birch - Oak Talus Woodlands and Ironwood - Oak - Ash Woodlands) are dominated by deciduous trees, not conifers. Large talus slopes at high elevations habitats (e.g., Mt. Abram) will likely be classified as one of the alpine or sub-alpine types.

Location Map

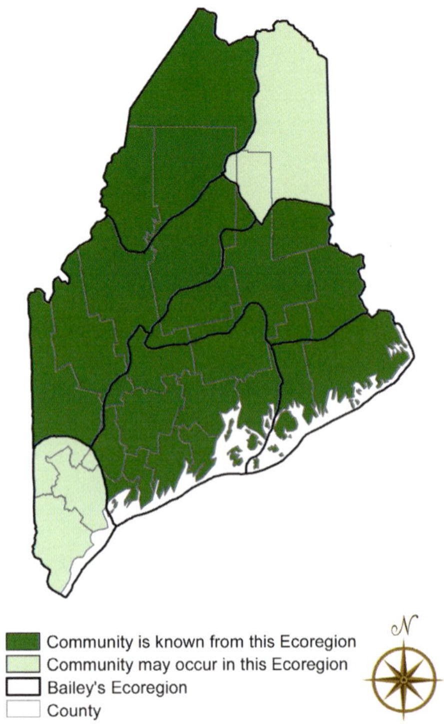

Spruce Rocky Woodland

Characteristic Plants

These plants are frequently found in this community type. Those with an asterisk are often diagnostic of this community.

Canopy
Paper birch
Red spruce*

Sapling/shrub
Mountain holly*
Red raspberry*
Red spruce

Dwarf Shrub
Black huckleberry
Lowbush blueberry*
Sheep laurel*

Herb
Common hairgrass
Rock polypody

Bryoid
Dicranum moss
Fringed Ptilidium liverwort
Rock-tripe lichen
Tufted reindeer-lichen*

Conservation, Wildlife, and Management Considerations

Most sites have little potential economic use. Sites are generally small and embedded within a coniferous or mixed forest hill slope.

The cool, moist crevices in these talus slopes provide excellent habitat for the rock vole and long-tailed shrew. This community may also host the Gaspé shrew, a species not currently known from Maine that occurs just to the north in Canada.

Spruce Rocky Woodland

Distribution

New England - Adirondack Province and Laurentian Mixed Forest Province, extending westward (and probably eastward and northward) from Maine.

Landscape Pattern: Small Patch

Examples on Conservation Lands You Can Visit

- Acadia National Park – Hancock Co.
- Deboullie Ponds Public Lands – Aroostook Co.

Subalpine Fir Forest

State Rank S3

Community Description

Balsam fir, or mixtures of fir and heart-leaved birch, form a dense canopy of somewhat stunted trees. Patches of heart-leaved birch and mountain ash are common where wind, fire, or landslides have created openings, along with a dense shrub layer of mountain ash, hobblebush, and regenerating fir. Herbs may be sparse, or may form locally dense patches in openings; wood ferns and big-leaved aster in particular tend to be patchy. In some expressions of this type that have developed after fire, the canopy consists almost entirely of paper birch or heart-leaved birch. Fir waves, an unusual landscape pattern of linear bands of fir dieback and regeneration, are another variant of this community.

Soil and Site Characteristics

These forests are commonly found above 2700' on level ridgetops and steep, upper slopes. The mineral soil layer is thin, typically 10-30 cm, and rocky. Natural disturbances such as landslides, wind, fire, and spruce-budworm can exert lasting influences on community dynamics. Recurrent landslides can keep some areas in birch - mountain-ash dominance.

Fir Waves on Crocker Mountain

Diagnostics

Fir or heart-leaved birch (occasionally paper birch) are dominant in a subalpine setting.

Similar Types

One form of the Maritime Spruce - Fir Forest type is compositionally very similar but occurs at sea level in the extreme environment of the Downeast coast. Decreasing in elevation, this type can grade into Spruce - Fir - Wood-sorrel - Feather-moss Forest or Spruce - Fir - Broom-moss Forest, which are distinguished by their higher proportion of spruce in the canopy and by less stunted trees.

Conservation, Wildlife, and Management Considerations

Although subalpine forests are naturally dynamic as they cycle through periods

Location Map

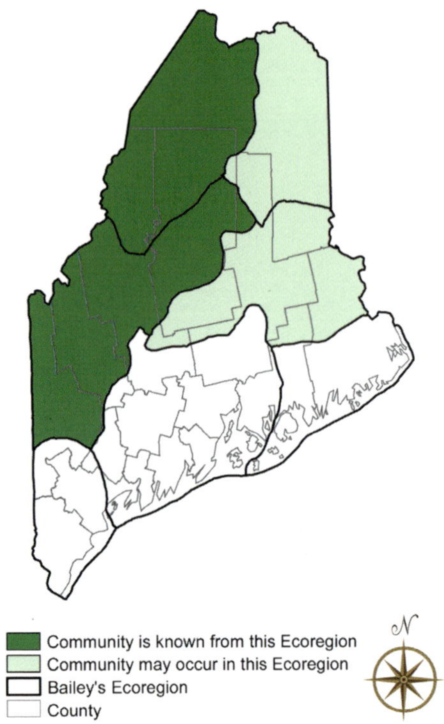

Subalpine Fir Forest

of weather and insect damage and regeneration, they appear to be relatively stable in overall extent and are extensive on Maine's higher mountains. Many major occurrences are well protected within public lands or private conservation lands. On the few remaining sites on private lands, timber harvesting, recreation, and windpower development could cause lasting impacts. At some sites, past harvesting has resulted in prolific growth of hay-scented and mountain wood fern, inhibiting tree regeneration.

This high-elevation forest community type may be used as nesting habitat by a number of high elevation and/or coniferous forest specialist bird species, such as the spruce grouse, dark-eyed junco, bay-breasted warbler, black-backed woodpecker, white-throated sparrow, and blackpoll warbler. The rare Bicknell's thrush inhabits structurally complex forests above 2500'. The rock vole and long-tailed shrew both inhabit cool moist crevices in rocky habitat at high elevations. Northern bog lemmings may inhabit wet sub-alpine spruce - fir forests in which peat moss is present.

Distribution

Western and central Maine westward (New England - Adirondack Province); likely extends northeasterly to the Gaspé Peninsula.

Landscape Pattern: Large Patch

Characteristic Plants

These plants are frequently found in this community type. Those with an asterisk are often diagnostic of this community.

Canopy

Balsam fir*
Heart-leaved paper birch
Paper birch*
Red spruce

Sapling/shrub

Balsam fir*
Black spruce*
Heart-leaved paper birch*
Mountain ash*
Wild-raisin

Herb

Balsam fir*
Big-leaved aster*
Bluebead lily
Mountain wood fern*
Northern wood-sorrel
Spinulose wood fern*
Starflower

Bryoid

Common broom-moss
Three-lobed bazzania

Associated Rare Plants

Northern comandra

Examples on Conservation Lands You Can Visit

- Baxter State Park – Piscataquis Co.
- Big Squaw Mountain Public Lands – Piscataquis Co.
- Bigelow Preserve Public Lands – Somerset Co.
- Crocker Mountain, Appalachian Trail – Franklin Co.
- Mahoosuc Mountain, Mahoosuc Public Lands – Oxford Co.
- Sugarloaf Mountain, Appalachian Trail – Franklin Co.

Sugar Maple Forest

State Rank: S4

Community Description

The canopy is dominated by sugar maple, with white ash a frequent associate in southern or central Maine. Other associated hardwood species include yellow birch and striped maple. Beech may be present but is less abundant than in Beech - Birch - Maple Forests. Conifers are usually sparse. The closed-canopy forest has little to moderate shrub cover, moderate to dense herb cover, and may have local carpets of sugar maple seedlings. Characteristic shrubs include alternate-leaved dogwood and red-elderberry. Typical herbs of this type, which are scarce or absent from beech-birch-maple forests, include jack-in-the-pulpit, doll's eyes, round-leaved violet, grape fern, zig-zag goldenrod, spikenard, false Solomon's seal, and long-stalked sedge.

Soil and Site Characteristics

These northern hardwood forests occur at moderate elevations on slightly enriched, moderately drained soils – often silt loams derived from pelite or other subacidic bedrock. They may occur as inclusions within typical northern hardwood forests, or, in northern Maine, they may occur over larger areas (hundreds of acres) and be the locally dominant northern hardwood forest.

Diagnostics

Sugar maple is dominant or co-dominant (>50% cover) with yellow birch or white ash. Conifers and beech each form less than 25% cover. The herbaceous layer includes at least several species indicative of moderate soil enrichment, as noted above, but lacks richer site species characteristic of Maple - Basswood - Ash Forest.

Similar Types

These forests are intermediate in nutrient regime and composition between Maple - Basswood - Ash Forests and Beech - Birch - Maple Forests. Maple - Basswood - Ash Forests are typically small-patch types in ravines or coves, with more restrictive rich-soil indicators such as basswood, maidenhair fern, blue cohosh, broad-leaved sedge, plantain-leaved sedge, and Braun's holly fern. Beech - Birch - Maple Forests typically contain as much if not more beech than sugar maple and lack the diversity and abundance of rich site indicators. Beech - Birch - Maple Forests

Location Map

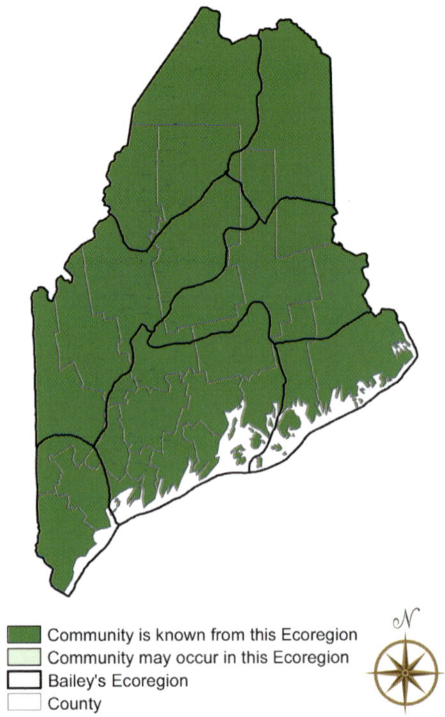

Sugar Maple Forest

often occur on drier sites (e.g., ridgetops) and/or more acidic soils (e.g., granitic types in Downeast Maine).

Conservation, Wildlife, and Management Considerations

This forest type has been extensively harvested and managed in northern Maine. Most management techniques diverge from the natural gap pattern, which is at the scale of single trees or small groups of trees. Large (>100 acres) examples reflecting only natural disturbance are scarce statewide, and intact examples in central and southern Maine tend to be smaller and more isolated.

This type provides nesting habitat for a large number of passerine bird species, such as the black-throated green warbler, rose-breasted grosbeak, scarlet tanager, ovenbird, and a large proportion of the global population of black-throated blue warblers.

Distribution

New England - Adirondack Province and Laurentian Mixed Forest Province. Extends east, west, and north from Maine.

Landscape Pattern: Small to Large Patch

Characteristic Plants

These plants are frequently found in this community type. Those with an asterisk are often diagnostic of this community.

Canopy

American beech
Sugar maple*
White Ash*
Yellow birch

Sapling/shrub

Alternate-leaved dogwood
Red Elderberry
Striped maple*
Sugar maple*
Yellow birch

Herb

Christmas fern
Doll's eyes*
Jack in the pulpit
Round-leaved violet
Sarsaparilla
Shining clubmoss*
Spikenard*
Striped maple

Associated Rare Plants

Autumn coral-root
Broad beech fern
Cut-leaved toothwort
Nodding pogonia
Tall white violet

Associated Rare Animals

Early hairstreak

Examples on Conservation Lands You Can Visit

- Deboullie Ponds Public Lands – Aroostook Co.
- Squa Pan Public Lands – Aroostook Co.
- White Mountain National Forest – Oxford Co.

White Cedar Woodland

State Rank S2

Community Description

This type is a partial canopy upland woodland (30-85% closure) in which northern white cedar is the dominant tree. Cedar may make up almost the entire canopy, or it may be mixed with lesser amounts of white pine, balsam fir, or other species. Lower layers all tend to be sparse (<25% cover each), with only scattered plants. Species composition varies according to the site's moisture regime. Where seeps maintain somewhat mesic conditions, one finds plants such as bush-honeysuckle and meadowsweet, while under more xeric conditions mountain holly and black huckleberry are representative. The sparse bryoid layer contains mostly bryophytes rather than lichens.

Soil and Site Characteristics

Sites occur on rocky hillslopes with 10-70% slope. The substrate is bedrock, talus, or coarse glacial till, with a thin (<15 cm) layer of sandy soil or poorly decomposed organic duff developing in pockets or over portions of the rock. Sites with seepage moisture tend to have deeper soils, but soil depth is <30 cm. Known sites are either on acidic substrates near the coast or on somewhat calcareous rocks inland.

Lowbush Blueberry

Diagnostics

This type is a partial canopy woodland dominated by northern white cedar (>30% cover); heath shrubs are not prominent in the herb layer (<20% cover).

Similar Types

Some Red Spruce - Mixed Conifer Woodlands may be co-dominated by northern white cedar, but they also have a heath shrub layer. Other northern

Location Map

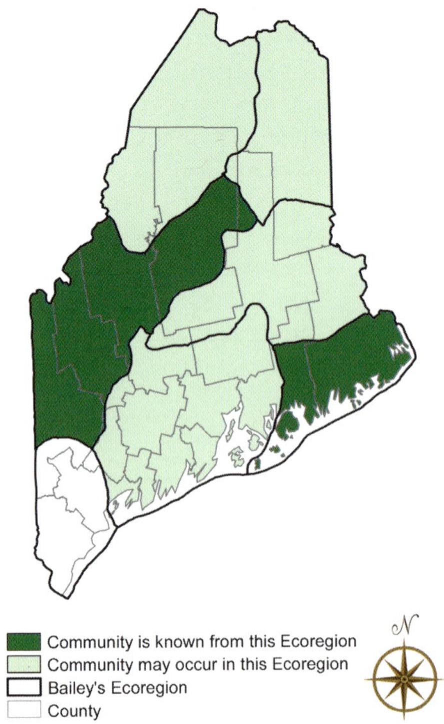

Fragrant Wood Fern

white cedar dominated community types are wetlands or seepage forests on flatter, deeper, mineral or organic soils.

Conservation, Wildlife, and Management Considerations

Little is known about the distribution and landscape setting of these woodlands. Documented at sites either on the Downeast Coast (e.g., Acadia National Park) or at higher elevations inland. Sites are generally inoperable from a forestry standpoint, and recreational use may be the biggest conservation concern. There is evidence of fire at some sites, but the role of fire in creating or perpetuating this type is not clear.

Examples of this community type that occur on bedrock or talus may provide cool, moist crevices that offer suitable habitat for the long-tailed shrew.

Distribution

Laurentian Mixed Forest and New England - Adirondack Provinces, known from east-coastal and north-central Maine; distribution elsewhere not well documented.

Landscape Pattern: Small Patch

Characteristic Plants

These plants are frequently found in this community type. Those with an asterisk are often diagnostic of this community.

Canopy

Balsam fir

Northern white cedar*

Paper birch

Sapling/shrub

Balsam fir*

Northern white cedar

Dwarf Shrub

Lowbush blueberry

Herb

Starflower

Bryoid

Dicranum moss

Associated Rare Plants

Fragrant wood fern

Narrow-leaved reed grass

Examples on Conservation Lands You Can Visit

- Bubble Pond, Acadia National Park – Hancock Co.
- Pemetic Mountain, Acadia National Park – Hancock Co.

White Oak - Red Oak Forest

State Rank S3

Community Description

This deciduous forest type is dominated by red oak with white oak as a canopy associate. White pine is occasionally present, but conifers comprise only a small proportion (<20%) of the canopy. Sugar maple and beech may be present in minor amounts. Shrubs occur as well spaced patches; typical species include striped maple and ironwood. The forest floor is characterized by low heath shrubs such as lowbush blueberry. Common herbs include woodland sedge, bracken fern, whorled loosestrife, and Canada mayflower. Bryoids are very sparse.

Soil and Site Characteristics

Sites of this type are on well drained gentle slopes (up to 20%) below 600' elevation. The soil is generally well drained, stony, sandy loam, fairly acidic (pH 4.8-5.0), and 20-50 cm deep. These forests are usually on somewhat sheltered sites.

Diagnostics

Forests dominated by a mixture of red oak and white oak, without a strong white pine or hickory component. White oak forms at least 25% of the canopy.

White Oak –Red Oak Forest

Similar Types

Oak - Pine Forests lack white oak and may have white pine co-dominant with red oak. Red Oak - Northern Hardwoods - White Pine Forests lack white oak and have a larger component of beech or sugar maple. Shagbark hickory is dominant in Oak - Hickory forests.

Conservation, Wildlife, and Management Considerations

The few mature sites of White Oak - Red Oak Forest known to remain in Maine are all on land that was once cleared. The known sites are subject to fragmentation

Location Map

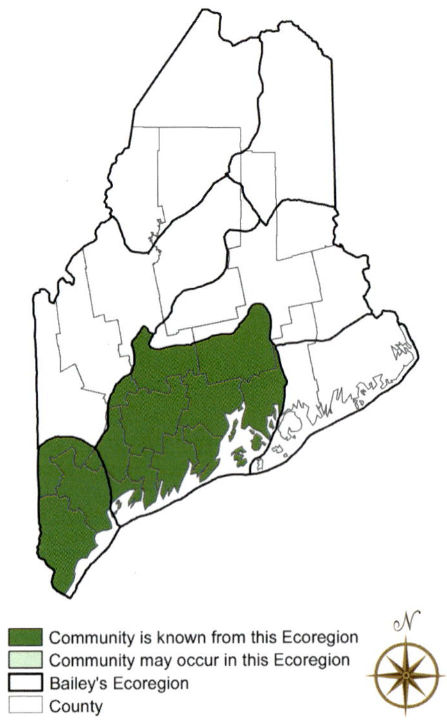

Scarlet Oak Leaf held against Oak Bark

Characteristic Plants

These plants are frequently found in this community type. Those with an asterisk are often diagnostic of this community.

Canopy

Red oak*
Sugar maple
White oak*

Herb

Rough-leaved ricegrass
Wild sarsaparilla
Wild-oats
Wintergreen
Woodland sedge*

Associated Rare Plants

Bitternut hickory
Chestnut oak
Flowering dogwood
Scarlet oak

Associated Rare Animals

Early hairstreak
Red-winged sallow
Whip-poor-will

by timber harvesting, clearing for agriculture, and residential development, uses that have reduced this naturally rare type even further. Community dynamics are not well known, but there are some indications that red oak regenerates more strongly than white oak at some sites and may replace it over time. Fire may also play a role in natural regeneration. Most occurrences of this type are on private lands.

This type offers habitat for a variety of birds, including scarlet tanager and ovenbird. Mature occurrences of this community type offer excellent potential sites for cavity dwellers such as the southern flying squirrel. The rare red-winged sallow moth uses red oak as one of its host plants and may be found in this community.

White Oak Leaf

Distribution

Restricted primarily to southern Maine, characteristic of the Eastern Broadleaf Forest Province.

Landscape Pattern: Small to Large Patch, generally 100 acres or less.

Examples on Conservation Lands You Can Visit

- Mt. Agamenticus – York Co.
- Sebago Lake State Park – Cumberland Co.

White Pine Forest

State Rank S4

Community Description

This is a closed canopy forest type in which white pine is dominant. Occasionally red spruce, red pine, hemlock, or (coastally) northern white cedar may be nearly co-dominant with the white pine; in fact, because the pine trees tend to be larger and the other trees smaller, the smaller trees may be more numerous. In many of these forests, the dense and strongly coniferous canopy limits understory growth. Shrub cover is rarely >20% and the herb layer rarely exceeds 30%. The herb layer can include a spotty mixture of dwarf shrubs such as lowbush blueberry, forbs, or ferns, but graminoids are very uncommon. Canada mayflower is frequent. The ground layer is mostly conifer litter, with bryoid cover <25%; large hair-cap moss and red- stemmed moss are common species.

Soil and Site Characteristics

This type occurs on sandy to loamy mesic soils (usually well drained, occasionally imperfectly drained or very well drained), often with a slowly decomposing duff layer of conifer needles. Soils are generally shallow (<40 cm) and moderately acidic (pH 5.0-6.0). These forests are usually at low elevations (<900') on slopes or coarse-textured flats.

White Pine

Diagnostics

White pine is dominant (>33% cover); red oak and northern hardwood species (beech, sugar maple, yellow birch) total <25% cover. The heath shrub and herbaceous layers are sparse.

Similar Types

Red Pine - White Pine Forests have >33% cover of red pine. Hemlock

Location Map

White Pine Forest

Forests have more hemlock than white pine. Where red spruce is co-dominant, this type can grade into Spruce - Fir forest types. Oak - Pine Forests and Red Oak - Northern Hardwoods - White Pine Forests have red oak at > 33% cover.

Conservation, Wildlife, and Management Considerations

Demand for white pine has considerably reduced mature, undisturbed examples of this type. Most sites known to be of high ecological quality lack formal protection. Maintaining the surrounding lands as forest is important in conserving particular stands of this type, particularly given that many known examples are small (<50 acres).

This community type may be used as nesting habitat by a number of coniferous or mixed forest specialist bird species such as the sharp-shinned hawk, pine warbler, black-throated green warbler, blackburnian warbler, and red crossbill.

Distribution

Statewide, less common northward; extends in all directions from Maine.

Landscape Pattern: Large Patch

Characteristic Plants

These plants are frequently found in this community type. Those with an asterisk are often diagnostic of this community.

Canopy
Eastern hemlock*
Northern white cedar
Red maple
Red spruce
White pine*

Sapling/shrub
Balsam fir
Beaked hazelnut*
Eastern hemlock
Red maple*
Red spruce
White pine*
Wild-raisin*

Dwarf Shrub
Lowbush blueberry*
Swamp dewberry*

Herb
Balsam fir
Canada mayflower*
Eastern hemlock
Starflower
White pine*

Bryoid
Dicranum moss*

Examples on Conservation Lands You Can Visit

- Bearce Lake, Moosehorn National Wildlife Refuge – Washington Co.
- Bigelow Preserve Public Lands – Franklin Co.
- Chamberlain Lake Public Lands – Piscataquis Co.
- Gero Island Public Lands – Piscataquis Co.
- Scientific Forest Management Area, Baxter State – Piscataquis Co.

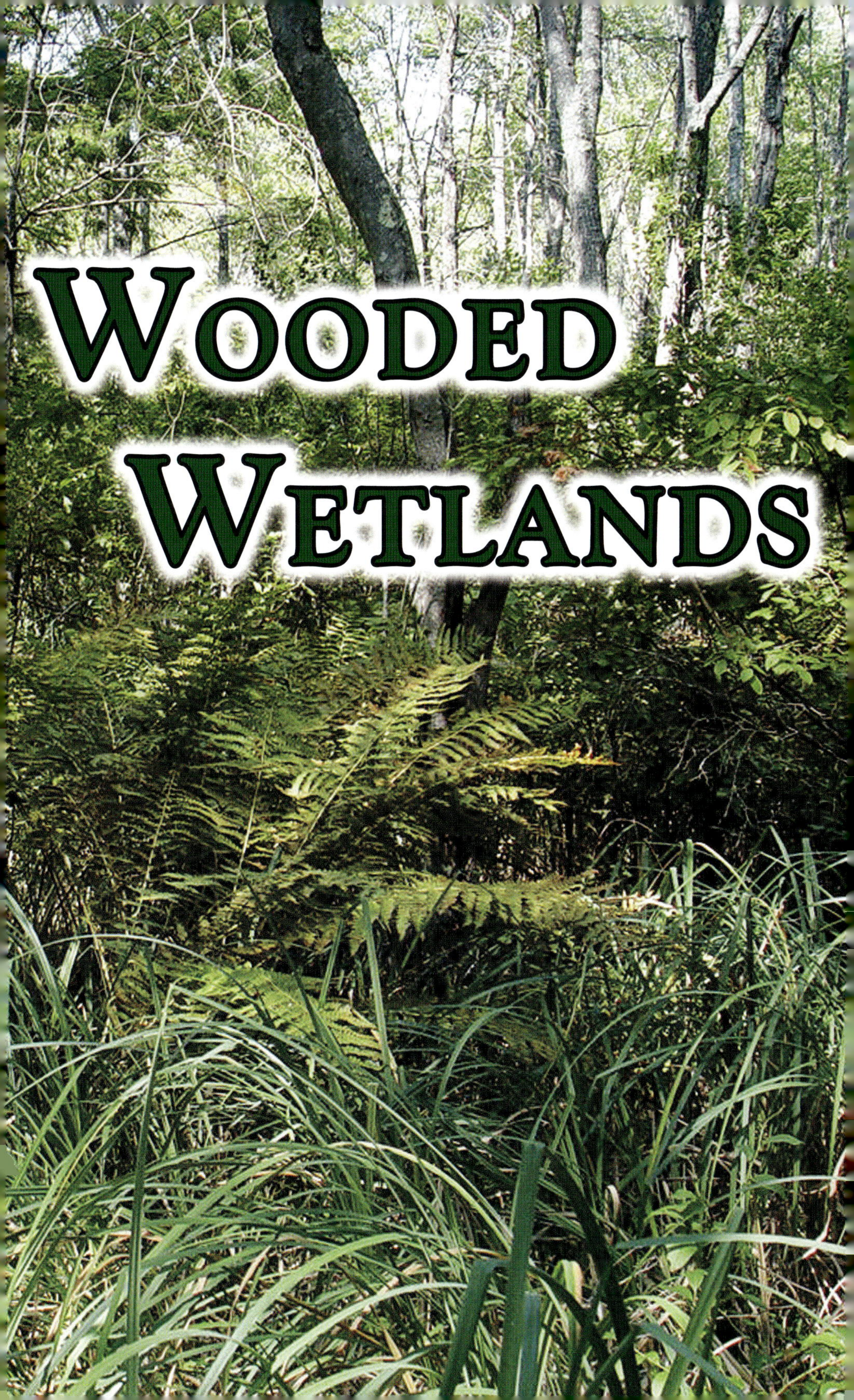
Wooded
Wetlands

Atlantic White Cedar Bog

State Rank S1

Community Description

These peatlands are dominated by dwarf heath shrubs with a sparse tree layer of Atlantic white cedar. Sheep laurel, Labrador tea, dwarf huckleberry, and other heath shrubs can form an almost continuous carpet beneath the stunted cedars. Leatherleaf is a common shrub. Herbs are sparse. The bryoid layer is well developed with peat moss covering the ground and forming the substrate.

Soil and Site Characteristics

Sites occur in basin wetlands of the southwestern coast, usually in areas transitional between fen and bog. They generally maintain contact with the groundwater but lack indicators of minerotrophic conditions. Peat and water are highly acidic.

Diagnostics

These peatlands possess abundant low heath shrubs and peat moss and are only sparsely forested by Atlantic white cedar.

Atlantic White Cedar Branch

Similar Types

Atlantic White Cedar Swamps occur on mineral soil or on thin peat (<30 cm) over mineral soil and typically have higher canopy closure (>50%). Other sparsely forested bog or fen communities, such as Sheep Laurel Dwarf Shrub Bog and Leatherleaf Boggy Fen, can have similar composition in the dwarf shrub, herb, and bryophyte layers, but lack Atlantic white cedar.

Conservation, Wildlife, and Management Considerations

These bogs occur as part of larger peatlands, and maintaining the hydrologic integrity of the entire

Location Map

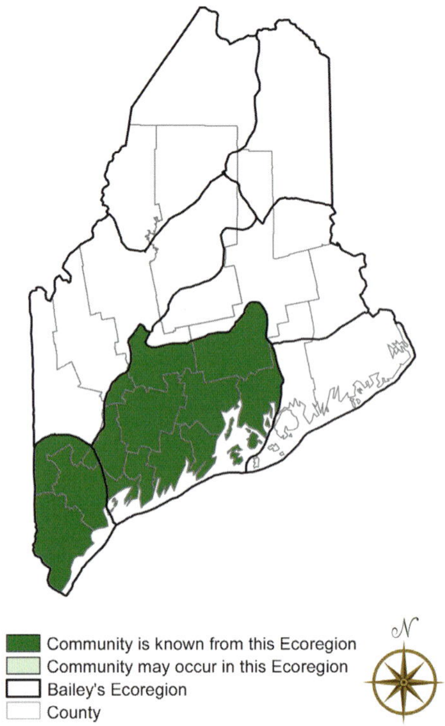

Atlantic White Cedar Bog

wetland with upland buffers is key. The cedars generally remain small, and therefore are of limited economic value. Most known sites in Maine are in conservation ownership. Frequent birds associated with this community include common yellowthroat and northern waterthrush. These wetlands provide habitat for the rare Hessel's hairstreak butterfly, which feeds on Atlantic white cedar in its larval stage.

Distribution

Eastern Broadleaf Forest Province, extending southward from Maine.

Landscape Pattern: Small Patch

Atlantic White Cedar Bark

Characteristic Plants

These plants are frequently found in this community type. Those with an asterisk are often diagnostic of this community.

Canopy
Atlantic white cedar*
Black spruce*
White pine

Sapling/shrub
Atlantic white cedar*
Black huckleberry
Mountain holly

Dwarf Shrub
Dwarf huckleberry*
Labrador tea*
Large cranberry
Leatherleaf*
Sheep laurel*

Herb
Pitcher plant
Tufted cotton-grass

Bryoid
Dicranum moss
Reindeer lichen
Sphagnum mosses*

Associated Rare Plants

Atlantic white cedar

Associated Rare Animals

Hessel's hairstreak

Examples on Conservation Lands You Can Visit

- Knight Pond, St. Clair Preserve – Waldo Co.
- Saco Heath Preserve – York Co.

Atlantic White Cedar Swamp

State Rank S2

Community Description

These densely forested communities allow little direct light to the forest floor. The canopy is usually a uniform cover of Atlantic white cedar with occasional black spruce and red maple; in some sites, red maple may be co-dominant. The ground is a mosaic of moss covered hummocks and hollows. Where light penetrates the canopy, shrubs such as highbush blueberry, black huckleberry, mountain holly, or winterberry may be prominent. The herb layer features dense patches of tree regeneration in some openings. Herbaceous species are typically more abundant than dwarf heath shrubs.

Soil and Site Characteristics

Atlantic white cedar swamps are underlain by shallow peat over mineral soil, or, less commonly, by mineral soil with little organic matter. They are flat, found in poorly drained depressions where standing water is present at least part of the year. All are at less than 300' elevation.

Atlantic White Cedar Branch with Fruit

Diagnostics

These wetlands have greater than 60% tree canopy cover. Atlantic white cedar is dominant or at least common. Evergreen shrubs are often present.

Similar Types

Northern White Cedar Swamps may be compositionally similar except for the dominant species. Other similar types lack Atlantic white cedar, except for Atlantic White Cedar Bogs, where the trees are stunted, the canopy is well under 60%, and the peat moss substrate is deep (>50 cm).

Conservation, Wildlife, and Management Considerations

Most occurrences of this community

Location Map

Atlantic White Cedar Swamp

type occur in rapidly growing southern Maine. Because of this and because the timber is valuable, these occurrences are vulnerable. All of Maine's stands have been harvested in the past; some are now mature. Their perpetuation appears to be partially dependent on disturbance events such as blowdowns or fire. In the absence of such events, cedar may be replaced by more shade tolerant species such as red maple and hemlock. The quality of Atlantic White Cedar Swamps may also be impaired by filling, water quality degradation, water level changes, and timber harvesting.

These forested wetlands provide habitat for a variety of conifer nesting birds and for the rare Hessel's hairstreak butterfly, which feeds on Atlantic white cedar in its larval stage.

Characteristic Plants

These plants are frequently found in this community type. Those with an asterisk are often diagnostic of this community.

Canopy
Atlantic white cedar*
Black spruce*
Red maple*

Sapling/shrub
Atlantic white cedar*
Mountain holly*

Dwarf Shrub
Labrador tea
Sheep laurel

Herb
Canada mayflower
Cinnamon fern*
Creeping snowberry
Goldthread*
Pitcher plant
Starflower
Three-seeded sedge
Wild sarsaparilla*

Bryoid
Dicranum moss
Sphagnum mosses*
Three-lobed bazzania

Associated Rare Plants

Atlantic white cedar
Smooth winterberry

Associated Rare Animals

Hessel's hairstreak

Distribution

Along the entire Atlantic coastal plain north to central Maine. They have been greatly reduced in extent and are one of the rarest forest types in the eastern US.

Landscape Pattern: Small to Large Patch.

Examples on Conservation Lands You Can Visit

- Appleton Bog Preserve – Knox Co.
- Massabesic Experimental Forest – York Co.
- Mt. Agamenticus – York Co.

Balsam Poplar Floodplain Forest

State Rank S2

Community Description

These partly open to closed canopy forests are dominated by balsam poplar and occur along medium-sized rivers. This type includes a wide range of early seral, wetland, and floodplain species. These typically include American elm, black or green ash, ostrich fern, lady fern, sensitive fern, bluejoint, red-osier dogwood, virgin's bower, speckled alder, dwarf raspberry, broad leaved goldenrod, inflated sedge, and wood nettle.

Soil and Site Characteristics

Sites occur on seasonally inundated floodplains or slightly elevated terraces flanking low gradient rivers in central and northern Maine. Sites are often embedded within a matrix of open oxbows and shrub thickets. These forests have lower frequency and duration of flooding than Silver Maple Floodplain Forests. Soils are alluvial fine sand or silt, usually with good drainage capacity.

Balsam Poplar Floodplain Forest

Diagnostics

Sites occupy a floodplain or river terrace setting with mineral soil. Balsam poplar is dominant or co-dominant; black or green ash and American elm may be present or codominant. The dense herb layer includes ferns, bluejoint, sedges, and shrubs such as red-osier dogwood and speckled alder.

Similar Types

Silver Maple Floodplain Forests and Hardwood River Terrace forests do not

Location Map

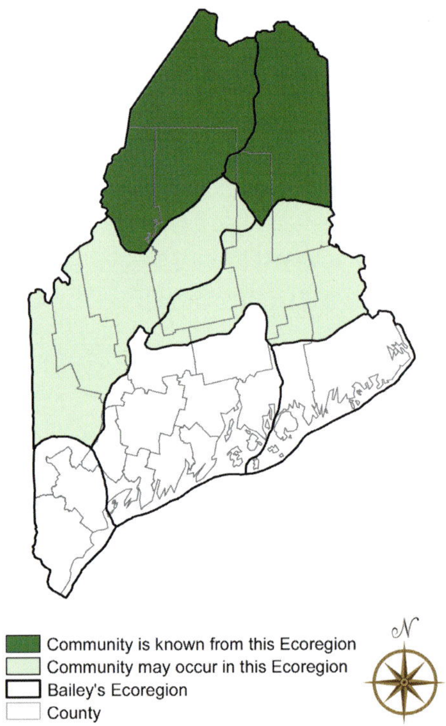

Balsam Poplar Leaves

have balsam poplar or black ash as a dominant or co-dominant.

Characteristic Plants

These plants are frequently found in this community type. Those with an asterisk are often diagnostic of this community.

Canopy

American elm
Balsam fir
Balsam poplar*
Black ash

Dwarf Shrub

Dwarf raspberry

Herb

Bluejoint
Ostrich fern*
Royal fern
Sensitive fern
Wood nettle

Conservation, Wildlife, and Management Considerations

The few known examples of this type have been influenced by past disturbances, including harvesting, beaver damage, and potentially fire. As a result, long term successional trends are not clear. Exotic plant species, which may displace those native to our area, may be easily transported to sites by river waters and represent a threat to the integrity of these forests. The biggest threat may be hydrologic alteration. All known sites in Maine are on private land.

Little data is available on wildlife use of this specific community type, but it is suspected that it would support many of those species associated with Hardwood River Terrace Forests in central and northern Maine.

Landscape Pattern: Small Patch

Sensitive Fern

Distribution

Narrow floodplains along the shores or islands of medium and smaller rivers in northern Maine, Quebec, and the Maritime Provinces.

Black Ash Swamp

State Rank S4

Community Description

This northern forested wetland type is characterized by a hardwood or mixed overstory and lush understory on shallow peat or muck soils. While black ash is characteristic in these small-patch communities, Black Ash Swamps often occur within or adjacent to larger expanses of Northern White Cedar Swamps or Red Maple Swamps and may be transitional between the two. Tree cover ranges from 25-80%. Common plants of forested wetlands are often dominant in the understory, including sensitive fern, royal fern, marsh fern, tussock sedge, fowl mannagrass, and spotted touch-me-not, as well as a scattering of rich-site wetland herbs including purple avens, Robbins' ragwort, foamflower, small enchanter's nightshade, and white turtlehead. Mosses and liverworts are usually abundant and may form a continuous carpet.

Soil and Site Characteristics

Black Ash Swamps occupy broad basins or seepage sites on lower gentle slopes, often in association with larger areas of cedar or red maple swamp. They typically occur in rolling to low terrain in central and northern regions of the state where higher pH soils or groundwater discharge occur. Saturated soils of well decomposed organic matter (peat or muck) are typical and surface rivulets and springs are occasional.

Showy Lady's-slipper

Diagnostics

Black ash is dominant or co-dominant (at least 40% cover) with northern white cedar or red maple in a wetland setting. Sensitive fern and royal fern each form more cover than cinnamon fern.

Similar Types

Northern White Cedar Swamps and

Location Map

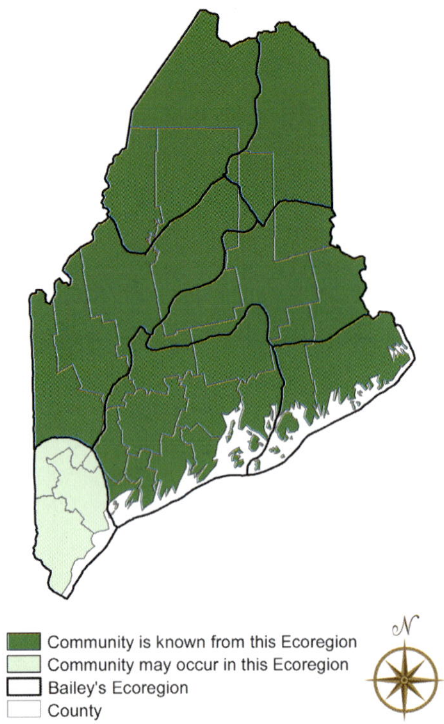

Black Ash Leaves

Cedar-Spruce Seepage Forests each have more conifer than hardwood cover in the canopy. Red Maple Swamps have more red maple than black ash in the canopy.

Conservation, Wildlife, and Management Considerations

Wind is apparently the primary form of natural disturbance in these swamps, as black ash is shallow-rooted. Although black ash does not grow as fast as other species, its ability to stump sprout may be an important reproductive strategy in response to disturbance. Black ash is fairly long lived (>250 years). Beavers have altered many black ash swamps, converting them into open or shrub-dominated wetlands.

Breeding birds found in black ash swamps may include great-crested flycatcher, brown creeper, veery, and northern waterthrush. Wood ducks may also breed in these swamps if there is open water available.

Characteristic Plants

These plants are frequently found in this community type. Those with an asterisk are often diagnostic of this community.

Canopy

Black ash*
Northern white cedar*
Red maple
Yellow birch

Sapling/shrub

Alderleaf buckthorn*
Black ash*
Mountain holly
Northern white cedar
Red maple
Speckled alder*
Winterberry
Witherod

Herb

Foamflower
Marsh fern
Purple avens
Robbins' ragwort
Royal fern
Sensitive fern*
Small enchanter's nightshade
Spotted touch-me-not

Associated Rare Plants

Bog bedstraw
Showy lady's-slipper
White adder's mouth

Distribution

New England - Adirondack Province and Laurentian Mixed Forest Province. Most common in northern and eastern Maine, and likely extends to the east, west, and north.

Landscape Pattern: Small patch.

Black Spruce Bog

State Rank S4

Community Description

This open canopy peatland type is characterized by black spruce and/or larch trees over typical bog vegetation of heath shrubs, graminoids, and peat mosses. It is the most common type of 'forested bog' in Maine. Canopy closure is usually 20-50% and occasionally ranges up to 85%. Black spruce is usually dominant, but in some cases larch (or rarely fir) may be more abundant. Red maple may be a component in somewhat more minerotrophic portions, and white pine may occur on hummocks. The shrub layer, including small trees, is usually well developed (>30% cover). Labrador tea and three-seeded sedge are characteristic species. The bryoid layer has close to 100% cover and is dominated by peat mosses; sparse reindeer lichens may occur.

Soil and Site Characteristics

Sites occur in a peatland setting, usually <1200' elevation, characteristic of nutrient poor or highly acidic peatlands (pH 4.2-5.2). These bogs may occur as part of fens, especially in kettleholes, and are standard constituents of raised (ombrotrophic) bogs.

Diagnostics

Canopy closure is at least 20%, with black spruce and/or larch strongly dominant, though red maple can also be present. Heath shrubs occur in the ground layer. Sites occur in a peatland setting (peat generally >30 cm deep).

Similar Types

Sheep Laurel Dwarf Shrub Bogs and Leatherleaf Boggy Fens, which are often adjacent on the ground, share many shrub and herb species but are considerably more open (<20% canopy cover). More minerotrophic sites tend to have wooded fens (Red Maple Wooded Fen, Northern White Cedar Woodland Fen), with black spruce less dominant.

Conservation, Wildlife, and Management Considerations

Black spruce bogs generally occur as part of larger peatlands. Maintaining the hydrologic integrity of the entire wetland with upland buffers is key. The trees mostly remain small and have limited economic use. Several known sites are in public ownership.

Bogs with scattered tall larch or snags provide suitable perching and foraging habitat for the rare olive-sided flycatcher. Similarly, the three-toed woodpecker

Location Map

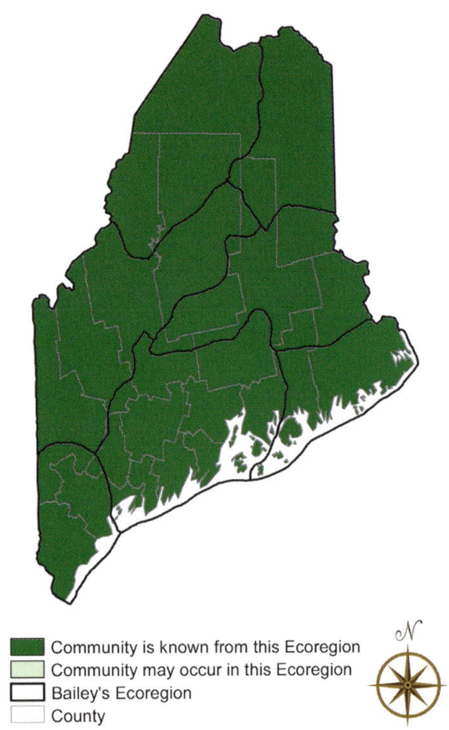

Black Spruce Bog

inhabits bogs with large numbers of dead trees. Palm warblers, common yellowthroats, and northern waterthrushes are specialists that breed primarily in this community type. The bog elfin butterfly uses black spruce as a larval host plant. The western pine elfin uses jack pine as its host plant in the Midwest but is associated with black spruce in Maine and may be found in this community in northwestern Maine. Thaxter's pinion moth uses larch as one of its larval hosts.

Distribution

Statewide, especially in the New England - Adirondack Province and Laurentian Mixed Forest Province. Extending eastward, westward, and northward from Maine, and as more isolated occurrences southward.

Landscape Pattern: Small to Large Patch

Examples on Conservation Lands You Can Visit

- Gassabias Lake, Duck Lake Public Lands – Hancock Co.
- Moose River – Somerset Co.
- Salmon Brook Lake Bog Public Lands – Aroostook Co.

Characteristic Plants

These plants are frequently found in this community type. Those with an asterisk are often diagnostic of this community.

Canopy

Balsam fir*
Black spruce*
Gray birch*
Red spruce*
White pine

Sapling/shrub

Balsam fir*
Black spruce*
Larch*
Mountain holly*
Rhodora*
Sheep laurel*

Dwarf Shrub

Black huckleberry*
Labrador tea*
Rhodora*
Sheep laurel
Velvet-leaf blueberry*

Herb

Balsam fir*
Black spruce*
Cinnamon fern*
Creeping snowberry
Lowbush blueberrry
Three-seeded sedge*

Bryoid

Dicranum moss
Red-stemmed moss
Reindeer lichen
Sphagnum mosses*

Associated Rare Plants

Northern comandra
Swamp birch

Associated Rare Animals

Olive-sided flycatcher
Rusty blackbird

Evergreen Seepage Forest

State Rank S4

Community Description

Northern white cedar and other conifers form a moderate to dense canopy cover (70-95%), allowing only patchy light to penetrate to the forest floor. Northern white cedar is the dominant tree, though red spruce, white spruce, or black spruce may be co-dominant on some sites. Balsam fir, red maple, or yellow birch may be present but not dominant. Shrubs and dwarf shrubs are typically sparse but may be more abundant in canopy gaps caused by harvesting or natural disturbance. The herb layer may be extensive, typically >50% cover, and comprised mostly of non-woody species with northern affinities such as bunchberry, twinflower, or creeping snowberry. The forest floor is characterized by a rich growth of mosses; generally, feather-mosses and liverworts are more abundant than peat mosses.

Soil and Site Characteristics

The substrate is shallow peat or organic material over mineral soil, generally saturated with cold groundwater. Water may emerge to form rivulets or small spring fed brooks, or it may remain under the thick layer of mosses. These forests are typically found on gentle, saturated slopes with groundwater seepage, often at the base of slopes near drainage outlets.

Diagnostics

These closed canopy (>70%) forests are dominated by northern white cedar on mineral soil (may have a thin organic layer on top, but not deep peat). The bryophyte layer is dominated by feather-mosses and leafy liverworts rather than by peat mosses. Groundwater seeps may be visible.

Similar Types

Most similar to and often grade into Northern White Cedar Swamps, which occur in saturated, level basins on organic rather than mineral soils with extensive cover of peat mosses. Northern White Cedar Woodland Fens are partially forested peatlands with a patchy canopy of cedar and a substantial cover of heath or other dwarf shrubs (usually >15% cover), on moderate to deep peat soils. Cedar - Spruce Seepage Forests with a substantial amount of red spruce could grade into Spruce - Fir - Cinnamon Fern Forests, which usually occur on flats.

Conservation, Wildlife, and Management Considerations

Most known occurrences of this community type in Maine have been harvested in the

Location Map

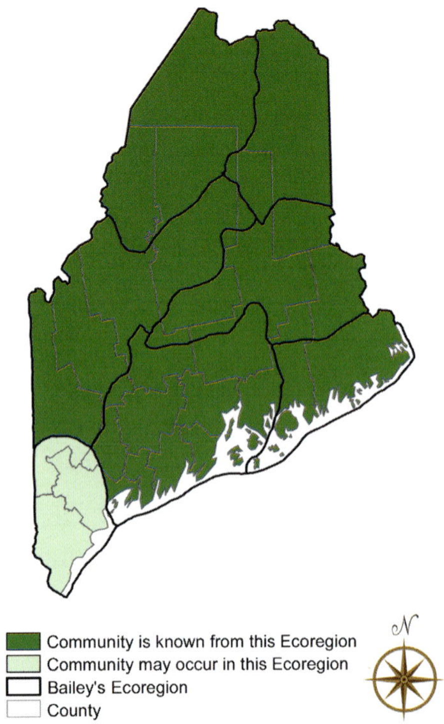

Northern White Cedar Branch

past, often targeting removal of spruce. Forest management generally does not result in permanent conversion of this type, although questions remain about how to most successfully regenerate cedar. Some high quality examples exist on public and private conservation lands.

Cedar - spruce seepage forests support numerous orchid species, including some that are rare. This community type may be used as nesting habitat by a number of coniferous forest specialist bird species, including black-backed woodpecker, palm warbler, yellow-bellied flycatcher, gray jay, boreal chickadee, Swainson's thrush, and northern waterthrush. Dead trees provide ideal habitat for three-toed woodpecker. Cool, well-oxygenated forested seeps provide habitat for the northern spring salamander.

Characteristic Plants

These plants are frequently found in this community type. Those with an asterisk are often diagnostic of this community.

Canopy
Balsam fir
Northern white cedar*
Red spruce*

Sapling/shrub
Fly honeysuckle

Herb
Bunchberry
Creeping snowberry
Dewdrop
Goldthread
Northern wood-sorrel
Oak fern
Three-seeded sedge
Twinflower

Bryoid
Mountain fern moss
Sphagnum mosses
Three-lobed bazzania

Associated Rare Plants
Giant rattlesnake-plantain
Showy lady's-slipper
Small yellow water crowfoot
Swamp fly honeysuckle

Associated Rare Animals
Northern spring salamander

Distribution
Primarily northern in distribution, extending westward and into Canada.

Landscape Pattern: Large Patch. Sites tend to be large (hundreds of acres) in northern Maine and smaller in central and southern Maine.

Examples on Conservation Lands You Can Visit
- Big Reed Pond Preserve – Piscataquis Co.
- Chamberlain Lake Public Lands – Piscataquis Co.
- Deboullie Ponds Public Lands – Aroostook Co.
- North Branch Inlet, Squa Pan Public Lands – Aroostook Co.
- Pollywog Gorge, Nahmakanta Public Lands – Piscataquis Co.
- Salmon Brook Lake Bog Public Lands – Aroostook Co.
- Sunkhaze Meadows National Wildlife Refuge – Penobscot Co.

Hardwood Seepage Forest

State Rank S3

Community Description

These closed canopy to partial canopy forests support a mixture of mostly deciduous overstory trees. Yellow birch, red maple, and/or green, black, or white ash are usually prominent species (35-85% cover each, sometimes lower). Hemlock or, less often, red spruce may create a mixed canopy (>25% conifer), with locally dense conifers. Sugar maple, beech, and red oak are occasional. The understory is usually open, with few shrubs and patches of tree regeneration. The herb layer is typically patchy, and reflects the underlying seepage gradients. Skunk cabbage, jewel weed, sensitive fern, and cinnamon fern occur in the wettest areas, and species less restricted by soil moisture occur elsewhere. Bryoids are sparse.

Soil and Site Characteristics

Sites occur on slight slopes (<15%) and adjacent bottoms where an impervious soil layer (~30 cm deep), such as marine clay or packed till, forces seepage water near the surface. Sites often occur at breaks in slope – either at the base of a slope, or on a slope bench. Soils are loamy, or grading to silty in flats, and moderately acidic to neutral (pH 5.2-7.0). Soils place this as a wetland type, but some sites may grade from wetland to upland as one moves upslope. Small sites, or 'forest seeps' (i.e., less than one acre) are frequent and are typically considered as inclusions within the broader forest rather than distinct natural communities.

Diagnostics

Ash and/or yellow birch are common in the canopy (red oak is prominent at some sites). Red maple may be present but is not dominant. Wetland species are common in the herb layer. Soils are saturated and often temporarily flooded.

Similar Types

Hemlock - Hardwood Pocket Swamps feature hemlock and/or red maple as dominants, have heaths or winterberry in the shrub layer, and typically occur in distinct basins. Some Maple - Basswood - Ash Forests may have areas of wet soils, but have very different herb layer composition. Red Maple - Sensitive Fern Swamps can occur in similar settings but have different canopy composition.

Location Map

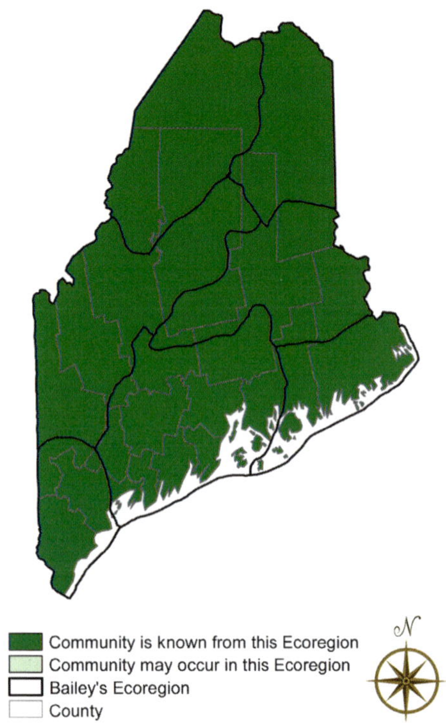

Hardwood Seepage Forest

Characteristic Plants

These plants are frequently found in this community type. Those with an asterisk are often diagnostic of this community.

Canopy

American beech
Eastern hemlock*
Green ash*
Red oak
Red spruce
Sugar maple
Yellow birch*

Sapling/shrub

American beech
Red spruce*

Herb

Bluejoint
Cinnamon fern*
Common speedwell
Goldthread
Jack-in-the-pulpit
New York fern
Sensitive fern*
Spinulose wood fern

Bryoid

Dicranum moss
Sphagnum mosses*

Associated Rare Plants

Spicebush
Swamp saxifrage

Associated Rare Animals

Northern spring salamander
Spicebush swallowtail

Conservation, Wildlife, and Management Considerations

Many sites are on land with a long settlement history and have been either cleared or harvested in the past. Because these tend to occur as small forest patches, their conservation depends in part on maintaining some surrounding forest cover (both upslope and downslope) as a buffer. Like vernal pools, recognition of this type is more difficult in the winter, when snow cover and plant senescence may make it difficult to distinguish these sites from upland forest. Seeps may remain unfrozen through the winter, making it difficult to operate logging equipment.

Birds using a variety of hardwood types may use these communities. Cool, well oxygenated forested seeps provide habitat for the northern spring salamander. Occurrences of this community type in southern Maine may host the rare spicebush swallowtail butterfly, whose larvae feed only on spicebush and sassafras.

Distribution

Statewide, though not well documented

Landscape Pattern: Small Patch. Occurrences less than one acre are generally considered inclusions rather than distinct natural communities.

Examples on Conservation Lands You Can Visit

- Dickwood IFW Lot – Aroostook County
- Sears Island – Penobscot Co

Northern White Cedar Swamp

State Rank S4

Community Description

This type is moderately to densely forested, often with little light penetrating to the forest floor. Northern white cedar is dominant (up to 95% cover), often forming a fairly uniform stand, but may be interspersed with various amounts of red maple (up to 25% cover), black spruce (up to 40% cover), or, less frequently, larch, yellow birch, or balsam fir. The variable shrub and ground layers form a lush mosaic of vegetated hummocks interspersed with moist hollows; alder may be frequent. The herb layer is well developed (>30% cover), with herbs more abundant than dwarf shrubs. Small cedar trees and an array of boreal herbs grow on the fallen logs and hummocks, including yellow lady's-slipper and several rare species. Sphagnum and other mosses blanket the hummocks, hollows, and fallen logs.

Soil and Site Characteristics

This community typically occupies level, poorly drained basins along stream flowages or the perimeter of ponds. The substrate is usually shallow peat (< 50 cm) over mineral soil; some sites are on deep peat accumulations. The characteristically alkaline conditions in this community type provide suitable habitat for a number of rare plant species.

Diagnostics

Sites are basin wetlands with >60% tree canopy cover and northern white cedar as the dominant tree. Peat mosses are the dominant bryophytes. The substrate is organic peat or muck.

Small Round-leaved Orchis

Similar Types

Northern White Cedar Seepage Forests are underlain by mineral soils, sometimes with a thin peat layer, and tend to occur on gentle slopes. They also have different mosses; feather-mosses (especially mountain-fern moss) and three-lobed bazzania will generally exceed peat mosses in total cover. Northern White Cedar Woodland Fens are partially forested peatlands with a patchy canopy of northern white cedar and a substantial cover of heath or other dwarf shrubs (usually >15% cover); open peatland vegetation types will usually occur with this type as part of the peatland. However, these three cedar types may grade into one another as part of large drainage basins.

Location Map

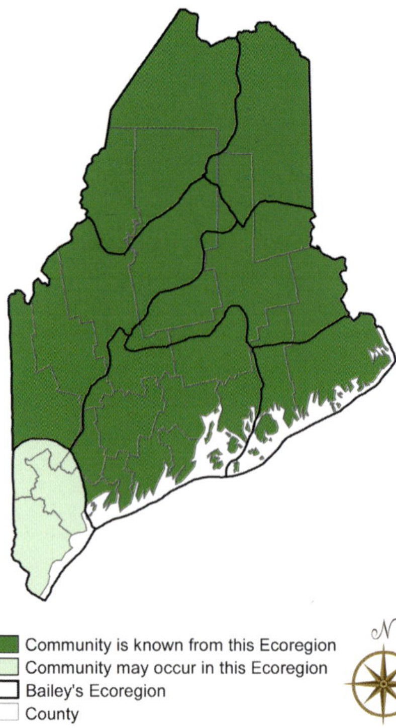

Community is known from this Ecoregion
Community may occur in this Ecoregion
Bailey's Ecoregion
County

Northern White Cedar Swamp

Conservation, Wildlife, and Management Considerations

Most examples of northern white cedar swamps have been logged at least once in the past. Cedar swamps in northern Maine can be very extensive, running into hundreds of acres. In southern Maine, they are often less than 50 acres in size. Although the overall extent of this community type appears stable, there are some indications that it is more difficult to regenerate cedar rather than fir through harvest practices. There has been little permanent conversion to other land uses or forest types. Some sites have been significantly altered by beaver activity.

This community type may be used as nesting habitat by a number of coniferous forest specialist bird species, including black-backed woodpecker, palm warbler, yellow-bellied flycatcher, gray jay, boreal chickadee, Swainson's thrush, and northern waterthrush. Northern White Cedar Swamps that have a large number of dead trees provide ideal habitat for the three-toed woodpecker.

Examples on Conservation Lands You Can Visit

- Deboullie Ponds Public Lands – Aroostook Co.
- Gott Brook, Dwinal Flowage Wildlife Management Area – Penobscot Co.
- Great Heath Public Lands – Washington Co.
- Petit Manan Point, Petit Manan National Wildlife Refuge – Washington Co.
- Salmon Brook Lake Bog Public Lands – Aroostook Co.

Characteristic Plants

These plants are frequently found in this community type. Those with an asterisk are often diagnostic of this community.

Canopy

Black spruce
Larch
Northern white cedar*
Red spruce
White spruce

Sapling/shrub

Black spruce*
Northern white cedar*
Red spruce
Speckled alder
Winterberry

Herb

Balsam fir*
Creeping snowberry
Northern white cedar*
Red spruce
Three-seeded sedge*

Bryoid

Mountain fern moss
*Sphagnum girgensohnii**
Other peat mosses

Associated Rare Plants

Bog bedstraw
Hoary willow
Lapland buttercup
Livid sedge
Northern bog sedge
Showy lady's-slipper
Small round-leaved orchis
Sparse-flowered sedge
White adder's-mouth

Distribution

Throughout the New England - Adirondack Province and Laurentian Mixed Forest Province, but most extensive in northern Maine. Extends north, west, and east from Maine.

Landscape Pattern: Large Patch (although in more southerly locations in the state, swamps are often small, < 50 acres).

Open Cedar Fen

State Rank S4

Community Description

These open canopy woodlands occur in a peatland setting with northern white cedar dominant. Canopy closure is 20-60%. Black spruce, red maple, balsam fir, black ash, or larch may be mixed with the cedar. The shrub layer may be locally dense with patches of trees and scattered shrubs of winterberry, alder, or mountain holly. The herb layer, usually with >50% cover, is variable in composition and may be predominantly heath shrubs or herbs with a prominent component of graminoids. Shrubby cinquefoil, alpine cotton-grass, sticky false-asphodel, and grass-of-parnassus may be at higher pH sites. The bryoid layer is mostly peat mosses, but the presence of mountain fern moss is indicative of this type.

Soil and Site Characteristics

Sites generally occur as part of a peatland (peat >30 cm deep) or occasionally along a peatland outlet stream where the peat substrate is shallower. Sites are in lowlands at elevations up to 1000', ranging from quite acidic to circumneutral (pH 4.6-7.2), and typical of somewhat minerotrophic (fen) conditions, not raised bog conditions.

Yellow Lady's-slipper

Diagnostics

Northern white cedar forms a partial canopy (<65%) and is frequently dominant in the shrub layer. Dwarf shrub peatland indicator plants are present. Sites often occur adjacent to an open peatland.

Similar Types

Northern White Cedar Swamps and Cedar - Spruce Seepage Forests occur on mineral soils or on thin to moderate peat over mineral soil, and not as part of a peatland with forested and non-forested areas (Northern White Cedar Swamps may occasionally occur in that setting). Cedar swamps and seepage forests also have very low cover of peatland dwarf shrubs. Red Maple Wooded Fens can be similar, but have far less northern white cedar in the

Location Map

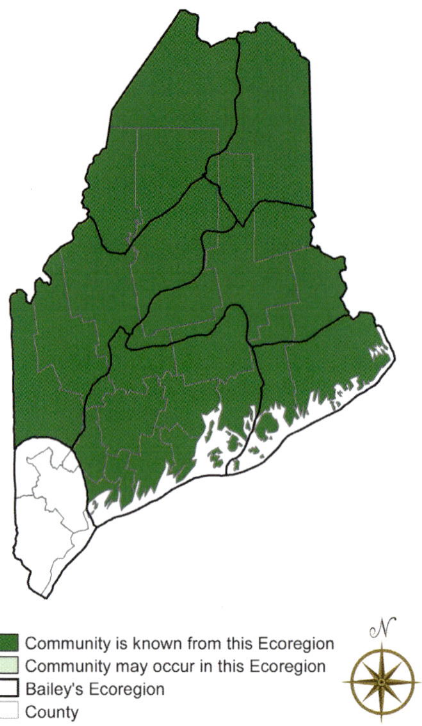

Northern White Cedar Woodland Fen

canopy. Gradations from cedar fens to Spruce - Larch Wooded Bogs also occur.

Conservation, Wildlife, and Management Considerations

These fens usually occur as part of larger peatlands, and maintaining the hydrologic integrity of the entire wetland is key. The cedars generally remain small, therefore this type is typically not a target for forest management. In some areas these fens have been altered by beaver activity.

Conifer-preferring birds that may use this partly open type include black-backed woodpecker, palm warbler, common yellowthroat, Lincoln's sparrow, and Swainson's thrush. Cedar fens that have a large number of dead trees provide habitat for the three-toed woodpecker.

Distribution

New England - Adirondack Province and Laurentian Mixed Forest Province, extending eastward, westward, and northward from Maine.

Landscape Pattern: Large Patch

Characteristic Plants

These plants are frequently found in this community type. Those with an asterisk are often diagnostic of this community.

Canopy

Black spruce*
Larch*
Northern white cedar*

Sapling/shrub

Balsam fir*
Black ash*
Northern white cedar*
Red maple*
Speckled alder

Dwarf Shrub

Black huckleberry*
Labrador tea
Leatherleaf*
Sheep laurel
Sweet gale*

Herb

Bluejoint*
Cinnamon fern*
Creeping snowberry*
Dwarf raspberry*
Northern white cedar*
Three-leaved false Solomon's seal
Tussock sedge*

Bryoid

Dicranum moss*
Mountain fern moss
Sphagnum mosses*

Associated Rare Plants

Bog bedstraw
Dioecious sedge
Horned beak-rush
Livid sedge
Marsh valerian
Showy lady's-slipper
Sparse-flowered sedge
Swamp birch

Examples on Conservation Lands You Can Visit

- Lake Umbagog National Wildlife Refuge – Oxford Co.
- Mattagodus Wildlife Management Area – Penobscot Co.
- Number Five Bog Public Lands – Somerset Co.

Pitch Pine Bog

State Rank S2

Community Description

Pitch pine is the dominant tree in these sparsely forested peatlands. The shrub layer likewise indicates the more southerly affinities of this type, with maleberry and highbush blueberry common along with the standard bog shrubs of huckleberry and mountain holly. The herb layer may be dense evergreen heath shrubs, especially leatherleaf, or it may be more sparse peat mosses covering the ground.

Soil and Site Characteristics

Sites occur in shallow basins on the coastal plain; typical acidic bog conditions predominate. Peat may be shallow, over sandy mineral soil, or deep (>50 cm) as is typical of peatlands. In some, but not all, cases these types are adjacent to pitch pine uplands.

Diagnostics

This is an organic soil wetland with abundant peat and low heath shrubs, sparsely forested by pitch pine.

Sheep Laurel

Similar Types

Several other peatland community types have very similar dwarf shrub, herb, and bryophyte composition, especially Spruce - Larch Wooded Bog, Sheep Laurel Dwarf Shrub Bog, and Leatherleaf Boggy Fen. The predominance of pitch pine in the tree layer makes this type unique in Maine.

Conservation, Wildlife, and Management Considerations

Strong development pressures in southern Maine may threaten unprotected sites and degrade their landscape surroundings. Four sites in southern Maine are in public or private conservation ownership.

Location Map

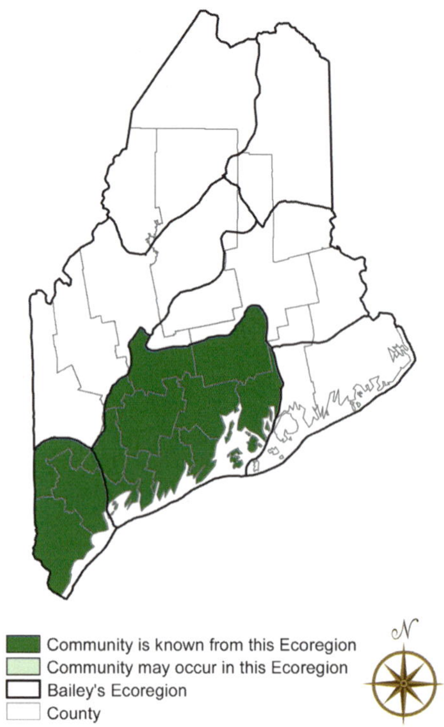

Pitch Pine Bog

Birds associated with this community include wetland species such as the common yellowthroat and northern waterthrush.

Distribution

Along the north Atlantic coastal plain (Eastern Broadleaf Forest Province) to southern and midcoast Maine.

Landscape Pattern: Small Patch, mostly 5-40 acres.

Large Cranberry

Characteristic Plants

These plants are frequently found in this community type. Those with an asterisk are often diagnostic of this community.

Canopy

Pitch pine

Sapling/shrub

Black chokeberry
Black huckleberry*
Highbush blueberry*
Maleberry
Mountain holly
Speckled alder

Dwarf Shrub

Large cranberry
Leatherleaf*
Sheep laurel

Herb

Bracken fern
Cinnamon fern
Three-seeded sedge
Wild sarsaparilla

Bryoid

*Sphagnum girgensohnii**

Associated Rare Plants

Smooth winterberry holly

Examples on Conservation Lands You Can Visit

- Brownfield Bog Wildlife Management Area - Oxford Co.
- East of Little River, Rachel Carson National Wildlife - York Co.
- Saco Heath Preserve - York Co.
- Scarborough Marsh Wildlife Management Area - Cumberland Co.

Pocket Swamp

Hemlock - Hardwood Pocket Swamp

State Rank S2

Community Description

These forested wetlands may be deciduous or mixed and typically occur as small depressions within an upland landscape. Red maple almost always dominates the canopy and occurs with hemlock and/or black gum. Black gum is an uncommon tree in Maine and is a good indicator of this community. Shrubs may be locally dense and include highbush blueberry and winterberry. The herb layer is variable in extent, and often features large clumps of ferns. In southern Maine, several rare plants are associated with these wetlands.

Soil and Site Characteristics

Sites occur in small isolated basins, sometimes perched on the sides of gentle hills, with a seasonal high water table. The soil may dry out during the summer, or pools of water may remain among the forested hummocks. Often these basins have no surface outlet, or they may drain only at high water. Soils are acidic, usually with a thin peat layer over mineral soil, occasionally with deeper peat. These wetlands typically occur as small patches (typically <3 acres) in otherwise well drained, forested uplands.

Diagnostics

Hemlock and/or red maple are dominant; black gum is often present, and high bush blueberry, winterberry, and/or maleberry occur in the understory. Wetland soils occur in isolated drainages, not along a stream or large drainage flat.

Spicebush Swallowtail Larvae

Similar Types

Red Maple - Sensitive Fern Swamps occur along alluvial soils of broad streams or lake basins and are more strongly dominated by red maple. Hardwood Seepage Forests have less red maple, more ash and yellow birch.

Conservation, Wildlife, and Management Considerations

Timber harvesting lowers the natural community values of these small and often isolated wetlands. Because these tend to occur as small forest patches, their conservation depends in part on maintaining some surrounding forest cover

Location Map

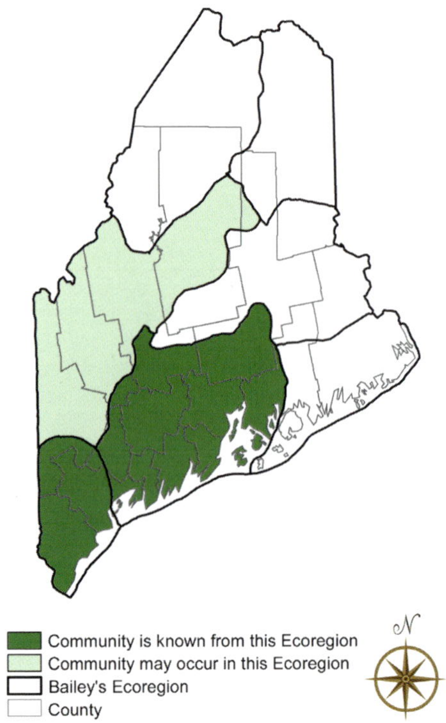

Spicebush

as a buffer. They should be buffered from direct impacts, such as physical disturbance to the soil, and indirect impacts, such as water quality degradation. Only a few examples are known on public lands or private conservation lands.

Many occurrences of this community type function as vernal pools, which are important breeding habitats for a variety of amphibians including wood frogs, spotted salamanders, and blue-spotted salamanders. Rare turtles such as Blanding's and spotted turtles may feed on amphibian egg masses present in such pools. If peaty hummocks are common, four-toed salamanders may breed in these wetlands. Occurrences of this community type in which spicebush is present may host the spicebush swallowtail butterfly, whose larvae feed only on spicebush and sassafras.

Distribution

Limited to the southern, central, and midcoast regions of the state, primarily in the Eastern Broadleaf Forest Province, and extending southward and southwestward from Maine.

Landscape Pattern: Small Patch; known sites are 2 - 30 acres.

Characteristic Plants

These plants are frequently found in this community type. Those with an asterisk are often diagnostic of this community.

Canopy

Black gum
Eastern hemlock
Red maple
Yellow birch

Sapling/shrub

Highbush blueberry
Maleberry
Mountain holly
Wild-raisin
Winterberry
Yellow birch

Dwarf Shrub

Lowbush blueberry

Herb

Bluejoint
Cinnamon fern*
Goldthread
Long sedge
Marsh fern
Northern water-horehound
Royal fern*
Three-seeded sedge*

Associated Rare Plants

Mountain laurel
Smooth winterberry
Spicebush

Associated Rare Animals

Blanding's turtle
Spicebush swallowtail
Spotted turtle

Examples on Conservation Lands You Can Visit

- Brownfield Bog Wildlife Management Area – Oxford Co.
- Long - Short Pond, Ferry Beach State Park – York Co.
- Mt Agamenticus – York Co.

Red Maple Fen

Red Maple Wooded Fen

State Rank S4

Community Description

Red maple dominates the canopy of this partly forested peatland, or it may be co-dominant with larch or black spruce. Canopy closure is usually <50%, sometimes to 65%. The shrub layer is locally dense, with small trees and thickets of winterberry, mountain holly, highbush blueberry, or maleberry. Sweetgale and heath shrubs that are typically dwarfed in bog settings grow taller (often >1 m) in this setting. Sedges are characteristic in the herb layer, and cinnamon fern and other wetland plants may be locally common. The bryoid layer is extensive (>60% cover) and dominated by peat mosses. A variant features larch as the dominant canopy tree, but vegetation is otherwise the same.

Soil and Site Characteristics

Sites occupy low basins (up to 1000' elevation) and are typically a peripheral portion of a larger wetland. The saturated soils are organic and the peat layer may be deep (>50 cm). The substrate is less acidic than most true bogs (pH 5.0-5.4).

Skunk Cabbage

Diagnostics

Sites occur in a peatland setting with red maple dominant in a partial canopy (<65%, usually <50%) or co-dominant with larch or black spruce. Heath shrubs and other characteristic peatland plants are present in shrub and herb layers. The substrate is dominated by Sphagnum moss.

Similar Types

Red Maple - Sensitive Fern Swamps can have similar overstory vegetation, but occur on mineral soils (perhaps with a thin peat layer) rather than on a peat substrate, and typically have taller trees with a more continuous canopy; they lack the heath shrubs characteristic of this

Location Map

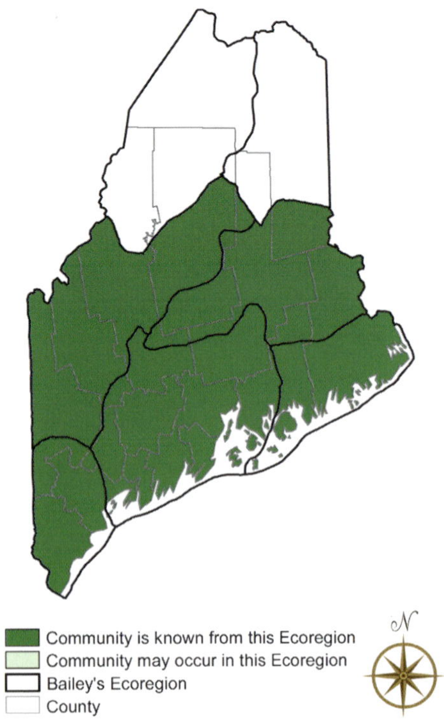

Red Maple Fen

type. Northern White Cedar Woodland Fens are similar but are dominated by northern white cedar, sometimes mixed with red maple. Spruce - Larch Wooded Bogs occur in similar settings, but generally in more nutrient poor conditions, and have spruce and/or larch much more abundant than red maple in the canopy.

Conservation, Wildlife, and Management Considerations

These fens usually occur as part of larger peatlands, and maintaining the hydrologic integrity of the entire wetland with upland buffers is key. This community type is widespread and apparently has few or no competing uses.

Birds associated with this community include wetland species such as the common yellowthroat and northern waterthrush. Thaxter's pinion moth uses larch and sweetgale as its larval host plants and may be found in this community.

Distribution

Statewide, less abundant northward; extending westward and southward (and perhaps eastward) from Maine

Landscape Pattern: Large Patch

Characteristic Plants

These plants are frequently found in this community type. Those with an asterisk are often diagnostic of this community.

Canopy
Black spruce
Larch*
Red maple*

Sapling/shrub
Balsam fir
Rhodora*
Sweetgale*
Wild-raisin*
Winterberry*

Herb
Beaked sedge
Cinnamon fern*
Goldthread*
Marsh fern*
Skunk cabbage*
Three-leaved false Solomon's seal*
Three-seeded sedge*
Tussock sedge*

Bryoid
Sphagnum mosses*

Associated Rare Plants

Bog bedstraw

Examples on Conservation Lands You Can Visit

- Appleton Bog Preserve – Knox Co.
- Middle Pond State Park – Oxford Co.
- The Heath, Massabesic Experimental Forest – York Co.

Red Maple Swamp

Red Maple - Sensitive Fern Swamp

State Rank S4

Community Description

Red maple dominates the somewhat open to nearly closed canopy (20-90% closure), sometimes with a relatively large component (up to 40% cover) of balsam fir, red spruce, or northern white cedar. Green ash and yellow birch are common, but rarely abundant, associates. The maples may be widely spaced with multiple trunks and arching crowns. The shrub layer is patchy; winterberry is common and various other shrubs may be locally abundant. The herb layer is well developed and dominated by herbs, with dwarf shrubs <20% of herb cover. Bluejoint and sensitive fern are characteristic herbs. The bryoid layer is usually <35% cover; peat mosses are typical but do not form extensive, deep carpets as they do in peatlands.

Soil and Site Characteristics

Sites occupy mineral soils or well decomposed organic material over mineral soil on flats or gentle slopes in small basins, often on floodplains of streams to small rivers. Soils are typically 30-60 cm deep, loamy to silty in texture, sometimes with well decomposed muck over the mineral fraction, and pH 4.8-5.4.

Diagnostics

These are mineral soil wetlands in which red maple dominates the canopy or is co-dominant with conifers other than black spruce or larch. The seasonally flooded soils usually remain saturated through the growing season.

Red Maple Flowers

Similar Types

Red Maple Wooded Fens are similar, but either occur in association with large peatlands or occupy small somewhat peaty basins; they do not occur on mineral soils. Some small Northern White Cedar Swamps and Spruce - Fir - Cinnamon Fern Forests, particularly along the coast, include a fair amount of red maple but have cedar or spruce/fir, respectively, as the most abundant canopy species. Silver Maple Floodplain Forests are dominated by silver maple and generally occur along larger

Location Map

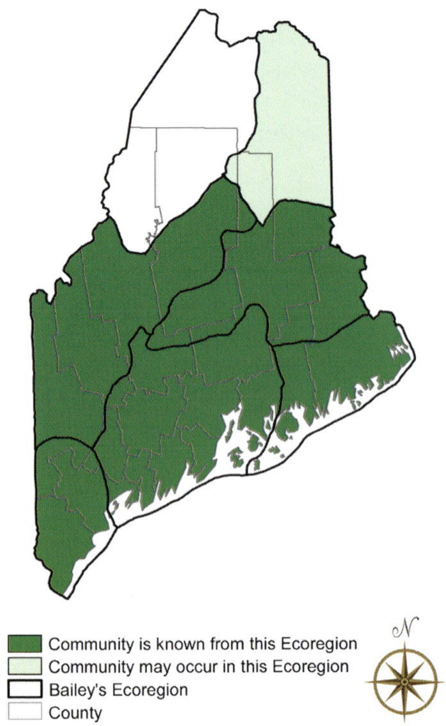

Red Maple Swamp

rivers, but the two types can intergrade on some floodplains.

Conservation, Wildlife, and Management Considerations

Maintaining the hydrologic integrity of these stream drainages with upland buffers is key. These swamps typically have had few conflicting uses, although some have been recently harvested. ATV use has been observed at some sites.

Red maple swamps often provide habitat in which spotted turtles hibernate. If wet Sphagnum hummocks are present, four-toed salamanders may breed in this community. Examples that occur on floodplains of streams and small rivers may contain wood turtles, which overwinter in the stream channel and forage in the floodplain. The silver-haired bat often roosts in riparian habitats in trees with loose bark. The northern waterthrush is a common associate of this community type. In the southern part of the state, the Louisiana waterthrush and yellow-throated vireo may be associates if the canopy is closed or nearly so.

Distribution

Statewide, but most common in the southern half of state. Extends southward and southwestward from Maine; eastward distribution unknown.

Landscape Pattern: Large Patch

Characteristic Plants

These plants are frequently found in this community type. Those with an asterisk are often diagnostic of this community.

Canopy

Balsam fir
Gray birch
Northern white cedar
Red maple*
Red spruce

Sapling/shrub

Arrowwood*
Balsam fir
Gray birch*
Red spruce
Speckled alder*
Winterberry*

Herb

Bluejoint*
Flat-topped white aster*
Interrupted fern
Tussock sedge
Royal fern*
Sensitive fern*

Bryoid

Sphagnum mosses*

Associated Rare Plants

Smooth winterberry holly
Spicebush
Swamp saxifrage
Swamp white oak
Sweet pepper-bush

Associated Rare Animals

Spotted turtle
Wood turtle

Examples on Conservation Lands You Can Visit

- Kennebunk Plains Preserve – York Co.
- Mt Agamenticus – York Co.
- Steep Falls Wildlife Management Area – Cumberland Co.
- Waterboro Barrens Preserve – York Co.

Silver Maple Floodplain Forest

State Rank S3

Community Description

These forests are dominated by silver maple (>60% cover). Associates include red maple and American elm (up to 30% cover) or, in a few locations, bur oak (up to 25% cover). Widely spaced trees, many with multiple trunks, give a park like feeling. The understory is open and shrubs are sparse. Musclewood may be present and is a good indicator. The lush carpet of herbs changes from spring ephemerals such as trout lilies and bloodroot to dense fern cover in summer. Bryoid cover is minor. Some forests have a berm adjacent to the river channel, and herbaceous species composition here is different from the lower elevation interior of the floodplain.

Soil and Site Characteristics

Sites occur on plains flanking low gradient rivers, within the reach of seasonal floods, at elevations <700'. Soils are fine sand or silt, usually with good drainage capacity; the water table fluctuates. Relatively high nutrient levels result from sediment deposition of annual floods; pH is typically 5.0-6.2.

Diagnostics

Sites occur in a floodplain setting with mineral soil. Silver maple is the dominant tree. There is a dense herb layer with sensitive fern and, locally, ostrich fern. Spring ephemerals are frequent.

Similar Types

Hardwood River Terrace Forests may be adjacent to this type on the higher floodplain, but these have a much smaller proportion of silver maple in the canopy. Instead, the canopy is dominated by sugar maple, red oak, or green ash, and herb diversity is higher. Red Maple - Sensitive Fern Swamps lack the dominance of silver maple, and have surface water or saturated soil throughout the growing season. Hardwood Seepage Forests occur along small drainages, usually sloping, rather than in extensive floodplains.

Conservation, Wildlife, and Management Considerations

Although a number of sites have been cleared or pastured in the past, current shoreland regulations provide increased protection to a number of these sites. Exotic plant species such as Japanese knotweed, which may displace those native to our area, also represent a threat to the integrity of these forests and have degraded some Maine examples. Several of the known examples are formally protected from conversion.

Location Map

Silver Maple Floodplain Forest

Northern waterthrush, barred owl, belted kingfisher, bank swallow, and green heron are associates of this community type. In the southern part of the state, the Louisiana waterthrush and yellow-throated vireo are likely associates if the canopy is closed or nearly so. Rare turtles like wood, spotted, and Blanding's turtles may feed on amphibian egg masses present in isolated pools within such forests. Wood turtles overwinter in river channels and forage in floodplain forests. The silver-haired bat often roosts in riparian habitats in trees with loose bark.

Distribution

Long and narrow floodplains along the shores or islands of large rivers and streams throughout Maine, New England, and New Brunswick.

Landscape Pattern: Large Patch (remaining sites mostly 20-200 acres, up to 1000 acres)

Characteristic Plants

These plants are frequently found in this community type. Those with an asterisk are often diagnostic of this community.

Canopy

American elm
Basswood*
Black ash
Black cherry*
Bur oak*
Green ash*
Red oak*
Silver maple*

Sapling/shrub

Arrowwood*
Buttonbush*
Common blackberry*
Gray birch*
Meadowsweet*
Musclewood*
Nannyberry*
Winterberry holly*

Dwarf Shrub

Swamp dewberry*

Herb

Bluejoint
Cinnamon fern*
Green ash
Jack-in-the-pulpit
Ostrich fern*
Royal fern*
Sensitive fern*
Tall meadow-rue
Wood-nettle*

Associated Rare Plants

Swamp white oak
Wild garlic
Wild leek

Associated Rare Animals

Wood turtle

Examples on Conservation Lands You Can Visit

- Brownfield Bog Wildlife Management Area - Oxford Co.
- Saco River Preserve - York Co.
- Sunkhaze Meadows National Wildlife Refuge - Penobscot Co.
- The Oxbow, East Branch Penobscot River - Penobscot Co.
- Trout Brook, Baxter State Park - Piscataquis Co.
- Wassataquoik Public Lands - Penobscot Co.

Spruce - Fir Wet Flat

State Rank S4

Community Description

This natural community is a fairly homogeneous forest type in which red spruce, black spruce, or red-black spruce hybrids grow on poorly drained, level to gently sloping sites. Balsam fir may be present in regenerating patches or stands but tends to give way to the longer-lived spruces over time. Stands often form even-aged blocks hundreds to thousands of acres in size. The even-aged structure likely results from the past influences of spruce-budworm, fire, harvesting, blowdowns, or a combination of multiple factors.

Cinnamon fern and three-seeded sedge are typical in these types statewide. In northern Maine, understory herbs and shrubs are sparse, and the forest floor is dominated by a dense carpet of mosses – typically Sphagnum species, three-lobed bazzania, and red-stemmed moss. Dwarf heath shrubs may be abundant at St. John Valley sites, which approach boreal 'muskeg'. In southern Maine, red maple may be sub-dominant. At sites near the coast, skunk cabbage may be a prominent understory species.

Soil and Site Characteristics

Sites usually occur along drainages or low flats where soil remains moist throughout the growing season and may be saturated or temporarily flooded in the springtime. The substrate is acidic mineral soil and may be very stony, with or without an organic layer (<30 cm) on top. More information is needed statewide to determine if this type should be split into two separate types, reflecting northern and southern Maine variants.

Diagnostics

Sites occur on moist to saturated mineral soils, usually with a dense carpet of mosses and liverworts. Closed canopies are dominated by spruce (>40% cover), or are rarely more open where red maple or northern white cedar mixes with spruce. Wetland plants occur in the herb layer, usually including cinnamon fern and three-seeded sedge.

Similar Types

Other spruce - fir types occur on better-drained upland soils and gentle to steeper slopes. Red Maple - Sensitive Fern Swamps can be similar but will have more red maple and less spruce and fir. Spruce - Larch Wooded Bogs can have similar species composition (especially where black spruce is dominant) but occur on peat deposits (>30 cm) rather than on mineral soils.

Location Map

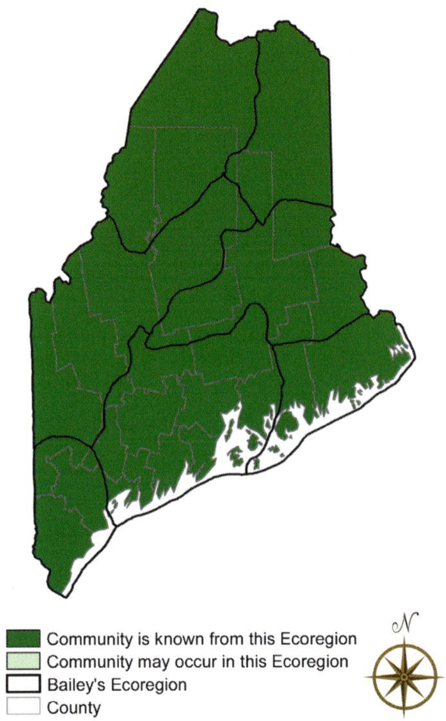

Spruce – Fir Wet Flat

Conservation, Wildlife, and Management Considerations

Nearly all known occurrences of this community type in Maine have been harvested in the past, and many have a history of natural disturbance such as fire or spruce-budworm. Large (>1000 acres) examples free from human disturbance are scarce. Forest management with natural regeneration generally does not with result in conversion of this type. Studies on some examples on public and private conservation lands may provide further information on the natural dynamics in these systems.

These stands may serve as deer wintering areas and may also provide habitat for pine marten and Canada lynx, depending on the age and successional stage. This community type may be used as nesting habitat by a number of coniferous forest specialist bird species, including the yellow-bellied flycatcher, sharp-shinned hawk, black-backed woodpecker, pine grosbeak, green heron, black-throated green warbler, Blackburnian warbler, common yellowthroat, Wilson's warbler, spruce grouse, blackpoll warbler, and the rare rusty blackbird.

Distribution

Statewide, more common and extensive northward. Characteristic of the Laurentian Mixed Forest Province and New England - Adirondack Province.

Landscape Pattern: Large Patch

Characteristic Plants

These plants are frequently found in this community type. Those with an asterisk are often diagnostic of this community.

Canopy

Black spruce*
Larch
Northern white cedar
Red maple
Red spruce*
White pine

Sapling/shrub

Alder*
Balsam fir
Black huckleberry*
Mountain holly*
Red maple
Wild-raisin

Dwarf Shrub

Leatherleaf*
Lowbush blueberry*
Rhodora*
Sheep laurel*

Herb

Bunchberry
Cinnamon fern*
Dwarf raspberry*
Goldthread
Skunk cabbage
Three-seeded sedge*

Bryoid

Red-stemmed moss*
Sphagnum mosses*
Three-lobed bazzania

Associated Rare Animals

Rusty blackbird

Examples on Conservation Lands You Can Visit

- Chamberlain Lake Public Lands – Piscataquis Co.
- Round Pond Public Lands – Aroostook Co.

Upper Floodplain Hardwood Forest

State Rank S2

Community Description

An almost complete canopy is dominated by sugar maple, red oak, or yellow birch, with red maple and ash often common and basswood or black cherry occasional. The understory is open and shrubs are sparse. The lush carpet of herbs changes from spring ephemerals such as trout lily and bloodroot to variable cover of mixed graminoids and forbs in summer. Bryoid cover is minor.

Soil and Site Characteristics

Sites occur on slightly elevated terraces flanking low-gradient rivers at elevations typically <1,000'. Flooding is occasional, sometimes less frequent than annually. These forests have lower frequency and duration of flooding than silver maple floodplain forests. Most known examples are along medium to larger rivers. Soils are fine sand or silt, usually with good drainage capacity and relatively high nutrient levels; pH is 5.0-6.2.

Diagnostics

Sites occupy floodplain or river terrace settings with mineral soil. The canopy is dominated by sugar maple, red oak or yellow birch. Silver maple and red maple may be present. A dense herb layer includes species not typical of wetlands (e.g. starflower, zig-zag goldenrod, big-leaved aster, silvery spleenwort). Sensitive fern is often present but not dominant. Spring ephemerals are often abundant.

Floodplain High Water Marks

Similar Types

On large rivers, Silver Maple Floodplain Forests are often adjacent to these forests, occurring between them and the channel, and are dominated by silver maple. Red Maple - Sensitive Fern Swamps have red maple dominant, and have soils that are flooded or saturated throughout the growing season. Beech - Birch - Maple Forests, Maple - Basswood - Ash Forests, and Red Oak - Northern Hardwoods - White Pine Forests can be similar in canopy composition, but are not in the floodplain or terraces of a river. Hardwood Seepage Forests occur along small stream

Location Map

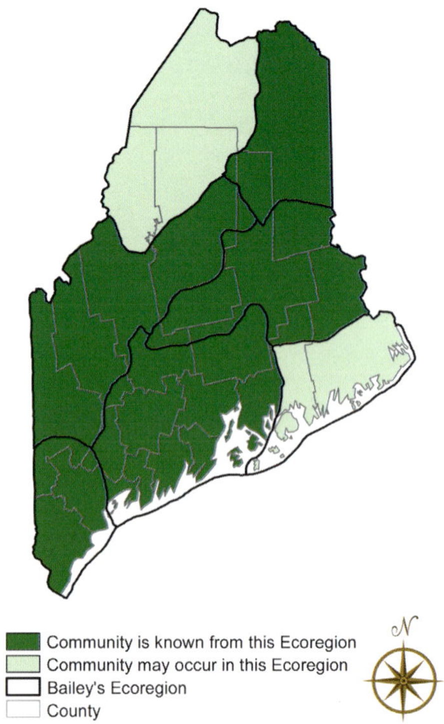

Upper Floodplain Hardwood Forest

drainages, usually sloping with steeper gradients.

Conservation, Wildlife, and Management Considerations

Virtually all of these forests have been harvested, and many have been converted to agriculture. Non-native plant species such as Japanese knotweed and Asiatic bittersweet, which may displace plants native to our area, represent a threat to the integrity of these forests and have degraded some Maine examples.

The northern waterthrush, barred owl, belted kingfisher, bank swallow, scarlet tanager, and green heron are associates of this community type. Wood turtles overwinter in river channels and forage in floodplain forests where they may feed on amphibian egg masses in vernal pools. The silver-haired bat often roosts in riparian habitats in trees with loose bark. Fairy shrimp may also occur in isolated vernal pools.

Distribution

Throughout Maine, New England, and New Brunswick.

Landscape Pattern: Large Patch

Characteristic Plants

These plants are frequently found in this community type. Those with an asterisk are often diagnostic of this community.

Canopy

Basswood*
Black cherry*
Green ash*
Red oak*
Sugar maple*
Yellow birch*

Sapling/shrub

American elm
Choke cherry
Musclewood*
Nannyberry*
White ash

Herb

Bloodroot
Blue cohosh
Jack-in-the-pulpit
Lady fern*
Ostrich fern*
Sensitive fern
Silvery spleenwort*
Tall meadow-rue

Associated Rare Plants

Bottlebrush grass
Pubescent sedge
Wild garlic
Wild ginger
Wild leek

Associated Rare Animals

Wood turtle

Examples on Conservation Lands You Can Visit

- Trout Brook, Baxter State Park – Piscataquis Co.
- Wassataquoik Public Lands – Penobscot Co.
- West Branch Piscataquis River, Appalachian Trail – Piscataquis Co.

OPEN
UPLANDS

Acidic Cliff

State Rank S4

Community Description

Sparse vegetation occurs on steep outcrops or cliffs of granitic or other acidic rock. Marginal wood fern and rock polypody are characteristic ferns; fragrant wood fern can be found on cooler sites. Rock tripe lichens may form extensive patches.

Soil and Site Characteristics

Sites occupy nearly vertical to vertical outcrops of non calcareous, erosion resistant rocks. Most are dry, with large unvegetated areas; a moist microclimate is maintained over local areas by runoff or seeps from higher elevations, or, in gorges, by the flowing streamwater. Smaller ledges and outcrops (e.g., less than 5,000 square feet of rock exposure) should be considered as inclusions in the surrounding forest rather than distinct natural communities.

Diagnostics

Sparsely vegetated cliffs occur below treeline, without circumneutral indicator species.

Rock Tripe and Rock Polypody

Similar Types

Boreal Circumneutral Outcrops have circumneutral indicator species such as shrubby cinquefoil or certain uncommon herbs. Three-toothed Cinquefoil - Blueberry Low Summit Balds are on summits, not cliffs, and usually have heath shrubs mixed with the herbs.

Conservation, Wildlife, and Management Considerations

Many sites are relatively inaccessible and minimally affected by either forestry or recreational activities. Several are within public lands or conservation ownership.

Location Map

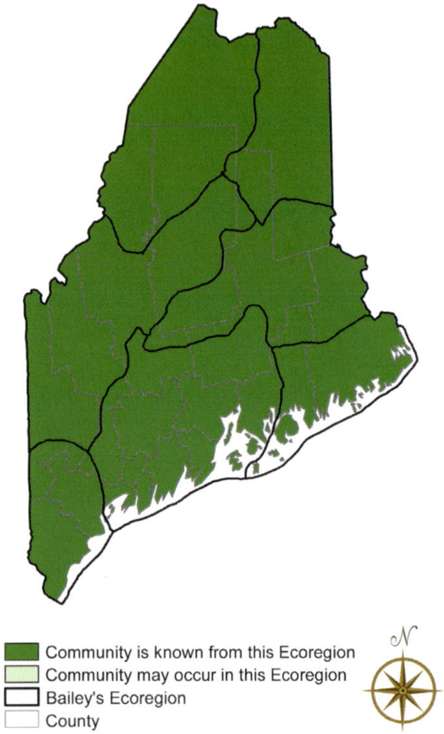

Acidic Cliff - Gorge

Common ravens, peregrine falcons, and golden eagles may nest on cliffs in western, northern and coastal Maine.

Distribution

Essentially statewide except for extreme southern Maine, more common northward.

Landscape Pattern: Small Patch. The minimum mapping unit is 5,000 square feet of rock exposure; smaller ledges and outcrops should be considered as inclusions in the surrounding forest rather than distinct natural communities.

Acidic Cliff - Gorge

Characteristic Plants

These plants are frequently found in this community type. Those with an asterisk are often diagnostic of this community.

Herb

Brownish sedge
Common hairgrass
Marginal woodfern
Rand's goldenrod
Rock polypody

Bryoid

Rocktripe lichen

Associated Rare Plants

Fragrant wood fern

Associated Rare Animals

Golden eagle
Peregrine falcon

Examples on Conservation Lands You Can Visit

- Deboullie Ponds Public Lands – Aroostook Co.
- Dunn Falls, Appalachian Trail – Oxford Co.
- Grindstone Falls – Penobscot Co.
- Mount Kineo Public Lands – Piscataquis Co.
- Tunk Mountain, Donnell Pond Public Lands – Hancock Co.

Alpine Cliff

State Rank S1

Community Description

Herbaceous - bryoid vegetation occurs with a mix of forbs and graminoids in patches over bare rock. Low heath shrubs including alpine bilberry and Labrador tea are present, but the distinctive feature of the vegetation is the herbaceous component. Several montane herbs, such as star saxifrage, hairy arnica, and alpine bistort, may be locally abundant on these seepy sites. The bryoid layer is extensive and may include large patches of the liverwort *Scapania nemorea.*

Soil and Site Characteristics

Sites occur near or above treeline in seeps or rivulets, often associated with vertical to nearly vertical rock faces. The known locations are in cirques, with constant moisture, at elevations above 3000'.

Common Butterwort

Diagnostics

Sites are at or above treeline, on sheer cliffs or cirque walls, usually with seepage; vegetation is a mixture of herbs, bryophytes, and dwarf shrubs; bulrush sedge is usually present along with other herbaceous species that do not occur at lower elevations.

Similar Types

Other alpine community types occur either on drier substrates (lacking extensive bryophytes) or in boggy situations with peat mosses, cotton-grasses, and heath shrubs dominant. Mountain Alder - Bush-honeysuckle Subalpine Meadows may occur just

Location Map

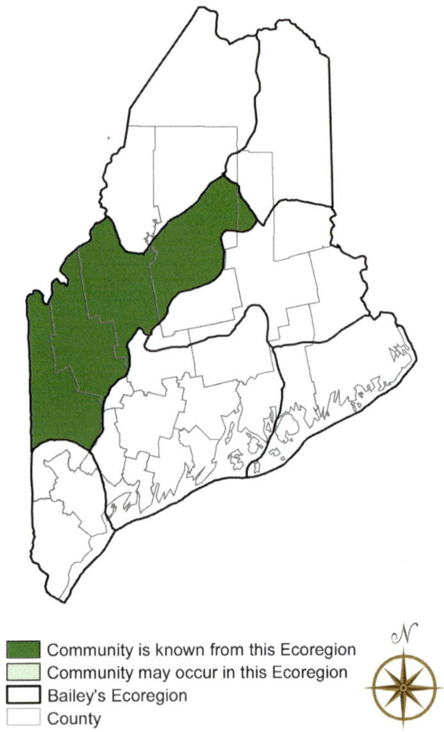

Alpine Cliff

below this community, but these are denser and the dominants much taller. Circumneutral Riverside Seeps may share some of the unusual species but are at lower elevations along major rivers.

Conservation, Wildlife, and Management Considerations

The known sites in Maine are on public lands or private conservation lands and somewhat removed from hiking trails. This vegetation is extremely sensitive to foot traffic, but given that access ranges from difficult to dangerous, the sites appear to be protected fairly well by their topography.

Alpine cliffs may provide nesting habitat for common ravens, peregrine falcons, and golden eagles.

Distribution

Restricted in Maine to a few of the state's highest mountains (New England - Adirondack Province). Distribution outside of Maine unknown.

Landscape Pattern: Small Patch

Characteristic Plants

These plants are frequently found in this community type. Those with an asterisk are often diagnostic of this community.

Sapling/shrub

Mountain alder
Shrubby cinquefoil*
Squashberry

Dwarf Shrub

Alpine bilberry*
Labrador tea

Herb

Bluejoint
Bulrush sedge*
Deer-hair sedge*
Highland rush
Low rough aster*

Associated Rare Plants

Alpine bistort
Alpine bitter-cress
Arctic red fescue
Bulrush sedge
Hairy arnica
Hornemann's willow-herb
Mountain timothy
Star saxifrage

Associated Rare Animals

Golden eagle
Peregrine falcon

Examples on Conservation Lands You Can Visit

- Mt. Katahdin, Baxter State Park – Piscataquis Co.

Alpine Snowbank

State Rank S1

Community Description

Dwarfed alpine shrubs form low mats, with herbs interspersed. Alpine bilberry and/or dwarf bilberry are dominant, but moss-plant and mountain-heath can form extensive patches among the bilberry. The relatively protected habitat amid otherwise exposed alpine vegetation allows some lower elevation species to exist, including Canada mayflower, bunchberry, mountain wood fern, bluebead lily, large-leaved goldenrod, and common hairgrass. Bryoids may be locally abundant, and include hair-cap mosses and red-stemmed moss.

Soil and Site Characteristics

Sites occupy protected upper mountain slopes above treeline, such as upper cirque walls. Snow lingers here after it has melted from the rest of the alpine zone.

Mountain Cranberry, Labrador Tea, and Reindeer Lichens

Diagnostics

Sites occur in protected habitats above treeline; bilberries are dominant; tundra dwarf birch, mountain-heath, and/or moss-plant are characteristic.

Similar Types

Subalpine Heath - Krummholz occurs at lower elevation and lacks the strictly alpine species typical of this type. Dwarf Heath - Graminoid Alpine Ridge is similar (and often adjacent) but lacks the characteristic snowbank species (moss-plant and mountain-heath), as well as those typical of lower elevations. Diapensia Alpine Ridge can also be adjacent, but is dominated by diapensia. Alpine Cliff vegetation has a higher

Location Map

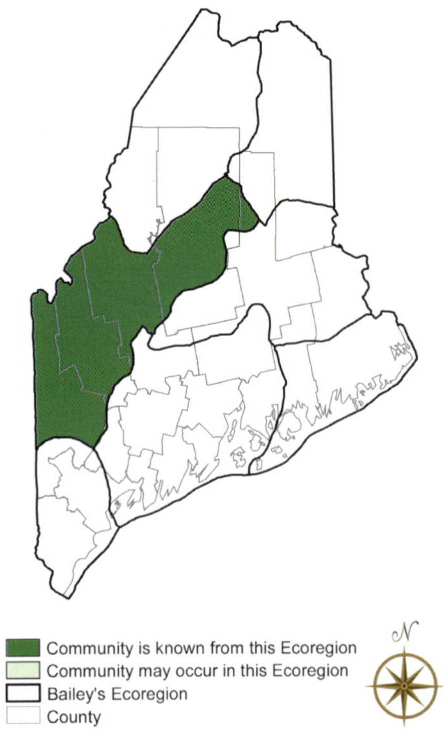

Alpine Azalea

proportion of herbaceous species and occurs on seeps, with a well developed bryophyte layer.

Conservation, Wildlife, and Management Considerations

The major documented occurrence of this type in Maine is on Katahdin. Hiker impacts to this type appear minimal, because most of its area is away from trails. Monitoring of impacts where alpine snowbank vegetation occurs near trails could help determine management needs.

Mt. Katahdin is the only known nesting area in the state for the rare American pipit, which utilizes a variety of natural community types above treeline.

Distribution

Restricted to Maine's highest mountains (New England - Adirondack Province), extending west to New Hampshire and along the Appalachians.

Landscape Pattern: Small Patch

Characteristic Plants

These plants are frequently found in this community type. Those with an asterisk are often diagnostic of this community.

Dwarf Shrub

Alpine bilberry*
Dwarf bilberry*
Moss-plant
Mountain cranberry
Mountain-heath
Pale laurel

Herb

Bunchberry
Canada mayflower
Common hairgrass*
Stiff clubmoss

Bryoid

Awned hair-cap moss
Juniper hair-cap moss

Associated Rare Plants

Alpine azalea
Alpine bearberry
Bearberry willow
Bigelow's sedge
Boreal bentgrass
Lapland rosebay
Mountain-heath
Northern comandra

Associated Rare Animals

American pipit

Examples on Conservation Lands You Can Visit

- Mt. Katahdin, Baxter State Park – Piscataquis Co.

Blueberry - Lichen Barren

State Rank S2

Blueberry - Lichen Barren

Community Description

These barrens are wide, flat to hummocky expanses of dwarf shrub vegetation punctuated by sparse pine or spruce trees. Lowbush and/or velvet-leaf blueberry is the predominant shrub, forming a fairly even carpet. Herbs are sparse. In patches among the shrubs, reindeer lichens may form extensive carpets. The characteristic expression of this community is as an opening within woodland barrens such as Pitch Pine - Scrub Oak Barrens or Spruce - Heath Barrens; smaller openings (e.g., less than an acre) would be considered inclusions, but larger ones should be segregated as this non-wooded type.

Soil and Site Characteristics

Coarse textured glacial outwash deposits form a flat to undulating substrate that can encompass a wide moisture gradient. Xeric conditions on hummocks or raised areas can grade into bog-like vegetation in depressions. Soils are highly acidic and nutrient poor. Sites are typically found in areas where fire has been frequent. Lichen carpets may be characteristic after particularly hot fires.

Gray Birch Leaves

Diagnostics

Lowbush blueberry is dominant, lichens are abundant, and conifers and other trees are sparse.

Similar Types

Little Bluestem - Blueberry Sandplain Grasslands are more common southward, lack the abundant lichens, and feature little bluestem grass as a local dominant. Pine and spruce barrens (Pitch Pine - Scrub Oak Barrens, Pitch Pine - Heath Barrens, Spruce - Heath Barrens) are closely related but have more tree cover.

Location Map

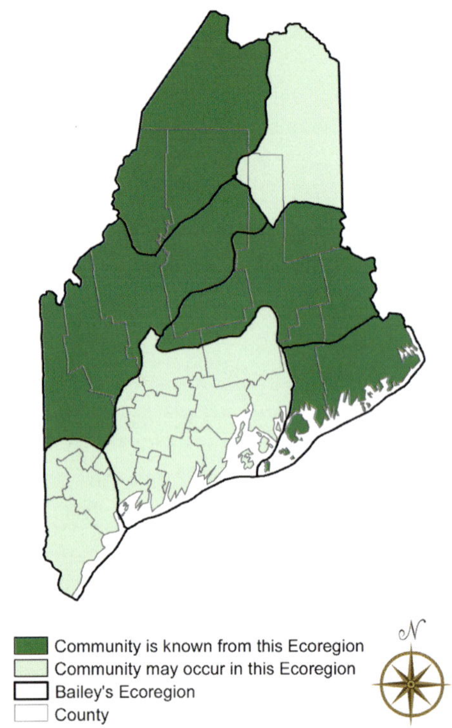

Burned Blueberry Barren

Conservation, Wildlife, and Management Considerations

Many former natural occurrences of this type have been converted to actively managed blueberry barrens in Downeast Maine. While these managed barrens maintain some superficial resemblance to natural barrens, pesticide use may have changed their species composition. Both of the documented occurrences of this type are on state-owned or private conservation land.

Managed blueberry barrens provide some of the best habitat in the northeast for the rare upland sandpiper and other ground-nesting species like the rare short-eared owl. The whimbrel, a non-breeding migrant shorebird, uses near-coastal barrens for foraging. The graceful clearwing, a butterfly that feeds on blueberries in its larval phase, is formerly known from Maine and may have inhabited examples of this community within pitch pine barrens.

Distribution

Natural occurrences are known only from Downeast and extreme northwestern Maine; poorly documented.

Landscape Pattern: Small Patch

Characteristic Plants

These plants are frequently found in this community type. Those with an asterisk are often diagnostic of this community.

Sapling/shrub
Gray birch
Red pine
White pine

Dwarf Shrub
Black chokeberry
Lowbush blueberry*
Meadowsweet
Sheep laurel
Sweetfern

Herb
Bracken fern
Wintergreen

Bryoid
Awned hair-cap moss
Juniper hair-cap moss
Red-stemmed moss
Reindeer lichen*

Associated Rare Plants

Canada mountain-ricegrass

Associated Rare Animals

Short-eared owl
Upland sandpiper
Whimbrel

Examples on Conservation Lands You Can Visit

- Nicatous Public Lands – Hancock Co.
- St. John River Preserve – Aroostook Co.

Circumneutral Outcrop

State Rank S2

Community Description

These rock faces are sparsely vegetated with herbs and scattered meadowsweet and paper birch. Composition varies from site to site, but includes at least some species indicative of higher pH conditions, such as shrubby cinquefoil, ebony sedge, rock whitlow-cress, lance-leaved draba, Laurentide primrose, smooth woodsia, etc. Graminoids, including brownish sedge, and ferns, such as fragile fern and bulblet fern, may be locally abundant in moist pockets.

Soil and Site Characteristics

This community occurs on low- to mid-elevation (<2700') outcrops of limestone, dolomite, or other rock where weathering produces circumneutral to calcareous substrates. Sites are usually below the hill summit, on side slopes or cliffs rather than ridges. Slope varies, often with vertical faces and near horizontal shelves within the same area. Sites are typically dry but may have moist spots where seepage occurs.

Fragrant Wood Fern

Diagnostics

A sparsely vegetated cliff/outcrop setting occurs below treeline, with herbs dominant; circumneutral indicator species are present.

Similar Types

Acidic Cliff - Gorge vegetation is also on bare rock faces, but on more acidic rock, and lacks any circumneutral indicator species. Three-toothed Cinquefoil - Blueberry Low Summit Balds likewise lack circumneutral indicators and typically occur on summits. Alpine Cliff vegetation can have some circumneutral indicators but is found at higher elevations (>3000').

Location Map

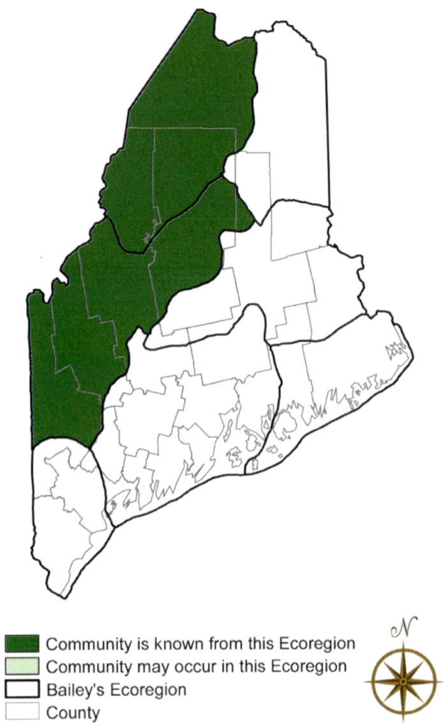

Circumneutral Outcrop

Conservation, Wildlife, and Management Considerations

Most known sites are relatively inaccessible and minimally affected by either forestry or recreational activities. Retaining a wooded buffer, where one would naturally occur, is important in maintaining the natural microclimate of the cliff and outcrop areas. Several known sites are within public lands or private conservation ownership.

Distribution

Hills of northwestern and north-central Maine (mostly the New England - Adirondack Province, extending into the Laurentian Mixed Forest Province). Distribution outside of Maine is poorly understood.

Landscape Pattern: Small Patch

Bulblet Fern

Characteristic Plants

These plants are frequently found in this community type. Those with an asterisk are often diagnostic of this community.

Sapling/shrub

Meadowsweet

Dwarf Shrub

Shrubby cinquefoil

Herb

Brownish sedge
Bulblet fern
Fragile fern
Fragrant wood fern
Rusty cliff fern

Bryoid

Juniper hair-cap moss

Associated Rare Plants

Ebony sedge
Intermediate sedge
Lance-leaved draba
Northern woodsia
Rock whitlow-grass
Slender cliffbrake
Smooth rockcress
Smooth woodsia
White bluegrass

Examples on Conservation Lands You Can Visit

- Deboullie Ponds Public Lands – Aroostook Co.
- Little Kineo Mountain Public Lands – Piscataquis Co.
- Mount Kineo Public Lands – Piscataquis Co.
- Trout Brook Mountain, Baxter State Park – Piscataquis Co.

Cold-air Talus Slope

State Rank S2

Community Description

This boreal assemblage is characterized by patchy mats of dwarf shrubs among the rocks, with scattered and stunted spruce (usually black spruce). Labrador tea and black crowberry are the most characteristic shrubs. Herbs are very sparse. Foliose and fruticose lichens are often extensive.

Soil and Site Characteristics

Sites occur near the base of somewhat sheltered or north-facing talus slopes where cold air collects. They are usually steep, though sometimes approaching level. The substrate is large talus blocks with pockets of organic duff among them. Ice blocks may persist into the summer in rock crevices.

Black Spruce

Diagnostics

Sites occupy the base of talus slopes; mats of Labrador tea and other heath shrubs are characteristic, as are abundant fruticose and foliose lichens.

Similar Types

Spruce Talus Woodlands lack the abundant Labrador tea characteristic of this community and also have fewer reindeer lichens. Other spruce woodland types (Red Spruce - Mixed Conifer Woodland and Black Spruce Woodland) have higher tree cover (>25%) and lower bryoid cover. Heath - Lichen Subalpine Slope Bogs are similar, but have more continuous vegetation, peat accumulation, and

Location Map

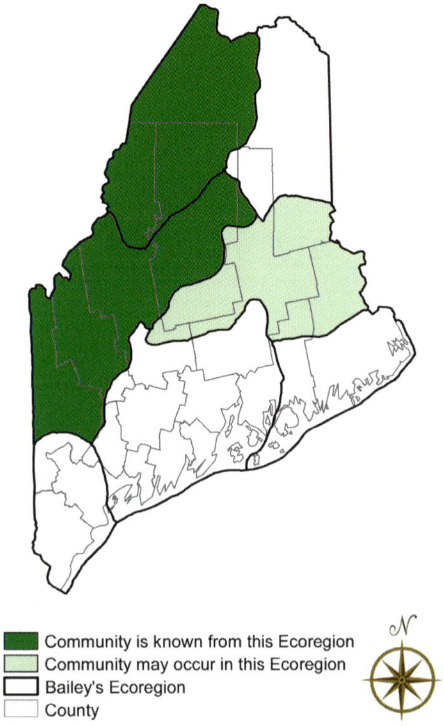

Cold-Air Talus Slope

occur on upper slopes at higher elevations.

Conservation, Wildlife, and Management Considerations

Known sites are on public lands or private conservation lands. One of the better known sites, Mahoosuc Notch, is traversed by the Appalachian Trail. Heavy foot traffic has caused some damage to the vegetation, but the community remains generally intact.

The long-tailed shrew and rock vole may show a preference for these rocky habitats.

Distribution

Restricted in Maine to northern and montane areas; New England - Adirondack Province extending west and likely northeastward to the Gaspé Peninsula.

Landscape Pattern: Small Patch

Characteristic Plants

These plants are frequently found in this community type. Those with an asterisk are often diagnostic of this community.

Sapling/shrub
Black spruce

Dwarf Shrub
Black crowberry*
Labrador tea*
Lowbush blueberry
Mountain cranberry
Sheep laurel

Herb
Rock polypody

Bryoid
Reindeer lichen*

Examples on Conservation Lands You Can Visit

- Deboullie Ponds Public Lands – Aroostook Co.
- Mahoosuc Public Land Unit – Oxford Co.

Lowbush Blueberry

Downeast Maritime Shrubland

State Rank S3

Community Description

Mat forming vegetation occurs on exposed maritime headlands. Most of the cover is dwarf heath shrubs and black crowberry growing only a few inches tall, punctuated by islands of taller shrubs such as bayberry or small spruce. Crowberry may carpet large areas. Creeping juniper or the more upright common juniper are often present. In the herb layer, three-toothed cinquefoil is characteristic. The bryoid layer is minor, but may feature reindeer lichens.

Soil and Site Characteristics

Sites occur on extremely exposed, usually granitic headlands. Sites may contain pockets of peaty soil developing in rock hollows. Salt spray, fog, and wind are near constant elements.

Diagnostics

Sites are characterized by the presence (and often dominance) of black crowberry, forming mats of vegetation with three-toothed cinquefoil and other species on exposed rocky coastal headlands.

Similar Types

Rose - Bayberry Maritime Shrublands occur in similar settings, but are dominated by taller shrubs without a strong component of mat-forming shrubs. Seaside Goldenrod - Goosetongue Open Headlands share many species and often co-occur with Crowberry - Bayberry Headlands, but the former are much more sparsely vegetated (vegetation often covers <25% of the rock surface overall), and without extensive mats of dwarf shrub vegetation.

Conservation, Wildlife, and Management Considerations

Several known occurrences are on public or private conservation lands. While this protects them from conversion to other uses, it also increases the opportunity for foot traffic. Because the vegetation is so low to the ground, it is easily trampled and some sites have been degraded. The primary management consideration is

Location Map

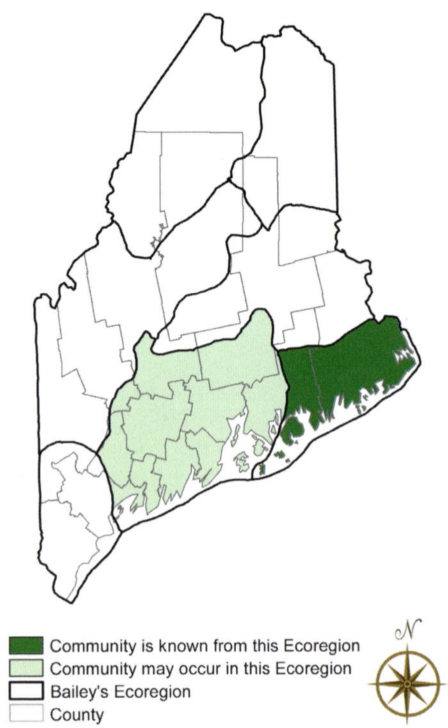

Downeast Maritime Shrubland

to keep visitors on established trails as much as possible.

Rocky headland communities, especially those that occur on uninhabited islands, may provide nesting habitat for some pelagic bird species, such as the Atlantic puffin, razorbill, and Leach's storm petrel. Coastal breeders such as the common eider, American black duck, herring and great black-backed gulls, and Atlantic puffin may also use this habitat in appropriate settings. The crowberry blue butterfly is restricted to coastal heaths in east-coastal Maine. It is typically found in peatlands, where it uses black crowberry as a larval host plant, but could occur in this community as well.

Distribution

Coastal Maine (often on islands) from Mount Desert Island eastward; extending into the Canadian Maritimes (Laurentian Mixed Forest Province).

Landscape Pattern: Small Patch

Characteristic Plants

These plants are frequently found in this community type. Those with an asterisk are often diagnostic of this community.

Dwarf Shrub

Black crowberry*
Creeping juniper
Large cranberry*
Lowbush blueberry*
Mountain cranberry
Swamp dewberry*

Herb

Bluebell
Canada mayflower
Red fescue
Three-toothed cinquefoil

Bryoid

Reindeer lichen

Associated Rare Plants

Nova Scotia false-foxglove

Associated Rare Animals

Crowberry blue

Examples on Conservation Lands You Can Visit

- Cutler Public Lands – Washington Co.
- Little Bois Bubert Island, Petit Manan National Wildlife Refuge – Washington Co.
- Mistake Island Preserve – Washington Co.
- Schoodic Peninsula, Acadia National Park – Hancock Co.
- West Quoddy Head State Park – Washington Co.

Dune Grassland

State Rank S2

Community Description

These sand dunes are strongly dominated by beach grass, with patches of beach-pea, red raspberry, bristly gooseberry, or poison ivy. In depressions or blowouts, grass cover is replaced by a mat of beach heather, golden heather, little bluestem, and pinweed, punctuated by tufts of reindeer lichens. Beach grass has stout rhizomes that allow it to colonize new places in its shifting substrate. Downeast of Acadia National Park, beach grass may be replaced by the ecologically and morphologically similar Virginia wild rye.

Soil and Site Characteristics

Sites occur on coastal sand dunes between the high tide line and back dune marshes abutting uplands; this type is best expressed along the southwest coast.

Diagnostics

Grass cover of beach grass or Virginia wild rye, which may include pockets of dwarf shrub - herb vegetation, characterizes these coastal dunes.

Piping Plover

Similar Types

Rose - Bayberry Maritime Shrublands can occur with these dune grasslands, but are dominated by shrubs rather than herbs. Beach Strands often occur on the shoreward margin of Dune Grasslands and are dominated by forbs rather than grasses.

Conservation, Wildlife, and Management Considerations

Beach dune systems along Maine's southwest coast have been reduced in extent by development. The intact dune systems that remain are almost all in public or private conservation ownership. Even with this protection, however, dunes are subject

Location Map

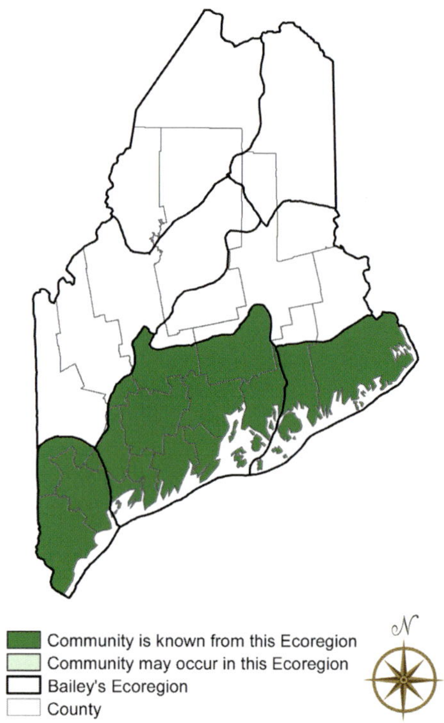

Dune Grassland

to degradation from uncontrolled recreational use. Even light foot traffic can initiate blowouts that will persist for years. Attempts over the last decade to educate people about the importance of staying out of dune areas are helping, but constant vigilance will be required in the face of increasing beach use. The prospect of sea level rise may also put these systems at greater risk.

Dune grasslands and adjacent upper beaches are critical breeding habitat for two species of endangered shorebirds: least terns and piping plovers. Nesting habitat for these birds has been reduced by human use of beaches, and the known nesting sites in Maine are carefully monitored. Other ground nesting bird species that may utilize these habitats include laughing gulls and short-eared owls.

Distribution

Mostly southwest of Merrymeeting Bay, occasionally eastward; extends southward along the Atlantic Coast to New Jersey

Landscape Pattern: Small Patch, linear.

Characteristic Plants

These plants are frequently found in this community type. Those with an asterisk are often diagnostic of this community.

Sapling/shrub
Bristly gooseberry
Red raspberry

Dwarf Shrub
Beach heather*

Herb
Beach grass*
Beach-pea*
Jointweed
Little bluestem*
Pinweed*
Virginia wild rye*

Associated Rare Plants

Beach plum
Coast-blite goosefoot

Associated Rare Animals

Laughing gull
Least tern
Piping plover
Short-eared owl

Examples on Conservation Lands You Can Visit

- Crescent Beach State Park – Cumberland Co.
- Popham Beach State Park – Sagadahoc Co.
- Rachel Carson National Wildlife Refuge – York Co.
- Reid Beach, Reid State Park – Sagadahoc Co.
- Seawall Beach, Morse Mountain Preserve – Sagadahoc Co.

Heath Alpine Ridge

State Rank S2

Community Description

This is the one of the most common community types above treeline. Vegetation can be quite variable and is dominated by a mixture of dwarf evergreen shrubs and herbs. Total vegetation cover is usually 35-65%. Shrubs usually make up 40-75% of the vegetation cover. The most abundant herbs are Bigelow's sedge and highland rush. Several rare species that occur only above treeline are found in this type. In one variant of this type, Bigelow's sedge is dominant and forms carpets, with heath shrubs and other herbaceous species far more sparse.

Soil and Site Characteristics

Sites occupy exposed, windswept ridges above treeline. Moisture conditions range from fairly xeric in exposed areas to moist patches in protected spots. Substrate varies from gravelly flats to stone pavements to rugged fellfields.

Diagnostics

Sites are above treeline and dominated by mixed dwarf shrubs, Bigelow's sedge, and highland rush. Alpine bilberry is prominent. Other strictly alpine species, such as boreal bentgrass, alpine sweetgrass, Lapland rosebay, bearberry willow, or alpine bearberry, are frequent but often at low cover.

Bigelow's Sedge

Similar Types

Diapensia Alpine Ridge vegetation has higher cover of diapensia and usually lower cover of graminoids. Subalpine Heath - Krummholz and Crowberry - Bilberry Summit Bald vegetation lack the strictly alpine species (although alpine bilberry and highland rush may occur in both). Bilberry - Mountain-heath Alpine Snowbank vegetation features tundra dwarf birch, moss-plant, or mountain-heath. Cotton-grass - Heath Alpine Bogs also lack the strictly

Location Map

Heath Alpine Ridge

alpine shrubs and have peat mosses and other wetland plants.

Conservation, Wildlife, and Management Considerations

Most Maine occurrences of this type occur on public or private conservation land; however, hiker traffic has caused degradation in heavily used areas. As with all above treeline vegetation in Maine, careful trail siting and efforts to minimize off-trail use are the important management considerations.

On the tablelands of Mt. Katahdin, this community type provides the only known habitat in the world for the Katahdin arctic butterfly. Mt. Katahdin is also the only known nesting area in the state for the American pipit.

Distribution

Upper-elevation ridges of Maine's western and central mountains (mostly in the New England - Adirondack Province), extending westward and southward along the Appalachians. Likely extends northeastward to the Gaspé Peninsula.

Landscape Pattern: Small Patch

Characteristic Plants

These plants are frequently found in this community type. Those with an asterisk are often diagnostic of this community.

Dwarf Shrub

Alpine bilberry*
Diapensia*
Mountain cranberry

Herb

Bigelow's sedge*
Highland rush*
Three-toothed cinquefoil

Associated Rare Plants

Alpine blueberry
Alpine sweet-grass
Bigelow's sedge
Boott's rattlesnakeroot
Boreal bentgrass
Cutler's goldenrod
Dwarf rattlesnake root
Mountain sandwort

Associated Rare Animals

American pipit
Katahdin arctic

Examples on Conservation Lands You Can Visit

- Bigelow Preserve Public Lands – Somerset Co.
- Goose Eye Mountain, Mahoosuc Public Lands – Oxford Co.
- Mt. Katahdin, Baxter State Park – Piscataquis Co.
- The Brothers, Baxter State Park – Piscataquis Co.

Low-elevation Bald

State Rank S3

Community Description

Patches of blueberry, lichens, low herbs, and bare rock form a mosaic on these summits. Vegetation may be sparse, but usually forms 10-50% cover overall, often comprised of only a few species. Three-toothed cinquefoil may be locally abundant. A few coastal sites feature broom-crowberry, an uncommon species. Bryoid cover may be low or high and usually is dominated by lichens rather than bryophytes. This is the typical habitat of the rare smooth sandwort.

Soil and Site Characteristics

Sites are on bald hilltops, mostly near the coast. Soils are patchy and usually consist of a thin, mostly organic layer, creating dry conditions. The substrate is acidic (pH 5.0-5.4). Elevations range from 600'-1500'.

Diagnostics

Tree cover is virtually zero. Total vegetation cover is 10-50%, rarely more, and is dominated by a mixture of dwarf ericads, forbs, and graminoids, with three-toothed cinquefoil often locally dominant. Vegetation height rarely exceeds 0.3 m.

Low-Elevation Bald on Mt. Cutler

Similar Types

Rocky Summit Heath is floristically similar and can occur with this type, but it has greater vegetation cover and height (patches > 0.5 m tall), and usually a sparse tree cover. Crowberry - Bilberry Summit Balds are structurally similar but feature crowberry and bilberry as locally dominant dwarf shrubs; they generally occur at higher elevations and more inland locations than this type.

Conservation, Wildlife, and Management Considerations

Many sites have evidence of intense, vegetation-clearing fire (e.g., Acadia National Park fires of 1947), but it is not clear whether past fires were of natural

Location Map

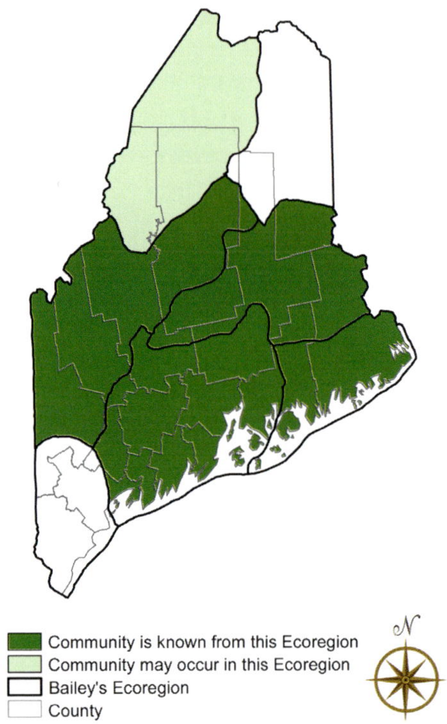

Low-Elevation Bald

Characteristic Plants

These plants are frequently found in this community type. Those with an asterisk are often diagnostic of this community.

Sapling/shrub
Black chokeberry

Dwarf Shrub
Broom-crowberry*
Lowbush blueberry*

Herb
Poverty oatgrass
Three-toothed cinquefoil*

Bryoid
Awned hair-cap moss
Reindeer lichen

Associated Rare Plants

Alpine blueberry
Mountain sandwort
Smooth sandwort

or human origin. Moreover, at some sites historic fires may have pushed the vegetation beyond a 'resilience threshold', and there is now so little vegetation that it is unlikely that fire could return for several hundred years.

This type is well represented on public lands and private conservation lands. However, because this community type is usually associated with nice views, many sites have moderate to heavy hiker or ATV use. Because the vegetation is rather sparse, it is easy for visitors to wander off the trail, and off-trail traffic can seriously degrade the vegetation and has done so at several sites.

Distribution

Mid-coast Maine eastward, along the immediate coast to a short distance inland.

Landscape Pattern: Small Patch

Examples on Conservation Lands You Can Visit

- Black Mountain, Donnell Pond Public Lands – Hancock Co.
- Mt Megunticook, Camden Hills State Park – Knox Co.
- Pemetic Mountain, Acadia National Park – Hancock Co.

Mountain Sandwort

Mid-elevation Bald

Crowberry - Bilberry Summit Bald

State Rank S3

Community Description

This patchy subalpine to alpine vegetation is dominated by a low mat of crowberry (usually black crowberry) and alpine bilberry. Fir, spruce, mountain holly, and/or heart-leaved birch occasionally grow 1-2 m high, but only at low cover (<25%). Total vegetation cover is usually 20-50% (excepting crustose lichens), with expanses of rock prominent among the vegetation. Crustose and foliose lichens are abundant on the exposed bedrock.

Soil and Site Characteristics

Sites are on very exposed bedrock summits or upper mountain slopes at moderate to high elevations (1800' and up). Typically sites are thin-soiled and well drained.

Diagnostics

Sparse and patchy vegetation occurs near or on mountain summits, with crowberry, bilberry, and/or highland rush prominent, and without strictly alpine species such as Bigelow's sedge, diapensia, Lapland rosebay, etc.

Mid-Elevation Bald

Similar Types

This type often co-occurs with other alpine or sub-alpine types. Rocky Summit Heath vegetation can occur at similar elevations, but is usually somewhat less exposed, with some tree cover (though stunted), and with typical lower-elevation heath shrubs such as Labrador tea, sheep laurel, black huckleberry, or rhodora. Dwarf Heath - Graminoid Alpine Ridge occurs at higher elevations and features some strictly alpine species such as Bigelow's sedge, boreal bentgrass, diapensia, etc.

Location Map

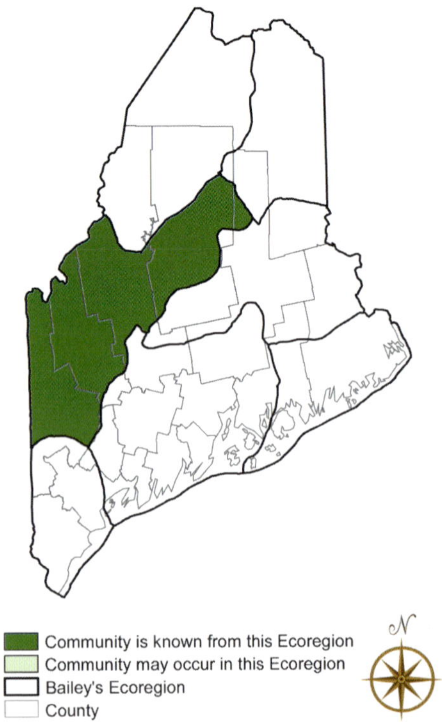

Mid-Elevation Bald

Conservation, Wildlife, and Management Considerations

Most occurrences are on public or private conservation land; however, hiker traffic can cause degradation even on these "protected" lands. As with all alpine and sub-alpine vegetation in Maine, careful trail siting and efforts to minimize off trail use are the important management considerations.

Birds of open habitats and grasslands, such as the savannah sparrow, may use this habitat. The rare crowberry blue butterfly is restricted to coastal heaths in east-coastal Maine. It is typically found in peatlands, where it uses black crowberry as a larval host plant, but could occur in this community as well.

Distribution

Mid- to upper elevation ridges of Maine's western and central mountains (mostly in the New England - Adirondack Province), extending westward and southward along the Appalachians.

Landscape Pattern: Small Patch

Characteristic Plants

These plants are frequently found in this community type. Those with an asterisk are often diagnostic of this community.

Sapling/shrub

Red spruce*

Dwarf Shrub

Alpine bilberry*
Alpine blueberry*
Black crowberry*
Sheep laurel*

Herb

Three-toothed cinquefoil*

Associated Rare Plants

Alpine blueberry
Alpine sweet-grass
Dwarf rattlesnakeroot
Mountain sandwort
Silverling

Associated Rare Animals

Crowberry blue

Examples on Conservation Lands You Can Visit

- Caribou - Haystack Mountain, White Mountain - Oxford Co.
- Mount Coe, Baxter State Park - Piscataquis Co.
- Moxie Bald Mountain, Appalachian Trail - Somerset Co.
- South Turner Mountain, Baxter State Park - Piscataquis Co.

Open Headland

Seaside Goldenrod - Goosetongue Open Headland

State Rank S4

Community Description

In this community patchy herbaceous vegetation of seaside goldenrod, goosetongue, bluebell, and yarrow covers coastline bedrock. Overall vegetation cover is usually 10-35% but may be locally more extensive. Crowberry or creeping juniper may be present, but they do not cover extensive areas. Downeast of Acadia National Park, species of the Canadian Maritimes may be present, such as roseroot, beachhead iris, marsh-felwort, and glabrous knotted pearlwort. The primary bryoids are crustose lichens on the rock, including the lime-green map lichen and bright orange Xanthoria lichen. Where the forest edge abuts, cover is more dense, and often includes shrubs such as meadowsweet, currants, or small white spruce.

Nova Scotia False-foxglove

Soil and Site Characteristics

Sites occur on bedrock areas exposed to salt spray and storm tides, typically between the high tide line and the upland forest. Sites are flat to sloping, and plants grow in the fissures of the rock or in small depressions with almost no soil.

Diagnostics

Sparse herb dominated vegetation occurs on oceanside bedrock.

Similar Types

Crowberry - Bayberry Headlands share many species, but these have more vegetation (usually >50% total cover and >25% shrub or dwarf shrub cover)

Location Map

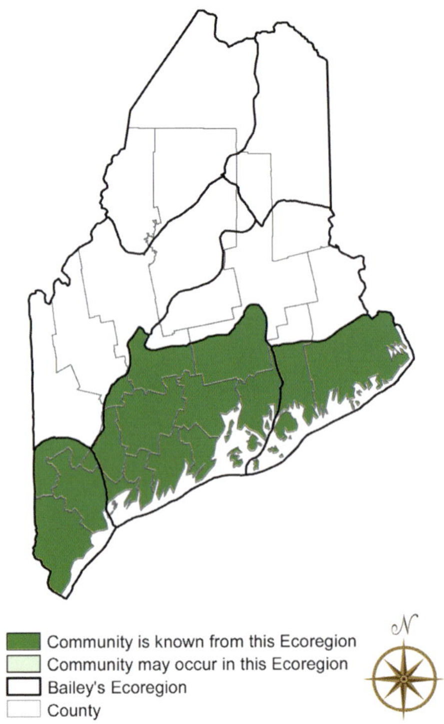

Open Headland

and are dominated by shrubs rather than herbs. Rose - Bayberry Maritime Shrublands are dominated by shrubs >1 m tall.

Conservation, Wildlife, and Management Considerations

This type is well distributed along the Maine coast. Many sites are on public lands or private conservation lands. Some areas receive moderate to heavy foot traffic. To the degree that users remain on the bare rock, impacts can be minimal.

Rocky headland communities, especially those that occur on uninhabited islands, may provide nesting habitat for some oceanic bird species.

Distribution

Coastwide, largest occurrences east of Penobscot Bay. Extends into the Canadian Maritimes and south to Massachusetts (Laurentian Mixed Forest Province and Eastern Broadleaf Forest Province).

Landscape Pattern: Small Patch

Characteristic Plants

These plants are frequently found in this community type. Those with an asterisk are often diagnostic of this community.

Herb

Bluebell
Bluejoint grass
Gall of the earth
Goosetongue
Red fescue
Roseroot*
Seaside goldenrod*
Yarrow

Associated Rare Plants

Birds-eye primrose
Blinks
Marsh-felwort
Nova Scotia false-foxglove

Examples on Conservation Lands You Can Visit

- Acadia National Park – Hancock Co.
- Cutler Public Lands – Washington Co.
- Great Wass Island Preserve – Washington Co.
- Petit Manan National Wildlife Refuge – Washington Co.
- Quoddy Head State Park – Washington Co.

Rivershore Outcrop

State Rank S3

Community Description

Sparse rivershore vegetation is dominated by herbs with occasional low shrubs. Total cover rarely exceeds 25%. Typical herbs include three-toothed cinquefoil, common hairgrass, hairy goldenrod, silverrod, bluebell, balsam ragwort, and narrow false oats. Shrubs include dwarf bilberry, lowbush blueberry, shrubby cinquefoil, and shadbush; poison ivy may be locally abundant. Where soil allows the growth of taller shrubs (e.g., at the upland transition into adjacent shrub vegetation), red osier dogwood, round-leaved dogwood, and willows may occur. The rare species associated with most of these ledges show an affinity to northern areas; in central Maine, one may find more temperate indicator species, such as Indian grass and little bluestem.

Soil and Site Characteristics

This type occurs on dry ledges and outcrops along rivershores. Substrate is typically circumneutral or calcareous slate, with plants growing in vertical fissures. Sites are subject to annual flooding and ice scour, which allows at least a small amount of silt to accumulate in the rock crevices.

Rivershore Outcrop

Diagnostics

Herb dominated sparse vegetation occurs on rivershore outcrops.

Similar Types

This is the only herb dominated rivershore ledge community type. Circumneutral Riverside Seeps share many species with moist pockets of these outcrops, but occur on gravelly (unconsolidated) substrates. The geographic transition seen from northern Maine rivers to those in central Maine may warrant splitting

Location Map

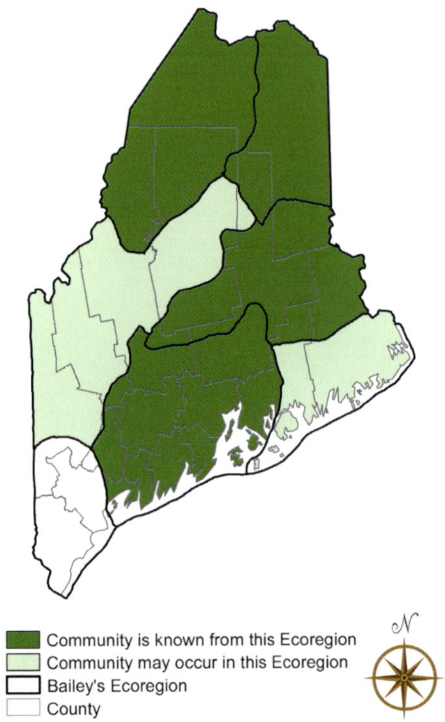

the central Maine occurrences into a "Bluestem Shoreline Outcrop" type, but more information is needed from both within and outside of Maine.

Conservation, Wildlife, and Management Considerations

Many sites are visited on foot for recreation. In sites with moderate to heavy foot traffic, some degradation of the vegetation is apparent. A few sites show some degradation by exotic species such as Japanese knotweed. Several sites are in public ownership or private conservation ownership; many are privately owned.

These rivershore shrublands provide habitat for common bird species that inhabit open shrublands such as common yellowthroat, alder flycatcher, Wilson's warbler, and Lincoln's sparrow.

Distribution

Along the major rivers from central Maine northward and eastward. Extends east and north into New Brunswick and west into New Hampshire and Vermont.

Landscape Pattern: Small Patch. Linear.

Examples on Conservation Lands You Can Visit

- Allagash Lake, Allagash Wilderness Waterway – Piscataquis Co.
- Allagash Public Lands – Aroostook Co.
- Coburn Park – Somerset Co.
- Rocky Island Preserve – Aroostook Co.

Characteristic Plants

These plants are frequently found in this community type. Those with an asterisk are often diagnostic of this community.

Sapling/shrub

Meadowsweet
Morrow's honeysuckle
Poison-ivy
Red osier dogwood
Round-leaved dogwood
Shining willow

Dwarf Shrub

Dwarf bilberry
Lowbush blueberry
Velvet-leaf blueberry

Herb

Balsam ragwort
Bluebell
Common pussytoes
Dwarf raspberry
Early goldenrod
Field pussytoes
Kalm's lobelia
Narrow false oats
Silverrod
Stiff aster
Tufted hairgrass
Wild chive

Associated Rare Plants

Alpine milk-vetch
Clinton's bulrush
Cut-leaved anemone
Indian grass
Mistassini primrose
New England violet
Pale green orchis
Purple clematis
Soft-leaf muhly
St. John oxytrope

Rocky Summit Heath

State Rank S4

Community Description

Dwarf shrubs and stunted spruce or fir are the dominant features of this patchy vegetation. The tree layer is sparse (<25% cover) and includes balsam fir, red or black spruce, and (especially near the coast) northern white cedar. Dwarf shrub dominants vary and include Labrador tea, blueberries, and mountain cranberries. Interspersed islands of taller shrubs and stunted, wind flagged trees (<2 m tall) may grade to krummholz form. Herbs are patchy and less extensive than shrubs; three-toothed cinquefoil is usually present and often prominent. The bryoid layer is usually sparse (<30% cover), but may be more extensive in moist bedrock depressions. In these low spots, peat mosses are typical; otherwise, lichens are the dominant bryoids.

Soil and Site Characteristics

Sites occupy upper slopes and ridges near the coast or mid-elevation balds (2,000'-3,000' inland). Soils are patchy, in bedrock pockets, consist of a thin layer (5-25 cm deep) of organic duff mixed with sand or rock fragments, and are acidic (pH 4.8-5.0) and excessively well drained, except in localized peaty pockets.

Diagnostics

Tree cover is sparse, stunted, and patchy (<25% cover); balsam fir and heart-leaved paper birch are typical; red oak may occur at some coastal locations. Shrub cover is typically 10-50% at lower elevation sites or in pockets at higher elevations. Dwarf shrub cover exceeds herb cover, with lowbush blueberry or mountain cranberry prominent.

Similar Types

Crowberry - Bilberry Summit Balds occur at higher elevations but lack tree cover and feature crowberry, bilberry, and/or highland rush. Three-toothed Cinquefoil - Blueberry Low Summit Balds occur at similar elevations and may co-occur with Rocky Summit Heaths, but the former usually have herbaceous species exceeding dwarf shrub species in the herb layer. Other open summit communities occur at higher elevations and feature at least some strictly alpine species such as Bigelow's sedge, highland rush, diapensia, Lapland rosebay, etc. Red Spruce - Mixed Conifer Woodlands are floristically similar and may co-occur but have >25% tree cover.

Location Map

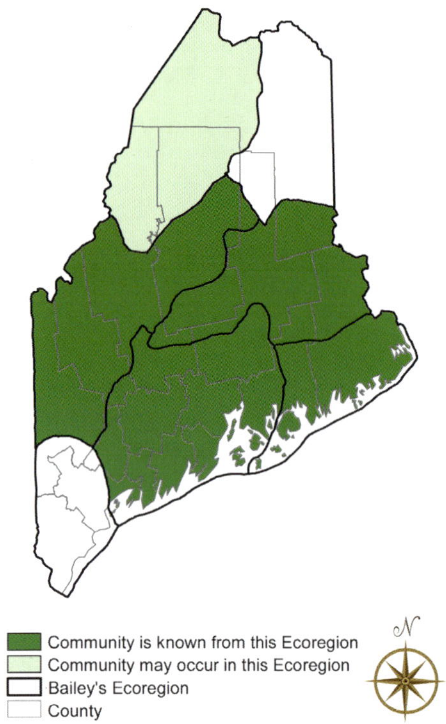

Rocky Summit Heath

Characteristic Plants

These plants are frequently found in this community type. Those with an asterisk are often diagnostic of this community.

Canopy
Balsam fir*

Sapling/shrub
Balsam fir*
Gray birch*
Mountain holly*
Northern white cedar*
Red spruce*
Shadbush*
Wild-raisin*
Winterberry holly*

Dwarf Shrub
Black huckleberry*
Labrador tea*
Lowbush blueberry*
Mountain cranberry*
Rhodora*
Sheep laurel*

Herb
Balsam fir*
Three-toothed cinquefoil

Bryoid
Reindeer lichen
Sphagnum mosses*

Associated Rare Plants

Alpine blueberry
Alpine sweet-grass
Smooth sandwort

Associated Rare Animals

Bicknell's thrush

Conservation, Wildlife, and Management Considerations

Because this community type is usually associated with nice views, many sites have moderate to heavy hiker use. Off trail traffic can seriously degrade the vegetation, but at most sites the relatively dense shrub and conifer vegetation does not invite off trail wanderings. This type is well represented on public lands and private conservation lands.

Coniferous forest specialists like blackpoll warblers and spruce grouse are common associates in this community.

Distribution

Western and coastal Maine (New England - Adirondack and Laurentian Mixed Forest Provinces), extending west into northern New England and New York and east into the Canadian Maritimes.

Landscape Pattern: Small Patch

Examples on Conservation Lands You Can Visit

- Bald Mountain, Little Concord Pond Public Lands - Oxford Co.
- Black Mountain, Donnell Pond Public Lands - Hancock Co.
- Blueberry Mountain, White Mountain National Forest - Oxford Co.
- Cadillac Mountain, Acadia National Park - Hancock Co.
- Cold Brook Trail, White Mountain National Forest - Oxford Co.
- Sargent Mountain, Acadia National Park - Hancock Co.
- Tunk Mountain, Donnell Pond Public Lands - Hancock Co.

Rose Maritime Shrubland

Rose - Bayberry Maritime Shrubland

State Rank S4

Community Description

Medium height shrubs (1-2 m) usually cover 30-60% but may form dense thickets. Bayberry and roses are characteristic; raspberry and poison ivy are frequent associates. On some islands, shrublands are dominated by raspberry or bush-honeysuckle, with little or no bayberry or rose. Wild-raisin and winterberry may occur in more protected or moist pockets. Lowbush blueberry and northern dewberry are occasional as dwarf shrubs. Herbs grow in sometimes extensive patches among the shrubs and include salt tolerant shore species such as beach grass, beach-pea, sea-beach sandwort, seabeach angelica, and seaside goldenrod. Bryoids are absent, except for small amounts of lichens in some areas.

Soil and Site Characteristics

The islands where this type occurs are exposed to onshore winds and salt spray; sometimes covering extensive areas on stabilized dunes or rocky islands. Except on dunes, soils are thin, usually less than 25 cm deep, and acidic (pH 4.8-5.5).

Virginia Rose

Diagnostics

This type is typified by a seaside setting and dominance of upland shrubs (> about 1 m tall), particularly bayberry and roses, with cover >20%.

Similar Types

Crowberry - Bayberry Headlands can share many species, but have dwarf shrub cover more abundant than taller shrub cover and lack poison-ivy. Other sandy beach community types can be adjacent to Rose - Bayberry Maritime Shrublands and share some species, but have different physiognomy: Pitch Pine Dune Woodlands have a tree canopy of >25%; Dune Grasslands are dominated by grasses, not shrubs; and Beach Strands are sparsely vegetated with herbs and no shrubs.

Location Map

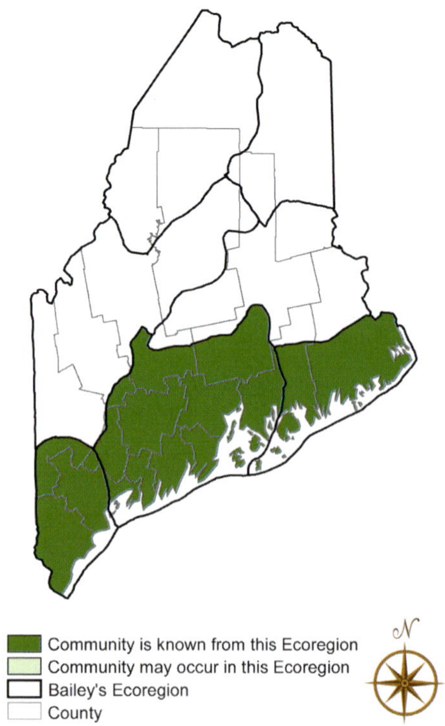

Bayberry Leaves

Conservation, Wildlife, and Management Considerations

Most known sites have been used historically for grazing sheep. Evidence of fire is common in most. Rugosa rose is an invasive species of these sites. The extent to which this community has developed as an artifact of clearing and grazing is unknown and would be an interesting study (along with successional dynamics). At least some small sites along the immediate coastline appear to have developed by and are persisting through natural processes. Several occurrences are on public lands or private conservation lands.

Maritime shrubland communities, especially those that occur on uninhabited islands, may provide nesting habitat for some oceanic bird species, such as the Atlantic puffin, razorbill, and Leach's storm-petrel. Coastal breeders such as the common eider, black duck, and herring and great black-backed gulls may also use this habitat in appropriate settings.

Distribution

Along Maine's immediate coastline and islands (Laurentian Mixed Forest Province).

Landscape Pattern: Small Patch

Characteristic Plants

These plants are frequently found in this community type. Those with an asterisk are often diagnostic of this community.

Sapling/shrub

Bayberry*
Bush-honeysuckle*
Meadowsweet*
Poison-ivy*
Red raspberry*
Rugosa rose*
Virginia rose*
Winterberry holly*

Dwarf Shrub

Large cranberry*
Lowbush blueberry*

Herb

Beach grass*
Canada mayflower*
Greene's rush*
New York aster
Rough-stemmed goldenrod
Sea-beach angelica
Wire rush
Yarrow

Associated Rare Plants

Beach plum
Small saltmarsh aster

Examples on Conservation Lands You Can Visit

- Little Duck Island– Hancock Co.
- North Libby Island Wildlife Management Area – Washington Co.
- Popham Beach State Park – Sagadahoc Co.
- Seawall Beach, Morse Mountain Preserve – Sagadahoc Co.
- Stratton Island – York Co.

Sandplain Grassland

State Rank S1

Community Description

These grassland barrens are expanses of graminoid and dwarf shrub vegetation dominated by poverty oatgrass, little bluestem, woodland sedge, and lowbush blueberry. Northern blazing star, a globally rare wildflower, is a dominant forb at the one viable Maine occurrence of this community. Local dominants include sweetfern and black chokeberry. Species allied with pine barrens are frequent, such as gray birch, whorled loosestrife, and wood lily. Bryoids are sparse, and up to 30% of the ground is unvegetated, usually with some graminoid or shrub litter.

Soil and Site Characteristics

These grasslands occur on flat sandy plains or deep outwash deposits. Soils are extremely well drained and acidic. Historically, fire was important in maintaining these communities and is now being used as a management tool.

Diagnostics

Canopy cover is less than 15%; tall shrub layer is patchy but less than 25%. Poverty oatgrass, little bluestem, and woodland sedge are the dominant graminoids, often over a layer of lowbush blueberry.

Similar Types

Pitch Pine - Scrub Oak Barrens and Pitch Pine - Heath Barrens share many species, but have well developed shrub and/or tree layers. Hudsonia River Beach also features little bluestem, but blueberry is scarce or absent and the riverside setting is distinctive.

Conservation, Wildlife, and Management Considerations

The one viable Maine site is in conservation ownership and is managed for the rare community, rare animals, and rare plants that occur there. Prescribed burning is a major management tool. This type appears to have been more common in southern

Location Map

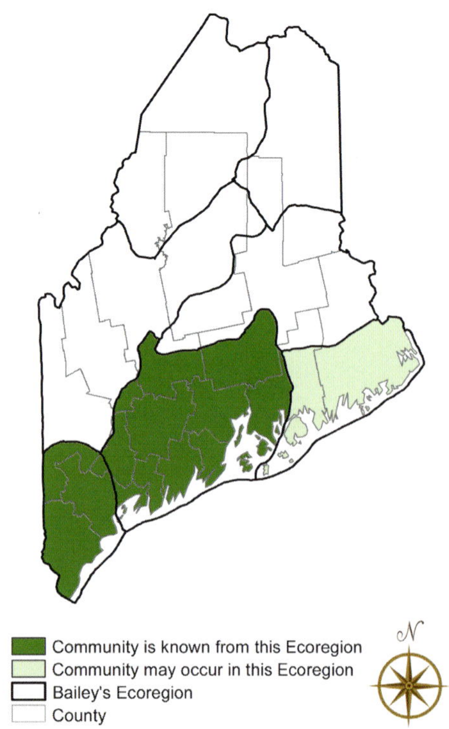

Main Photo: Sandplain Grassland
Inset: Black Racer

Maine historically, but most nearby sites that might have supported this community have either been developed or have transitioned to other types due to lack of fire.

This community provides nesting habitat for several rare ground nesting grassland birds including the grasshopper sparrow, upland sandpiper, and short-eared owl. These open grasslands also provide excellent habitat for a rare snake, the northern black racer. The cobweb skipper inhabits open dry fields with bluestem grasses, which it uses as its larval host plant. Several historical butterfly and moth species that may have inhabited formerly large tracts of this community include the graceful clearwing, persius dusky wing, and frosted elfin.

Distribution

Extreme southern Maine, extending southward along the Atlantic Coast to Cape Cod (Eastern Broadleaf Forest Province).

Landscape Pattern: Large Patch

Characteristic Plants

These plants are frequently found in this community type. Those with an asterisk are often diagnostic of this community.

Sapling/shrub

Gray birch
Pin cherry
Pitch pine
Sweetfern
Thicket shadbush

Dwarf Shrub

Black chokeberry
Lowbush blueberry*

Herb

Arrow-leaved violet
Little bluestem*
Northern blazing star*
Pinweed
Poverty oatgrass
Sharp-pointed ricegrass
Silverrod
Whorled loosestrife
Woodland sedge*

Associated Rare Plants

Clothed sedge
Dry land sedge
Northern blazing star
Upright bindweed
White-topped aster

Associated Rare Animals

Cobweb skipper
Grasshopper sparrow
Northern black racer
Short-eared owl
Upland sandpiper

Examples on Conservation Lands You Can Visit

Kennebunk Plains Preserve – York Co.

Spruce - Fir Krummholz

State Rank S3

Community Description

Krummholz refers to the zone between treeline and more open alpine vegetation, where tree species are limited by the harsh conditions to a dense shrub growth-form. Black spruce, balsam fir, and heart-leaved paper birch form masses of stunted and wind swept trees 0.5-2 m high. Mountain alder may be locally common, and mountain ash and mountain shadbush are occasional. Total shrub cover is often close to 100%, and these areas may be all but impenetrable. Boreal herbs, such as bluebead lily and Canada mayflower, grow with patches of mosses in small openings among the shrubs, but total herb cover is sparse. Bryoids may be extensive beneath the trees.

Soil and Site Characteristics

This type occupies upper mountain slopes above treeline, typically at elevations of 2700 - 3700'. The cool conditions, lingering snows, and frequent fog and clouds create a fairly moist microclimate, but the sites are very exposed to wind and storms.

Black Spruce Cones

Diagnostics

These are forests of the treeline zone in which dwarfed and matted trees form a dense shrub layer 0.5-2 m high; usually strongly coniferous.

Similar Types

Rocky Summit Heath can grade into or form a patchwork with this community, but it features lower tree cover (<25%) and more heath shrubs and open spaces. Fir - Heart-leaved Birch Subalpine Forest shares many overstory species and can grade into this community but is distinguished by having more upright trees and a fairly well developed herbaceous layer.

Location Map

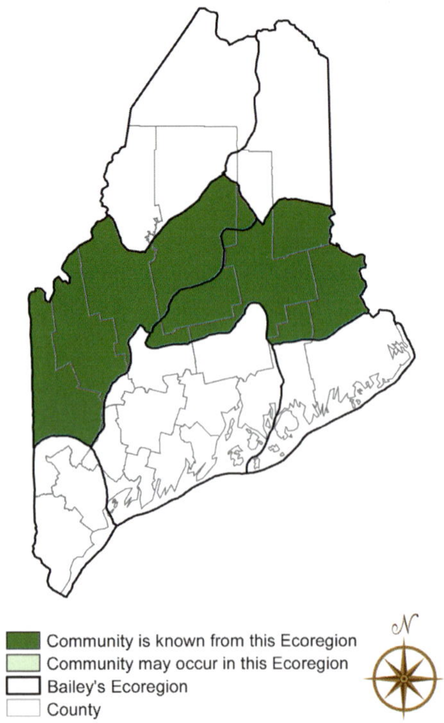

Spruce – Fir Krummholz

Conservation, Wildlife, and Management Considerations

Krummholz is extensive on Maine's higher mountains, and most major occurrences are well protected within public lands or private conservation lands. The historic extent has been somewhat reduced by the development of ski areas, and proposals for wind generators could impact other sites. Because traversing this vegetation is so miserable, off-trail impacts from hikers are minimal, in contrast to other alpine/subalpine vegetation types.

This high-elevation dwarfed forest community type provides habitat for Bicknell's thrush, which only inhabits structurally complex forests above 2500'. Coniferous forest specialists like blackpoll warblers and spruce grouse are common associates in this community.

Distribution

Upper-elevation ridges of Maine's western and central mountains (mostly in the New England - Adirondack Province), extending westward and southward along the Appalachians, and likely to the Gaspé Peninsula.

Landscape Pattern: Large Patch

Characteristic Plants

These plants are frequently found in this community type. Those with an asterisk are often diagnostic of this community.

Sapling/shrub

Balsam fir*
Black spruce*
Heart-leaved paper birch*

Dwarf Shrub

Alpine bilberry*
Labrador tea*

Herb

Black crowberry
Bluebead lily
Bunchberry
Canada mayflower
Creeping snowberry
Mountain cranberry
Stiff clubmoss

Bryoid

Common broom-moss
Fringed Ptilidium liverwort
Red-stemmed moss

Associated Rare Plants

Northern comandra

Associated Rare Animals

Bicknell's thrush

Examples on Conservation Lands You Can Visit

- Baldpate Mountain, Grafton Notch State Park – Oxford Co.
- Bigelow Preserve Public Lands – Somerset Co.
- Goose Eye Mountain, Mahoosuc Public Lands – Oxford Co.
- Mt. Abraham – Franklin Co.
- Mt. Katahdin, Baxter State Park – Piscataquis Co.
- Saddleback Mountain, Appalachian Trail – Franklin Co.

Subalpine Meadow

Mountain Alder - Bush-honeysuckle Subalpine Meadow

State Rank S1

Community Description

These shrub and graminoid dominated meadows occur near treeline. Dominants vary according to substrate moisture but often include mountain alder, bluejoint, and meadowsweet. In drier areas, bush-honeysuckle may be prominent. This vegetation is taller and more dense than typical alpine dwarf shrub vegetation. Openings around pondshores or other disturbed areas may support rare plants.

Soil and Site Characteristics

Sites occupy upper mountain slopes, flats, or basins near or above treeline. Slopes very from almost flat to quite steep, and substrate moisture varies from dryish to seepy soil conditions.

Diagnostics

Sites are in an alpine setting with dense mountain alder, bluejoint grass, and/or bush-honeysuckle; dwarf shrubs are absent or minor.

Bush-honeysuckle

Lowbush Blueberry

Similar Types

Other alpine and sub-alpine types have different characteristic species. However, this is not a well documented type, and its distribution and relationship to other alpine and subalpine vegetation types needs work.

Location Map

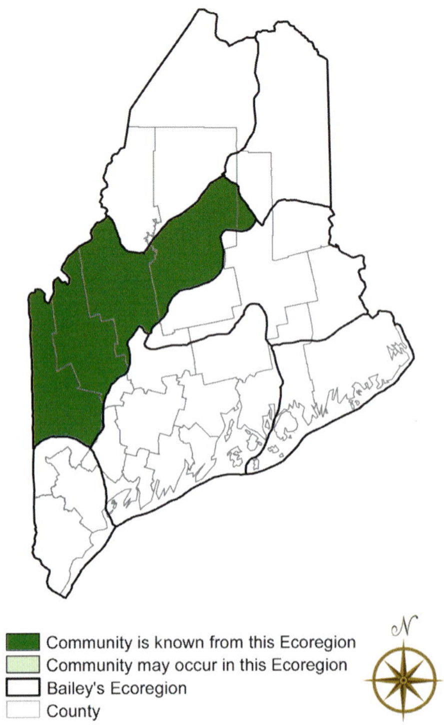

Subalpine Meadow

Characteristic Plants

These plants are frequently found in this community type. Those with an asterisk are often diagnostic of this community.

Sapling/shrub
Bush-honeysuckle*
Meadowsweet
Mountain alder*
Squashberry

Dwarf Shrub
Lowbush blueberry

Herb
Bluejoint*
Gall of the earth
Large-leaved goldenrod
Low rough aster
Three-toothed cinquefoil

Associated Rare Plants

Black sedge
Northern painted-cup
Russett sedge

Conservation, Wildlife, and Management Considerations

Both known occurrences in Maine are on protected lands. Hiker impacts have been minimal. Understanding the importance of fire and other disturbances to initiating and maintaining the vegetation in this community type warrants study.

Distribution

Upper-elevation ridges of Maine's western and central mountains (mostly in the New England - Adirondack Province); may extend to Gaspé Peninsula.

Landscape Pattern: Small Patch

Examples on Conservation Lands You Can Visit

- Mt. Katahdin, Baxter State Park – Piscataquis Co.
- North Traveler, Baxter State Park – Piscataquis Co.

State Rank S1

Community Description

Matted evergreen shrubs and a few herbs are scattered among boulders or in bedrock pockets in an open alpine setting. Cushions of diapensia are the dominant feature, and alpine bilberry is typically common as well. The cover of herb species is usually <20%, and graminoids such as Bigelow's sedge and highland rush are typically more abundant than forbs. Older mats of diapensia may exhibit dieback.

Soil and Site Characteristics

Sites occur on very exposed and windswept areas above treeline, with plants growing in gravelly substrate among fractured rocks. Slopes are flat to gentle.

Diagnostics

Diapensia is the dominant species with >20% cover.

Similar Types

Dwarf Heath - Graminoid Alpine Ridge can occur at similar elevations and can grade into this type, but diapensia, if present, is sparse (< 20% cover). Bilberry - Mountain-heath Alpine Snowbanks similarly lack extensive diapensia. Subalpine Heath Krummholz and Crowberry - Bilberry Summit Balds lack diapensia and other strictly alpine species and generally occur at lower elevations.

Diapensia

Conservation, Wildlife, and Management Considerations

This type is restricted even within the alpine zone and is very sensitive to hiker impacts. Because the vegetation is characteristically sparse and often occurs on flats, it is easy for hikers to wander off trail. The documented Maine sites occur on public lands or private

Location Map

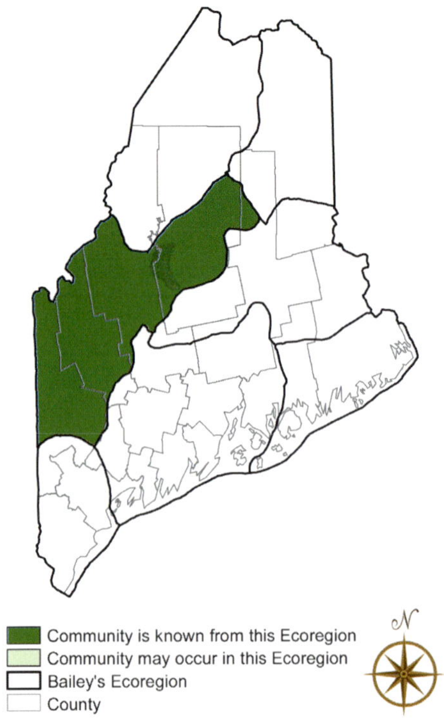

Windswept Alpine Ridge

conservation lands, where hiker impacts are the main management concern.

Mt. Katahdin is the only known nesting area in the state for the American pipit, which uses a variety of natural community types above treeline.

Distribution

Restricted to Maine's highest mountains (New England - Adirondack Province), extending west to New Hampshire and along the Appalachians, and possibly northeastward to the Gaspé Peninsula.

Landscape Pattern: Small Patch

Cutler's Goldenrod

Characteristic Plants

These plants are frequently found in this community type. Those with an asterisk are often diagnostic of this community.

Dwarf Shrub

Alpine azalea
Alpine bearberry
Alpine bilberry*
Black crowberry
Diapensia*

Herb

Bigelow's sedge
Boreal bentgrass
Highland rush
Mountain sandwort
Northern firmoss
Three-toothed cinquefoil

Bryoid

Cetraria lichen

Associated Rare Plants

Alpine bearberry
Alpine sweet-grass
Bearberry willow
Bigelow's sedge
Boreal bentgrass
Cutler's goldenrod
Diapensia
Lapland rosebay
Mountain sandwort
Northern firmoss

Associated Rare Animals

American pipit

Examples on Conservation Lands You Can Visit

- Mt. Abraham – Franklin Co.
- Mt. Katahdin, Baxter State Park – Piscataquis Co.
- Saddleback Mountain, Appalachian Trail – Franklin Co.

Open Wetlands

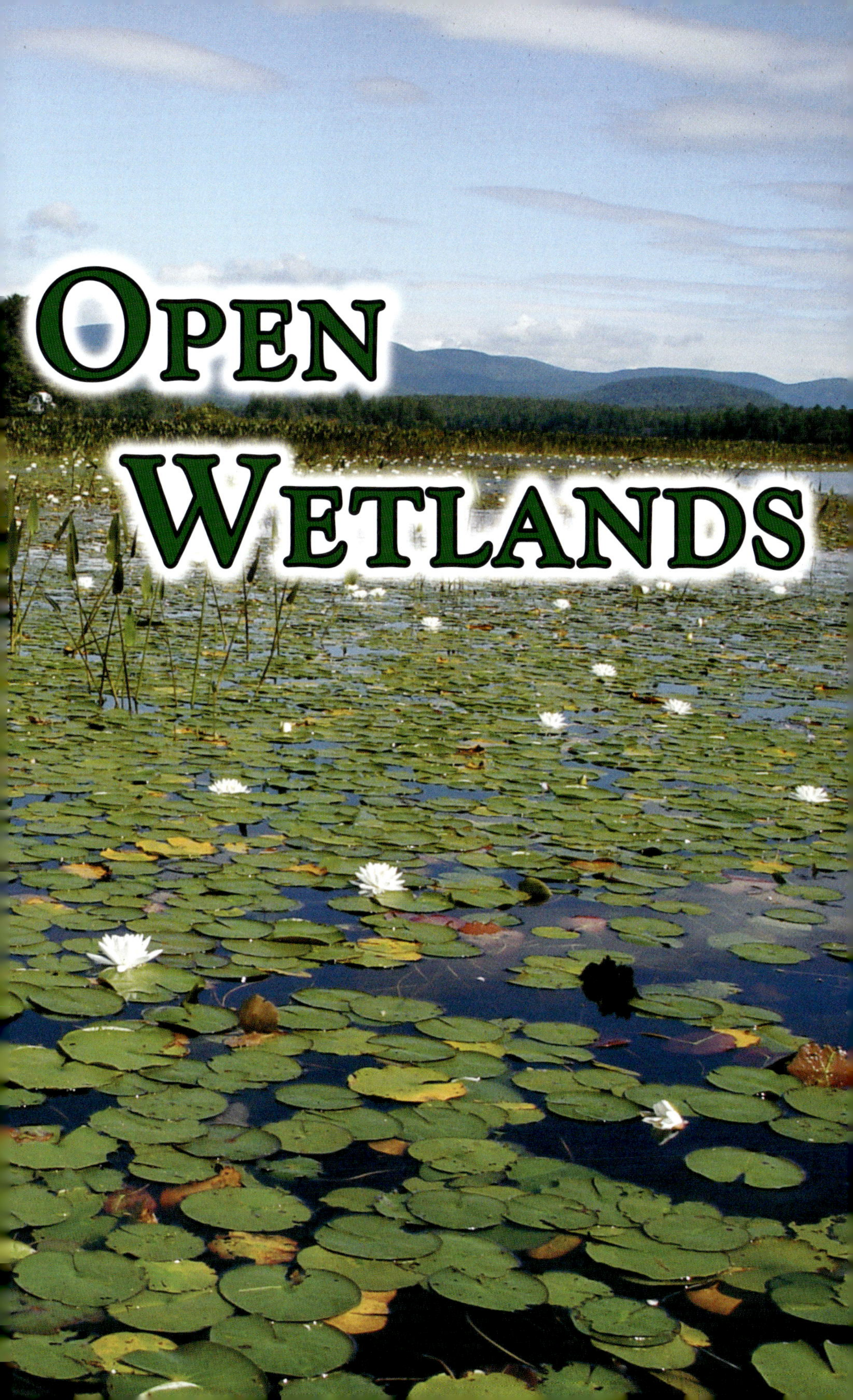

Alder Floodplain

State Rank S4

Community Description

These shrub dominated wetlands are characterized by speckled alder (or less commonly, mountain alder or smooth alder) and an assemblage of grasses and herbs indicative of alluvial conditions. Red maple, American elm, and red-osier dogwood may be co-dominant with alders, but other woody species are usually sparse in these moderate to high energy settings. Woody species may show signs of ice scour early in the season. Bluejoint is common, and frequent herbs include Virgin's bower, ostrich fern, flat-topped white aster, royal fern, sensitive fern, boneset, swamp dewberry, tall meadow-rue, and woodland horsetail.

Soil and Site Characteristics

This type occurs along medium to large moderate to high energy riverbanks that are seasonally flooded and subject to ice scour. It is usually on mineral soil.

Diagnostics

These are shrub-dominated floodplains along medium to large rivers. Alders dominate and comprise >30% cover. Heath shrubs and peat are sparse or absent.

Wood Turtle

Similar Types

Dogwood - Willow Shoreline Thickets occur in similar floodplain settings in northern Maine, but alder is not dominant. Alder Shrub Thickets typically occur in stagnant basins on deeper organic (muck or peat) soils, with peat mosses frequent.

Conservation, Wildlife, and Management Considerations

The rivershore habitat of this natural community suggests that direct threats from development are relatively low, though excessive

Location Map

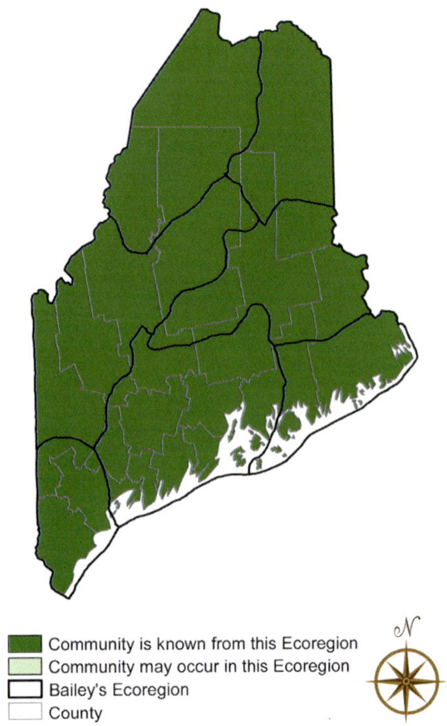

Alder Floodplain

watershed development would likely increase the volume and intensity of runoff and potentially alter riverine structure and community composition. Additionally, hydrologic alteration (i.e., impoundments) would compromise the disturbance regime, but new dams are unlikely on medium and large rivers.

These shrublands, especially when they occur in close proximity to open water, may provide habitat for bird species such as common yellowthroat, alder flycatcher, Wilson's warbler, and Lincoln's sparrow. Some occurrences of this community type support vernal pools, which are important breeding habitat for a variety of amphibians including wood frogs, spotted salamanders, and blue-spotted salamanders. In southern Maine, Blanding's and spotted turtles may feed on amphibian larvae and egg masses present in such pools. Wood turtles bask and forage in shrubby floodplains of moderate to large rivers statewide.

Distribution

Statewide; extending in all directions from Maine.

Landscape Pattern: Small to Large Patch

Characteristic Plants

These plants are frequently found in this community type. Those with an asterisk are often diagnostic of this community.

Sapling/shrub

Arrow-wood
Meadowsweet
Red maple*
Red osier dogwood
Speckled alder*

Herb

Bluejoint*
Boneset
False nettle
Flat-topped white aster*
Lady fern
Ostrich fern
Royal fern
Sensitive fern
Swamp dewberry*
Tall meadow-rue*
Virgin's bower*
Wild calla
Woodland horsetail

Associated Rare Plants

Hollow joe-pye weed

Associated Rare Animals

Wood turtle

Examples on Conservation Lands You Can Visit

- Moose River, Holeb Public Lands – Somerset Co.
- Narraguagus Wildlife Management Unit – Washington Co.
- St. John River – Aroostook Co

Alder Thicket

State Rank S5

Community Description

These tall (1-3 m) shrub dominated wetlands are characterized by dense growth of alder. Speckled alder is most typical, but rarely mountain alder or smooth alder may predominate. Red maple, gray birch, or other trees may be scattered sparsely above the shrubs. The herb layer is usually well developed (>35% cover), and is a variable mixture of forbs, graminoids, and ferns. The bryoid layer is patchy and dominated by peat mosses, especially *Sphagnum girghensonii*, *S. palustre*, and *S. magellanicum*.

Soil and Site Characteristics

This type occurs in basin wetlands that are usually saturated and may be seasonally flooded throughout the season. It is usually on muck or on peat. This type is very common as wet cleared areas revert to forest, such as old beaver meadows.

Diagnostics

These are shrub-dominated wetlands on peat or muck soils, often only temporarily flooded, in which alders dominate and comprise >20% cover, usually >40%.

Spotted Turtle

Similar Types

Alder Floodplains occur on mineral soils along medium to large rivers rather than in basins. Tussock Sedge Meadows, Bluejoint Meadows, and Mixed Graminoid - Shrub Marshes may have alder as a sub-dominant species, with lower cover than the graminoids. Mountain Holly - Alder Woodland Fens occur as part of a peatland and have mountain holly or heath shrubs mixed with the alder. Dogwood - Willow Shoreline Thickets often contain alder, but it is sub-dominant.

Conservation, Wildlife, and Management Considerations

Well distributed and well replicated. These shrublands, especially when they

Location Map

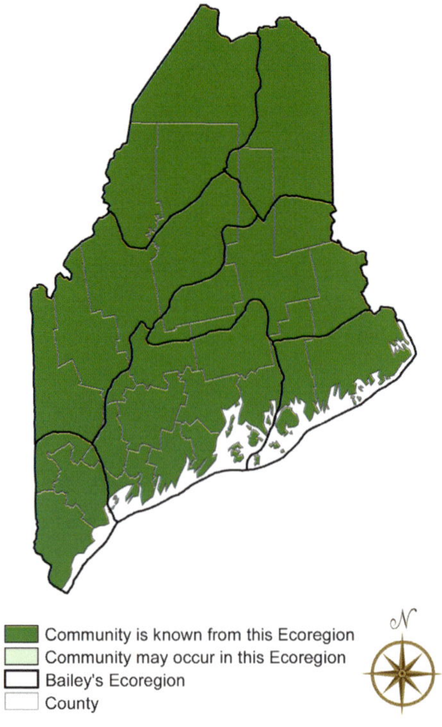

Alder Thicket

occur in close proximity to open water, may provide habitat for common bird species such as common yellowthroat, alder flycatcher, Wilson's warbler, and Lincoln's sparrow. Some occurrences of this community type support vernal pools, which are important breeding habitat for a variety of amphibians including wood frogs, spotted salamanders, and blue-spotted salamanders. Rare turtles like the wood turtle, or Blanding's and spotted turtles in southern Maine, may feed on amphibian egg masses present in such pools.

Distribution

Statewide; extending in all directions from Maine.

Landscape Pattern: Small to Large Patch

Examples on Conservation Lands You Can Visit

- Bigelow Preserve Public Lands – Somerset Co.
- Bradley Wildlife Management Area – Penobscot Co.
- Branch Lake Wildlife Management Unit – Hancock Co.
- Kennebunk Plains Preserve – York Co.
- Muddy River Wildlife Management Area – Sagadahoc Co.
- Narraguagus Wildlife Management Unit – Washington Co.
- Redington Pond Public Lands – Franklin Co.

Characteristic Plants

These plants are frequently found in this community type. Those with an asterisk are often diagnostic of this community.

Sapling/shrub
Balsam fir*
Black spruce*
Common blackberry*
Gray birch*
Hardhack*
Meadowsweet
Mountain alder*
Red maple*
Speckled alder*

Herb
Bluejoint
Cinnamon fern
Flat-topped white aster*
Royal fern
Sensitive fern*
Swamp dewberry*
Three-seeded sedge
Tussock sedge*
Wild calla

Bryoid
Sphagnum mosses*

Associated Rare Plants

Bog bedstraw
Northern bog sedge

Associated Rare Animals

Blanding's turtle
Spotted turtle
Wood turtle

Alpine Bog

State Rank S2

Community Description

Peat mosses form a carpet on which dwarf bog shrubs and herbs are scattered. The herb layer generally totals 25-40% cover and consists of patches of graminoids such as deer-hair sedge and tufted cotton-grass interspersed with patches of dwarf heath shrubs such as bilberries, crowberry, Labrador tea, leatherleaf, rhodora, and sheep laurel (the latter less abundant here than in other dwarf shrub bog vegetation types). A few islands of stunted balsam fir and black spruce may be present. Small cranberry and round – leaved sundew grow on the peat moss. Baked apple-berry is frequent and is restricted in Maine to these alpine bogs and Downeast peatlands.

Soil and Site Characteristics

These wetlands occur in alpine or subalpine settings near or above treeline. Sites are usually level, in small bedrock depressions, or slightly sloping on the edge of basins. The substrate is permanently saturated organic soil. Unlike its lower elevation counterparts, these bogs often have lenses of peat beneath the vegetation that remain frozen well into the growing season.

Cotton-Grass Tussock

Diagnostics

This type occurs near or above treeline on a saturated substrate. Peat mosses exceed cover of other vegetation, with reindeer lichens essentially absent. Shrubs include lower-elevation species as well as the more restricted bilberry and crowberry. Herbs are mostly graminoids.

Similar Types

The montane setting differentiates this from other peatland community types. The Sphagnum substrate differentiates it from other high-elevation types, except for Heath - Lichen Subalpine Slope

Location Map

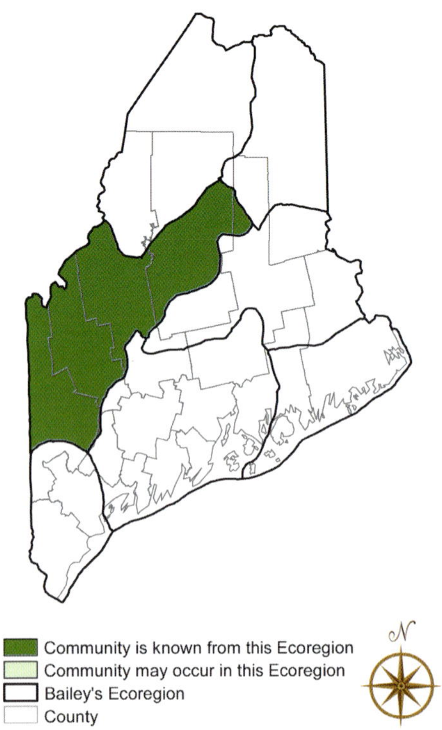

Alpine Bog

Bogs, which occur on steeper slopes (>20%), have the dwarf shrub - herb cover exceeding bryoid cover, have more krummholz (spruce, fir, or cedar) and fruticose lichens, and usually lack deer-hair sedge and cotton-grass. Both types are floristically similar to certain coastal raised peatlands and Heath - Crowberry Maritime Slope Bogs.

Conservation, Wildlife, and Management Considerations

The few known Maine occurrences of this type occur on public or private conservation land; hiker traffic has caused degradation in heavily used areas. As with all alpine vegetation in Maine, careful trail siting and efforts to minimize off trail use are the important management considerations. In these alpine wetlands, boardwalks or log crossings are helpful.

This high elevation bog community provides potential habitat for the rare northern bog lemming.

Distribution

Montane western Maine, extending westward into New Hampshire and possibly northeasterly to the Gaspé Peninsula.

Landscape Pattern: Small Patch

Characteristic Plants

These plants are frequently found in this community type. Those with an asterisk are often diagnostic of this community.

Dwarf Shrub

Alpine bilberry
Baked apple-berry
Black crowberry*
Labrador tea
Leatherleaf*
Pale laurel
Small cranberry*

Herb

Bog sedge*
Deer-hair sedge*
Few-flowered sedge*
Round-leaved sundew
Tufted cotton-grass*

Bryoid

Cetraria lichen*
Fen sickle-moss*
Sphagnum mosses*

Associated Rare Plants

Northern comandra

Associated Rare Animals

Northern bog lemming

Examples on Conservation Lands You Can Visit

- Mahoosuc Public Lands – Oxford Co.
- Saddleback Mountain, Appalachian Trail – Franklin Co.

Brackish Tidal Marsh

State Rank S3

Community Description

Brackish tidal marshes contain both freshwater and brackish water species, often in bands corresponding to tidal exposure. Tall rushes and bulrushes often predominate over extensive mid-elevation flats. At the lower elevations, rosette-forming herbs, such as lilaeopsis and tidal arrowhead, may be common on the mudflats. Near the high tide line, there may be a fairly narrow zone of muddy gravel or rock shore sparsely vegetated with low herbs, including some rare species such as Long's bitter-cress or water-pimpernel. Sweetgale and poison ivy are often present at the upper fringes of the marsh, at or above the tidal reach.

Soil and Site Characteristics

These marshes occupy intertidal reaches in larger tidal estuaries where freshwater and saltwater mix. Salinity ranges from 2-18 ppt, with gradients reflected in the species composition of particular sites. The substrate is usually mud rather than peat.

Diagnostics

These marshes occur in a brackish tidal setting. The vegetation is a mix of tall graminoids and rosette-forming forbs. Freshwater cordgrass, common arrow-grass, or wire-rush are present; obligate freshwater species, such as pickerelweed and common arrowhead, are absent.

Northern Blue Flag

Similar Types

Freshwater Tidal Marshes are most similar and can grade into this type as one moves downriver; they usually feature wild rice and/or softstem bulrush as dominants, and contain species such as pickerelweed and common arrowhead that do not tolerate brackish water. Some marshes are intermediate in character between the two types. Mixed Graminoid-Forb Saltmarshes and Spartina Saltmarshes develop a peat substrate and are dominated by obligate saltwater species such as saltmarsh cordgrasses, sea lavender, black-grass, or alkali bulrush.

Conservation, Wildlife, and Management Considerations

Tidal marshes provide valuable wildlife habitat and have received considerable

Location Map

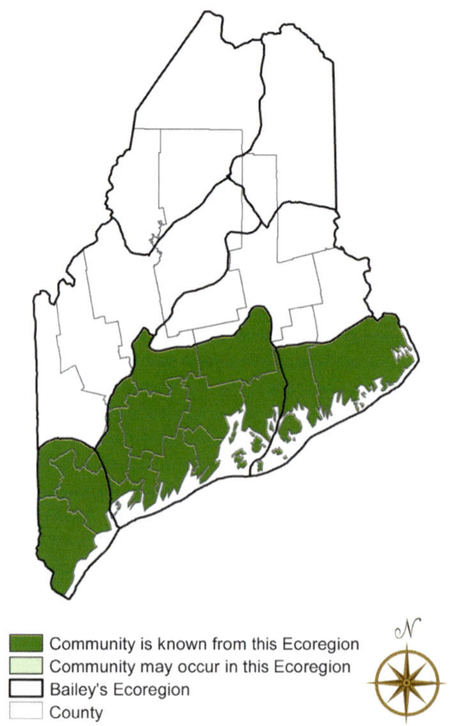

Brackish Tidal Marsh

conservation attention. Many occur on or adjacent to public lands or private conservation lands. With development of the uplands that border these marshes, maintenance of appropriate wetland buffers can help reduce degradation that could result from adjacent land uses. The prospect of sea level rise may also put these systems at greater risk.

Brackish marshes provide important nesting habitat for Nelson's sharp-tailed sparrow, seaside sparrow, and the rare saltmarsh sharp-tailed sparrow. These wetlands also provide foraging habitat for a large number of wading birds. The New England siltsnail inhabits coastal marshes and small tidal rivers where the water ranges from fresh to upper brackish. The spartina borer moth, whose historic range was along the immediate coast throughout New England, likely inhabited tidal marshes with sizeable populations of freshwater cordgrass, its larval host plant.

Distribution

Coastwide, extending in both directions from Maine (Laurentian Mixed Forest and Eastern Broadleaf Forest Provinces).

Landscape Pattern: Large Patch

Characteristic Plants

These plants are frequently found in this community type. Those with an asterisk are often diagnostic of this community.

Sapling/shrub

Poison-ivy*
Sweetgale*

Herb

Chaffy sedge*
Chair-maker's rush*
Common arrow-grass
Freshwater cordgrass*
Narrow-leaved cattail
Northern blue flag
Smooth cordgrass
Softstem bulrush
Wire rush*

Associated Rare Plants

Lilaeopsis
Parker's pipewort
Pygmyweed
Stiff arrowhead
Water-pimpernel

Associated Rare Animals

American oystercatcher
Black-crowned night-heron
Least bittern
Saltmarsh sharp-tailed sparrow
Short-eared owl

Examples on Conservation Lands You Can Visit

- Acadia National Park – Hancock Co.
- Hall Bay, Flying Point Preserve – Sagadahoc Co.
- Mendall Wildlife Management Area – Waldo Co.
- Morse Mountain Preserve – Sagadahoc Co.
- Rachel Carson National Wildlife Refuge – York Co.
- Scarborough Marsh Wildlife Management Area – Cumberland Co.

Bulrush Marsh

Bulrush Bed

State Rank S4

Community Description

Tall rushes and other non-persistent graminoids dominate this lakeshore or rivershore community. Hardstem bulrush, softstem bulrush, chair-maker's rush, and bayonet rush are common dominants. Aquatic macrophytes such as pickerelweed, pondweeds, and water-lilies may be present, but are not usually abundant. Species richness is often low.

Soil and Site Characteristics

These deepwater marshes have standing water over 15 cm deep all year, except during unusually prolonged low water levels. Bulrush beds are often near inlets and outlets of lakes or along slow moving portions of larger streams and rivers.

Diagnostics

This type is distinguished by a dominance of tall bulrushes and rushes to the exclusion of most other species. Standing water is present through most or all of season.

Softstem Bullrush

Similar Types

Pickerelweed - Macrophyte Aquatic Bed vegetation can share species but is dominated by pickerelweed and floating leaved macrophytes (water-lilies, pondweeds, etc.). Water-lily - Macrophyte Aquatic Bed vegetation has few emergent plants. Freshwater Tidal Marshes and Brackish Tidal Marshes may share some of the bulrush species, but occur in tidal situations and have different associated species.

Conservation, Wildlife, and Management Considerations

This type appears to be well distributed and secure in Maine. Anecdotal

Location Map

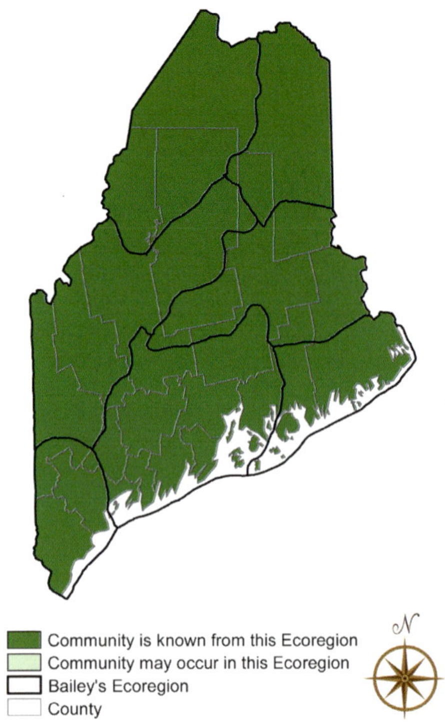

Bullrush Marsh

Characteristic Plants

These plants are frequently found in this community type. Those with an asterisk are often diagnostic of this community.

Herb

Bayonet rush*
Chair-maker's rush*
Hardstem bulrush*
Softstem bulrush*

Associated Rare Animals

American coot
Black-crowned night heron
Citrine forktail
Common moorhen
Least bittern
Scarlet bluet

information indicates that the waters of numerous public lands and private conservation lands include these marshes. Maintaining appropriate wetland buffers can help ensure that adjacent land uses do not result in degradation.

These deep emergent marshes provide foraging and nesting habitat for a large number of wading birds and waterfowl including rare species such as the least bittern, common moorhen, and American coot. They also offer excellent habitat for a number of insects including the New England bluet, pine barrens bluet, the rare scarlet bluet, and the rare citrine forktail.

Distribution

Statewide, although not well documented. Extends southward and westward from Maine, and presumably into Canada.

Landscape Pattern: Small Patch

Examples on Conservation Lands You Can Visit

- Acadia National Park – Hancock Co.

Bulrush Marsh

Aerial View of Bulrush Marsh

Cattail Marsh

State Rank S5

Community Description

Tall marsh vegetation is dominated by cattails and mostly deciduous shrubs. The cattails may be patchy, locally dominant, and grow taller than the other plant species. Common cattail, narrow-leaved cattail, or both may be present; the latter is common near the coast or in brackish settings and the former is more widespread. Shrubs include winterberry, meadowsweet, and others. The dense growth of shrubs and cattails leaves little room for other herbaceous species; additional occasional species include swamp milkweed, wild calla, and a few sedges and grasses. Bryophytes are usually sparse and occur on vegetation hummocks.

Soil and Site Characteristics

Cattail marshes often occur in impounded, semi-permanently flooded, and/or nutrient-rich waters. Documented sites are at low elevations (<500') and are generally associated with large basins and adjacent to open water. The substrate is muck or mineral soil rather than peat.

Variable Darner

Diagnostics

Standing water persists through all or most of season. Sites are non-tidal, and cattails are prominent above a mixture of low shrubs and sedges.

Similar Types

Mixed Graminoid-Shrub Marshes share many species, but lack cattails (rarely present at <5% cover) and typically occur in settings that are only temporarily flooded rather than with semi-permanent standing water. Brackish Tidal Marshes and Freshwater Tidal Marshes may have patches of narrow-leaved cattail but usually contain plants indicative of tidal habitats.

Conservation, Wildlife, and Management Considerations

This community is well distributed

Location Map

Cattail Marsh

statewide and apparently well represented (although not well documented) on public lands and private conservation lands. Maintaining appropriate wetland buffers can help ensure that adjacent land uses do not result in degradation.

These deep emergent marshes provide foraging and nesting habitat for a large number of wading birds and waterfowl, including rare species such as the least bittern, common moorhen, and American coot. Black terns may nest colonially in large inland occurrences of this community type with about 50% vegetative and 50% open water. Other birds commonly associated with cattail marshes include Virgina rails, red-winged blackbirds, and marsh wrens.

Distribution

Statewide, more common near the coast and at lower inland elevations. Extends southward and westward from Maine, and presumably into Canada.

Landscape Pattern: Small Patch

Characteristic Plants

These plants are frequently found in this community type. Those with an asterisk are often diagnostic of this community.

Sapling/shrub

Leatherleaf
Meadowsweet*
Sweetgale
Winterberry

Herb

Common cattail*
Narrow-leaved cattail
Pickerelweed
Small St. Johnswort

Associated Rare Plants

Tall beak-rush

Associated Rare Animals

American coot
Black tern
Common moorhen
Least bittern

Examples on Conservation Lands You Can Visit

- Alonzo Garcelon Wildlife Management Area – Kennebec Co.
- Hurds Pond Wildlife Management Area – Waldo Co.
- Jamie's Pond Wildlife Management Area – Kennebec Co.
- Sebago Lake State Park – Cumberland Co.

Circumneutral Fen

State Rank S2

Community Description

This peatland vegetation type is dominated by sedges or grades into dwarf shrubs. Dwarf shrub and graminoid cover each range from 10-75% and are inversely proportional to each other. Sparse cedar or larch may dot the fen. Shrubs may be patchy. Dominant sedges include deer-hair sedge and slender sedge; white beak-rush is locally common. Alpine cotton-grass, with its white wispy fruiting heads, is often obvious but not abundant. Shrubby cinquefoil and bog rosemary are characteristic. Northern bog aster and marsh muhly are good indicators, as are the calciphiles livid sedge, yellow sedge, sparse-flowered sedge, and northern bog sedge. The bryoid layer is extensive, with Campylium fen moss indicative.

Soil and Site Characteristics

These peatlands are influenced by calcium rich, circumneutral (rather than acidic) water. The substrate pH is 5.6 or higher, and remains saturated through the year. These peatlands occur in minerotrophic basins where contact with groundwater provides some nutrients to the plants. Sites are typically at lower elevations (<1000') and usually in areas underlain by limestone or other calcareous bedrock.

Sphagnum in Cedar

Diagnostics

Peatland vegetation is dominated by sedges or sedge/shrub mixtures including deer-hair sedge, slender sedge, and bog rosemary. Circumneutral indicators are present, such as shrubby cinquefoil, livid sedge, marsh muhly, grass-of-parnassus, and Kalm's lobelia.

Similar Types

Other peatland communities lack the circumneutral indicators typical of this type. Circumneutral Riverside Seeps have many similar species, but do not occur on peat substrate.

Conservation, Wildlife, and Management Considerations

This rare community type has been subject to few threats to date. Some examples occur on public lands and private conservation lands. Impoundment or draining would have negative

Location Map

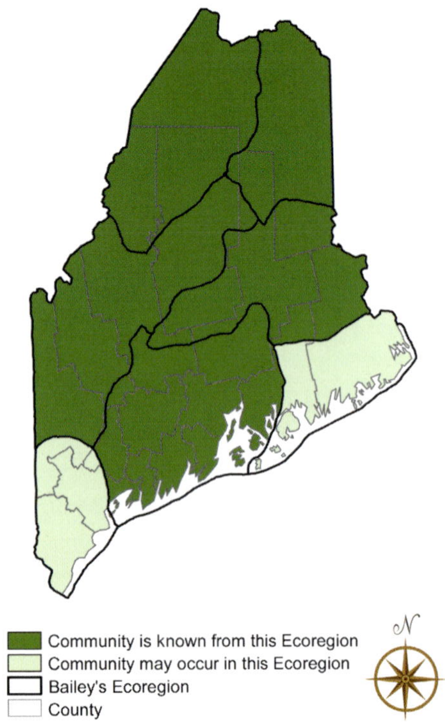

Clayton's Copper on Shrubby Cinquefoil

impacts on hydrology and consequently on vegetation. Maintaining appropriate wetland buffers is important in minimizing the effects of adjacent land use. Degradation from recreational use has not been an issue in most places, but if disturbance, such as foot traffic, is a necessity, traversing during frozen conditions or using boardwalks can reduce impacts.

This community is inhabited by the rare Clayton's copper butterfly, which uses shrubby cinquefoil as its sole larval host plant and primary adult nectar plant. This butterfly is found at only 14 sites worldwide, nine in Maine and five in New Brunswick. All known occurrences are in circumneutral fens with shrubby cinquefoil stands large enough to support a persistent population of the butterfly. Thaxter's pinion moth uses sweetgale and larch as larval host plants and may be found in this community as well.

Distribution

Most typically in the limestone regions of northern Maine, sporadically westward, eastward, and southward. (Laurentian Mixed Forest and New England - Adirondack Provinces.) Extends to northern New England and New York; Canadian distribution not well known.

Landscape Pattern: Small Patch

Characteristic Plants

These plants are frequently found in this community type. Those with an asterisk are often diagnostic of this community.

Sapling/shrub

Larch*
Northern white cedar*
Sweetgale*

Dwarf Shrub

Bog rosemary*
Leatherleaf*
Shrubby cinquefoil*
Sweetgale*

Herb

Deer-hair sedge*
Marsh muhly
Northern blue flag
Northern bog aster
Slender sedge*
Tussock sedge*

Bryoid

Campylium fen moss
*Sphagnum warnstorfii**

Associated Rare Plants

Capillary sedge
Dioecious sedge
Livid sedge
Low spike-moss
Prairie sedge
Slender-leaved sundew
Swamp birch

Associated Rare Animals

Clayton's copper

Examples on Conservation Lands You Can Visit

- Lake Umbagog National Wildlife Refuge – Oxford Co.
- Mattagodus Wildlife Management Area – Penobscot Co.
- Salmon Brook Lake Bog Public Lands – Aroostook Co.
- Woodland Bog Preserve – Aroostook Co.

Circumneutral Pond

State Rank S2

Community Description

The variable aquatic vegetation in this community type features floating leaved and submerged plants restricted or preferential to higher pH waters. Indicator species include tapegrass, common waterweed, water stargrass, white water crowfoot, Robbins' pondweed, alpine pondweed, Vasey's pondweed, and straight-leaved pondweed. Vegetation cover is usually 25-65%, occasionally higher. This type is broadly defined and heterogeneous; further sampling would be needed to discriminate subtypes.

Soil and Site Characteristics

This type occurs in alkaline lakes or ponds (conductivity usually >50 uMHOS/cm), in quiet waters at depths less than two meters.

Diagnostics

At least some of the indicator species listed above are present, and water-shield, pipewort, water lobelia, and pickerelweed are virtually absent. Yellow and white water-lilies, if present, are at low cover.

Circumneutral Fen Pond

Similar Types

This broad type can show similarities to other aquatic types, especially to the Water-lily - Macrophyte Aquatic Bed type. That type has higher cover of yellow and white water-lilies or of bigleaf pondweed, but more sampling of both types is needed to better distinguish them.

Conservation, Wildlife, and Management Considerations

This aquatic community type is sporadic in Maine, and further work is needed to document its distribution

Location Map

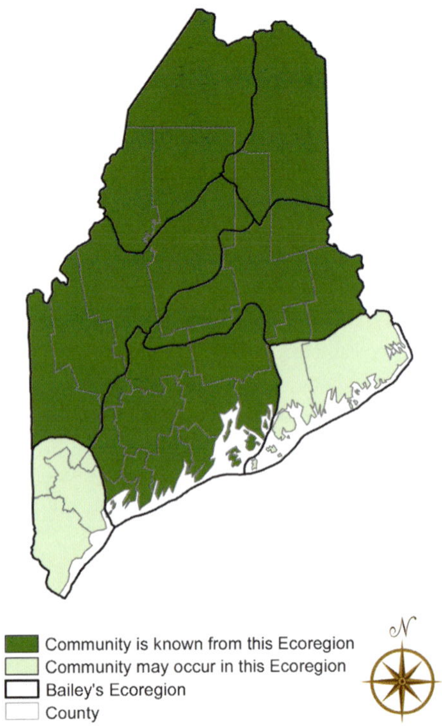

Aquatic Vegetation

Characteristic Plants

These plants are frequently found in this community type. Those with an asterisk are often diagnostic of this community.

Herb

Common waterweed*
Robbins' pondweed*
Slender water-milfoil*
Tapegrass*
Water-marigold*

Bryoid

Chara algae

Associated Rare Plants

Fries' pondweed
Pygmy water-lily
Straight-leaved pondweed
Vasey's pondweed
Water stargrass

and characteristics. A few examples occur on public lands and private conservation lands; many are in great ponds (>10 acres), which are a public resource. The major threats to this community are water quality degradation from excess nutrients in runoff and the spread of invasive aquatic plants, such as Eurasian water-milfoil and variable water-milfoil.

Alkaline ponds with aquatic vegetation may support many of the same wildlife species found in more acidic aquatic communities such as Water-lily Macrophyte Aquatic Bed or Pickerelweed - Macrophyte Aquatic Bed. It is unclear whether any wildlife species prefer the more alkaline conditions that this community provides.

Distribution

Central Maine northward (Laurentian Mixed Forest and New England - Adirondack Forest Provinces); distribution includes upstate New York; otherwise not well matched in other state classification documents.

Landscape Pattern: Small Patch

Examples on Conservation Lands You Can Visit

- Caswell National Guard Site – Aroostook Co.
- Number Five Bog Public Lands – Somerset Co.

Floating Pondweed

Coastal Beach

State Rank S4

Community Description

This type consists of sparsely vegetated upper beaches, where plants are adapted to the effects of salt spray, coarse sand and cobble, and drying winds. Characteristic plants include annuals such as sea-beach sandwort, sea-kale, beach-pea, and others. On Downeast gravel/cobble beaches, oysterleaf may be characteristic. This linear community is usually bordered landward by either sand dunes or upland forest vegetation.

Soil and Site Characteristics

Beach strands occur above the usual high tide line on sand or gravel coastal beaches. They are occasionally flooded at very high tides.

Diagnostics

Located just above the high tide line, beach strands are distinguished by the presence of beach-pea and sea-kale; forbs are more abundant than grasses.

Similar Types

Dune Grasslands may be adjacent, and species overlap between the two; but dune grasslands have a higher vegetative cover and are strongly dominated by graminoids rather than by forbs.

Arctic Tern

Conservation, Wildlife, and Management Considerations

This narrow band of vegetation usually lies landward of the most heavily walked portions of beaches and has likely been reduced or eliminated at popular beaches. Many examples exist on several public lands and private conservation lands. Raking seaweed from the beach to the dune fronts can affect the composition of this community. The prospect of sea level rise may also put these systems at greater risk.

Location Map

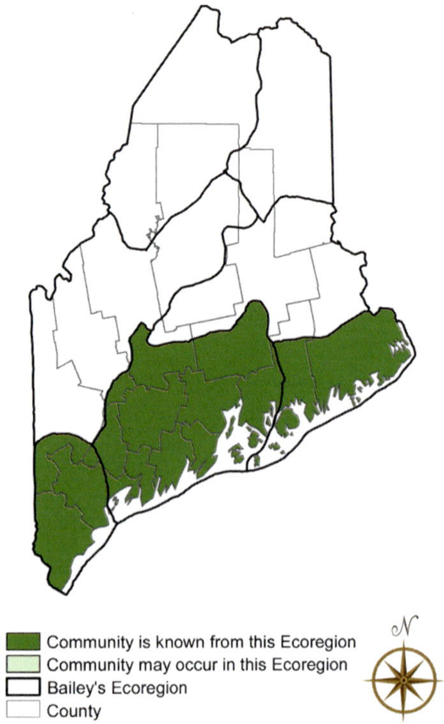

Beach Strand - Morse Mountain

Characteristic Plants

These plants are frequently found in this community type. Those with an asterisk are often diagnostic of this community.

Herb

American shore-grass
Beach-pea
Common saltwort
Sea milkwort
Sea-beach sandwort
Sea-kale
Spearscale
White sea-blite
Witch grass

Associated Rare Plants

Coast-blite goosefoot

Associated Rare Animals

American oystercatcher
Common tern
Least tern
Piping plover
Roseate tern

Upper beaches and adjacent dune grasslands are critical breeding habitat for two species of rare shorebirds: least terns and piping plovers. Nesting habitat for these birds has been reduced by human use of beaches and the known nesting sites in Maine are carefully monitored. Beach Strand communities also provide breeding habitat for common terns, though this species also breeds on inland lakeshore beaches. Rare roseate terns breed on sandy or gravelly beaches, especially those that occur on offshore islands. American oystercatchers are extremely rare breeders on coastal beaches of southern Maine. Beaches are also major roosting sites for migratory shorebirds.

Distribution

Coastwide, extending both south along the Atlantic seaboard and eastward into the Canadian Maritimes (Eastern Broadleaf Forest and Laurentian Mixed Forest Provinces).

Landscape Pattern: Small Patch, linear.

Examples on Conservation Lands You Can Visit

- Howard's Cove, Machiasport – Washington Co.
- Petit Manan National Wildlife Refuge – Hancock Co.
- Popham Beach State Park – Sagadahoc Co.
- Reid State Park – Sagadahoc Co.
- Roque Bluffs State Park – Washington Co.
- Scarborough Beach State Park – Cumberland Co
- Seawall Beach, Morse Mountain Preserve – Sagadahoc Co.

Coastal Sedge Bog

Deer-hair Sedge Bog Lawn

State Rank S3

Community Description
This raised bog type is dominated by carpets or patches of deer-hair sedge, often with very stunted (<0.3 m) heath shrubs such as black crowberry, dwarf huckleberry, or leatherleaf. Round-leaved sundew, pitcher plant, and small cranberry grow among the peat mosses, which form a dense and spongy ground layer. Reindeer lichens are scattered among the mosses.

Soil and Site Characteristics
This community is restricted to raised bogs along or near the coast, often forming expansive "lawns" on the raised portions. The substrate is saturated, acidic (pH ~4.5) peat moss. As with other bog vegetation, it occurs in nutrient poor, usually ombrotrophic settings.

Diagnostics
Sites are in a peatland setting, with a dominance of deer-hair sedge and a lack of other circumneutral indicators. Dwarf huckleberry is characteristic but not dominant.

Labrador Tea

Similar Types
Shrubby Cinquefoil - Sedge Circumneutral Fens can also be dominated by deer-hair sedge but are in fens, not raised coastal bogs, lack dwarf huckleberry, and have other circumneutral indicators present such as shrubby cinquefoil or certain sedges. Huckleberry - Crowberry Bogs can occur in similar settings to Deer-hair Sedge Bog Lawns, and can share many species, but will have dwarf shrubs more dominant than sedges; the two types may occur adjacent to each other with a continuous gradation from one type to the next.

Conservation, Wildlife, and Management Considerations
This community type is not widely

Location Map

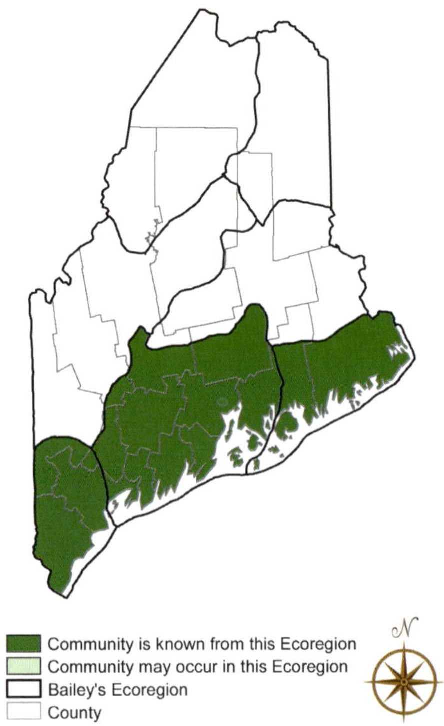

Deer-hair Sedge

distributed, but has been subject to few threats to date. Slow vegetation growth rates, due to the nutrient poor setting, mean slow recovery from physical disturbances, such as recreational use. If disturbance, such as foot traffic, is a necessity, traversing during frozen conditions or using boardwalks can minimize impacts. Peat harvesting could threaten some sites but is not currently much of a factor. Draining or other hydrologic changes would have negative impacts on bog vegetation. Several occurrences are on public lands or private conservation lands.

The rare crowberry blue butterfly is restricted to coastal heaths in east-coastal Maine. It uses black crowberry as a larval host plant.

Distribution

Downeast Maine, extending eastward into the Canadian Maritimes (Laurentian Mixed Forest Province).

Landscape Pattern: Small Patch, interspersed with other peatland types.

Characteristic Plants

These plants are frequently found in this community type. Those with an asterisk are often diagnostic of this community.

Sapling/shrub

Black chokeberry
Mountain holly

Dwarf Shrub

Black crowberry*
Dwarf huckleberry*
Labrador tea
Leatherleaf*
Pale laurel
Sheep laurel*
Small cranberry

Herb

Bog goldenrod
Coast sedge*
Deer-hair sedge*
Horned bladderwort
Pitcher plant
Round-leaved sundew

Bryoid

Bog broom-moss
Bog hair-cap moss
Little-tree reindeer-lichen
*Sphagnum rubellum**

Associated Rare Animals

Crowberry blue

Examples on Conservation Lands You Can Visit

- Acadia National Park – Hancock Co.
- Great Heath Public Lands – Washington Co.
- Great Wass Island Preserve – Washington Co.
- Larrabee Heath Preserve – Washington Co.
- Quoddy Head State Park – Washington Co.

Cobble Rivershore

State Rank: S4

Community Description

Vegetation on these rivershores may be sparse to dense depending on degree of flooding, length of exposure, and substrate type. Characteristic perennial species that tolerate inundation and flood scouring include twisted sedge and low-growing willow species. Associated species tend to vary widely from site to site and may be diverse; they include tufted hairgrass, red osier dogwood, sweetgale, water parsnip, water hemlock, cardinal flower, flat-topped aster, and smartweeds. Bryoids are usually sparse but where present may include Bryum species. Invasive, exotic species may be problematic in these areas, especially coltsfoot, purple loosestrife, and Japanese knotweed.

Soil and Site Characteristics

Herbaceous and shrub vegetation of this type occurs on coarse substrates deposited along medium- to high-energy river channels and, less frequently, exposed lakeshores with heavy wave action. Seasonal flooding and ice-scour maintain the open nature of these communities; generally, they develop in areas of the active channel that are exposed at low water or in drought years.

Cardinal Flower

Diagnostics

Herbs and graminoids are dominant or co-dominant with shrubs in a rivershore or streamshore setting.

Similar Types

Sand Cherry Tufted Hairgrass River Beaches are found along the St. John River and contain certain indicator species such as sand cherry and freshwater cordgrass. Dogwood-Willow Shoreline Thickets may be adjacent to (landward of) these types and are dominated by shrubs rather than herbs. Bluebell - Balsam Ragwort Shoreline

Location Map

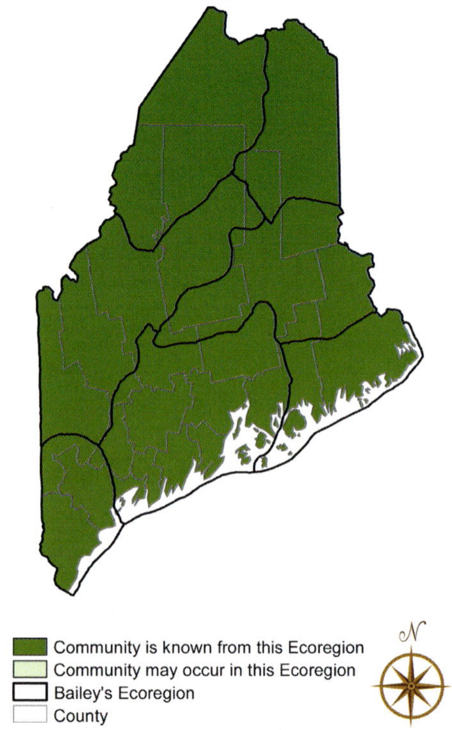

Cobble Rivershore

Outcrops occur on bedrock outcrops rather than cobble or gravel. The Hudsonia River Beach is restricted to sand and gravel bars of the Saco River and contains beach heather as a characteristic species.

Conservation, Wildlife, and Management Considerations

This community is linked to naturally fluctuating water levels and occasional ice scour. The rivershore habitat of this natural community suggests that threats from development are relatively low. Hydrologic alteration (i.e., impoundments) would compromise the disturbance regime, but new dams are unlikely on medium and large rivers. Exotic or agricultural species are common at some sites.

The rare White Mountain tiger beetle occurs in this habitat type. Another rare insect, the cobblestone tiger beetle, is currently under consideration for federal listing and in our area is known from Vermont, New Hampshire, and New Brunswick, but not Maine.

Characteristic Plants

These plants are frequently found in this community type. Those with an asterisk are often diagnostic of this community.

Herbs

Bluejoint
Boneset
Cardinal flower
Flat-topped aster
Goldenrods
Mad-dog skullcap
Purple-stemmed aster
Red osier dogwood
Smartweeds
Spotted joe-pye weed
Sweetgale
Tufted hairgrass
Twisted sedge
Virgin's bower
Water hemlock
Water parsnip
Wool-grass

Associated Rare Animals

White Mountain tiger beetle

Distribution

One of the predominant rivershore types in the New England - Adirondack Province and Laurentian Mixed Forest Province. Extends in all directions from Maine.

Landscape Pattern: Small to large patch, linear.

Dwarf Shrub Bog

State Rank S4

Community Description

A dense layer of dwarf heath shrubs dominates this common open peatland community. Stunted and scattered black spruce and larch trees form <25% cover. Heath shrubs carpet the hummocks and hollows of the peat substrate; sheep laurel or rhodora are typically dominant. Sedges contribute little cover (usually <15%, occasionally 20-25%); the most common is tufted cotton-grass, whose bright white tufts decorate the bog vegetation early in the summer. Insectivorous plants such as pitcher plant and sundew can be quite numerous. The ground surface is covered by a spongy carpet of peat mosses. The floristic composition varies depending upon bog morphology and nutrient availability.

Soil and Site Characteristics

This type occurs within raised portions of peatlands, where ombrotrophic conditions prevail (plant growth is raised above the water table, and virtually all nutrients come from precipitation). Although standing water may not be visible, the peat is commonly saturated with water throughout most of the year. The substrate is highly acidic, with pH 3.9-4.6.

Pitcher Plant Flower

Diagnostics

Open peatland vegetation is raised above the regional water table and dominated by sheep laurel, rhodora, and/or Labrador tea. Leatherleaf is less abundant. Tufted cotton-grass is present. Black spruce is common as scattered trees, but tree cover <25%.

Similar Types

Spruce - Larch Wooded Bogs have more trees (canopy >25%). Huckleberry - Crowberry Bogs also feature sheep laurel, but have patches of dwarf huckleberry and crowberry or deer-hair sedge. These types can grade into each other within a peatland. Other dwarf shrub peatland types occur in more minerotrophic conditions and are dominated by leatherleaf, bog rosemary, and/or shrubby cinquefoil. Heath - Lichen Subalpine Slope Bogs and Heath - Crowberry Maritime Slope

Location Map

Dwarf Shrub Bog

Bogs share similar vegetation but overlay rock rather than occurring as part of a peatland.

Conservation, Wildlife, and Management Considerations

This community type is well represented in Maine and is fairly stable in extent, with many examples on public lands and private conservation lands. Peat harvesting is a direct threat to a few sites. Changes to bog hydrology, through impoundment or draining, lead to vegetation changes. Slow vegetation growth rates, due to the nutrient poor environment, result in slow recovery from physical disturbances, such as recreational trail use. If disturbance such as foot traffic, is a necessity, traversing during frozen conditions or using boardwalks can minimize impacts.

Occurrences of this community type in northwestern Maine may include the bog fritillary butterfly, which uses small cranberry as its larval host plant. The bog elfin butterfly uses black spruce as a larval host plant and may be found in this community when black spruce is abundant. Several dragonfly species may be found in examples of this community where bog pools occur, including the zigzag darner, subarctic darner, and incurvate emerald.

Distribution

Common throughout the state, occurring as a large uniform community or as a small component of a complex bog. Extends south and west through northern New England and New York, and northward and eastward into Canada.

Landscape Pattern: Small to Large Patch

Characteristic Plants

These plants are frequently found in this community type. Those with an asterisk are often diagnostic of this community.

Sapling/shrub

Black spruce*
Mountain holly*

Dwarf Shrub

Black huckleberry*
Labrador tea*
Leatherleaf*
Pale laurel
Rhodora*
Sheep laurel*
Small cranberry

Herb

Black spruce*
Narrow-leaved cotton-grass*
Pitcher plant
Tufted cotton-grass*

Bryoid

Reindeer lichen*
Sphagnum mosses*

Associated Rare Plants

Swamp birch

Examples on Conservation Lands You Can Visit

- Gassabias Stream, Duck Lake Public Lands – Hancock Co.
- Harmon Heath, Cutler Public Lands – Washington Co.
- Number Five Bog Public Lands – Somerset Co.
- Second Lake, Rocky Lake Public Lands – Washington Co.
- Sunkhaze Meadows National Wildlife Refuge – Penobscot Co.
- The Heath, Massabesic Experimental Forest – York Co.

Freshwater Tidal Marsh

State Rank S2

Community Description

These tidal marshes are dominated by patchy stout herbs, typically a mixture of wild rice, softstem bulrush, and pickerelweed, often covering extensive areas. Mixed in with the tall herbs are lower forbs including several rare species. Some marshes may have mudflats dominated by forbs and low vegetation in patches among the graminoids; many have a very narrow band of low forbs near the high tide/upland interface. Brackish marsh species, such as chair-maker's rush, may be in these marshes as well, but at least some obligate freshwater plants such as pickerelweed, common arrowhead, sweet flag, and northern water-plantain will also be present. Bryophytes are essentially absent.

Soil and Site Characteristics

Freshwater tidal marshes are associated with major rivers, in low-gradient areas of the mid to upper tidal reaches. Freshwater inputs lower the salinity to <1 ppt. Substrate is usually mud, or mud mixed with gravel. The tidal regime affects substrate and plant zonation.

Torrey's Bulrush

Diagnostics

These graminoid dominated marshes occur along tidal rivers, with patches of forbs locally abundant. Obligate freshwater species are present, such as sweetflag, yellow water-lily, large yellow pond-lily, or pickerelweed.

Similar Types

Brackish Tidal Marshes are most similar and grade into this type as salinity decreases. Mixed Graminoid Shrub Marshes and Pickerelweed - Macrophyte Aquatic Bed types can contain several of the same species, but do not occur in tidal settings.

Conservation, Wildlife, and Management Considerations

Tidal marshes provide valuable wildlife habitat and have received considerable

Location Map

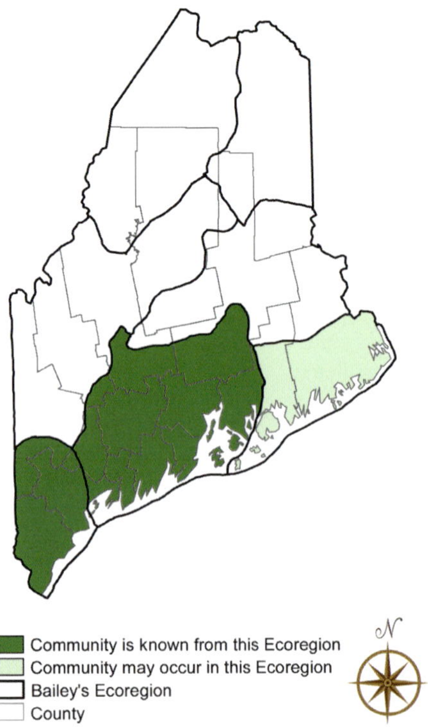

Freshwater Tidal Marsh

conservation attention. Heavy metals, sewage overflows, and other pollutants have degraded the substrate in many areas, but some have recovered as water quality has improved over the past decades. Many occur on or adjacent to public lands or private conservation lands. Some have been managed for waterfowl by planting wild rice. With development of the uplands that border these marshes, maintenance of appropriate wetland buffers can help reduce degradation that could result from adjacent land uses. Invasive species such as Japanese knotweed and purple loosestrife have invaded the upper reaches at some sites. The prospect of sea level rise may also put these systems at greater risk in the future.

The tidal marshes of Maine's larger estuaries, especially Merrymeeting Bay, are important pre-migration staging habitat for thousands of waterfowl and wading birds. The rare New England siltsnail inhabits coastal marshes and small tidal rivers where the water ranges from fresh to upper brackish.

Distribution

Upper tidal reaches of major rivers: most well known from the Kennebec and Penobscot Rivers (Laurentian Mixed Forest Province).

Landscape Pattern: Large Patch, often linear.

Characteristic Plants

These plants are frequently found in this community type. Those with an asterisk are often diagnostic of this community.

Herb

Chair-maker's rush*
Common arrowhead
Eaton's bur-marigold
Nodding beggar ticks
Northern water-plantain
Parker's pipewort
Pickerelweed*
Softstem bulrush*
Tidal arrowhead
Wild rice*

Associated Rare Plants

Beaked spikerush
Eaton's bur-marigold
Long's bitter-cress
Parker's pipewort
Pygmyweed
Spongy arrowhead
Stiff arrowhead
Water-pimpernel

Associated Rare Animals

American oystercatcher
Black-crowned night-heron
Least bittern
Short-eared owl

Examples on Conservation Lands You Can Visit

- Merrymeeting Bay Wildlife Management Area – Sagadahoc Co.
- Muddy River Wildlife Management Area – Sagadahoc Co.
- Swan Island Wildlife Management Area – Sagadahoc Co.

Grassy Shrub Marsh

State Rank S5

Community Description

This is a heterogeneous wetland type in which herbs and shrubs occur in various assemblages and proportions. Many examples are transitional to other open wetland types. A variant in southern Maine has buttonbush as a prominent shrub. The more typical expression is dominated by herbs, with a mixture of graminoids making up at least 50% of the cover, often with a sparse shrub layer containing meadowsweet or hardhack. Bluejoint is frequent, but not in large swards. Any of a variety of graminoids may be prominent at different sites. Three-way sedge and yellow loosestrife are indicators. Bryophytes are generally minor. This type is very broadly defined and could be subdivided into shrub versus herbaceous types using additional site data and analyses.

Soil and Site Characteristics

Sites are typically on mineral soils that are flooded early in the growing season and remain saturated (or occasionally flooded) throughout the season. Soil pH is typically 5.0-6.0. Beaver activity often affects these wetlands, and can cause dramatic (although sometimes temporary) changes in dominance.

Diagnostics

Wetland herb and shrub species are mixed (herbs 25-95%, shrubs 0-70% cover), without a dominance of tussock sedge, bluejoint, or alder. Sites occur on mineral soil, but may have a thick organic layer of muck. Red maple may be occasional in sapling form. Mountain holly, dogwoods, or alder are often dominant shrubs; sweetgale and leatherleaf are frequent but usually total less than 50% cover.

Similar Types

Sweetgale Mixed Shrub Fens are similar, but occur on organic soils and have plants more characteristic of peatlands, such as rhodora, sheep laurel, Labrador tea, bog rosemary, or pitcher plants. Those plants may be present in this type, but form <1% cover. Tussock Sedge Meadows and Bluejoint Meadows are strongly dominated by tussock sedge and bluejoint, respectively. Alder Shrub Thickets have >20% cover of alder or alder mixed with gray birch. Three-way Sedge - Goldenrod Outwash Plain Pondshores share diagnostic species but occur on sandy shores and contain other characteristic species.

Conservation, Wildlife, and Management Considerations

These marshes are well distributed throughout Maine and well represented on public lands and private conservation lands. Maintaining appropriate wetland buffers can help ensure that adjacent land uses do not degrade the marshes.

Several rare reptiles may be found in this community type. The ribbon snake seeks out prey in these wetlands. In Southern Maine, spotted turtles and Blanding's turtles may overwinter in marshes where water levels remain

Location Map

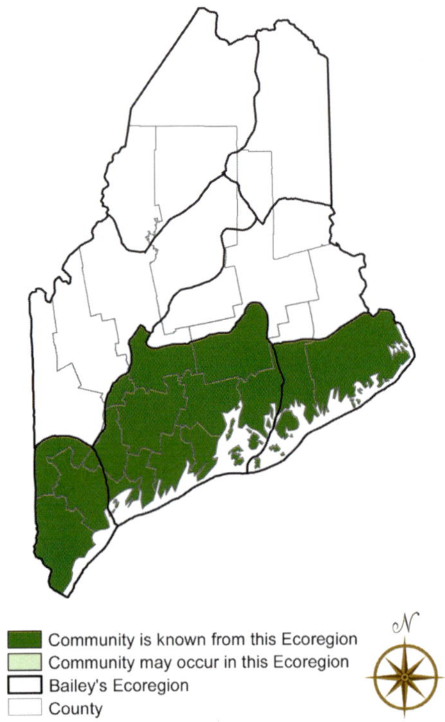

Mixed Graminoid Shrub Marsh

reliably deep throughout most years. Some occurrences may function as vernal pools, which provide important breeding habitat for a variety of amphibians including wood frogs, spotted salamanders, and blue-spotted salamanders. These wetlands provide nesting and foraging habitat for a number of wading birds including the green heron, American bittern, and the rare least bittern. The rare sedge wren nests in graminoid marshes and wet meadows.

Distribution

Statewide, extending southward, westward, and into Canada.

Landscape Pattern: Small to Large Patch

Examples on Conservation Lands You Can Visit

- Kennebec Crossing, Appalachian Trail – Somerset Co.
- Mt Agamenticus – York Co.
- Stratton Brook Pond, Bigelow Preserve – Franklin Co.
- Tunk Lake Area, Donnell Pond Public Lands – Hancock Co.

Characteristic Plants

These plants are frequently found in this community type. Those with an asterisk are often diagnostic of this community.

Sapling/shrub

Bog willow*
Buttonbush*
Meadowsweet*
Mountain holly
Red osier dogwood*
Speckled alder*
Sweetgale*
Winterberry

Dwarf Shrub

Leatherleaf*
Sweetgale*

Herb

Beaked sedge*
Black bulrush*
Bluejoint*
Expanded bulrush*
Few-seeded sedge*
Marsh St. Johnswort*
Royal fern*
Sensitive fern*
Three-way sedge*
Wool-grass*
Yellow loosestrife*

Bryoid

Sphagnum mosses*

Associated Rare Plants

Comb-leaved mermaid-weed
Featherfoil
Hollow joe-pye weed
Red-root flatsedge
Tall beak-rush

Associated Rare Animals

Black-crowned night-heron
Blanding's turtle
Least bittern
Ribbon snake
Sedge wren
Short-eared owl
Spotted turtle

Lakeshore Beach

State Rank S4

Community Description

These low sandy or gravelly lakeshores and associated berms are characterized by a variety of herbaceous plants that are tolerant of significant fluctuations in water level, ice scour, and wave action. Composition varies with substrate, with assemblages on sand possibly including large cranberry, switch-grass, black chokeberry, pinweed, poverty oat-grass, purplish northern panic-grass, and deertongue. On berms, a sparse canopy of trees or shrubs may be present, including common juniper, speckled alder, red maple, white pine, and gray birch. Typical assemblages on cobble or gravel include sweetgale, tufted hairgrass, water parsnip, beggar-ticks, spikerushes, common horsetail, boneset, creeping spearwort, and three square.

Soil and Site Characteristics

Beaches typically occur on the perimeters of larger lakes in landscapes of low to rolling terrain, where ice-push, wave action, and natural or anthropogenic drawdown (e.g., from impoundments) have created broad open bands and berms around portions of the lakeshore. Berms may consist of sand or gravel, with organic matter deposited by waves and forming wrack lines.

Water Parsnip

Diagnostics

Herbaceous plants are dominant or co-dominant with shrubs in a lakeshore setting.

Similar Types

Three-way Sedge - Goldenrod Outwash Plains may have similar soil and site characteristics but contain certain indicators that are not present on other lakeshores, including three-leaved goldenrod, golden pert, and meadow

Location Map

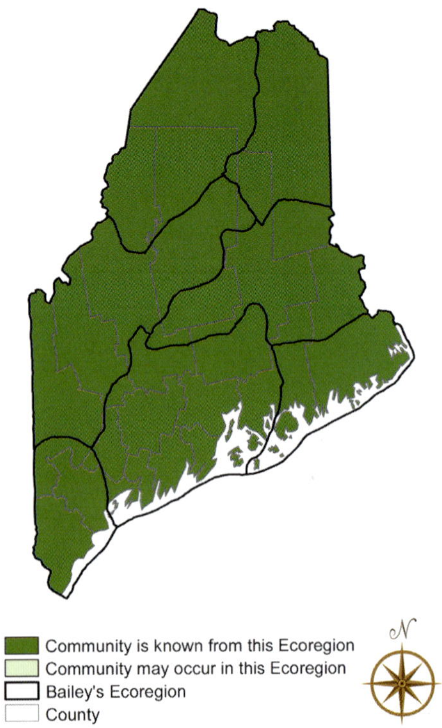

Lakeshore Beach

beauty. Aquatic community types are dominated by submerged plants.

Conservation, Wildlife, and Management Considerations

This is the most frequent lakeshore type for medium and large lakes in Maine. While many lakeshore beaches are protected by fee ownership, easement, or regulation, activities such as ATV use have significantly altered the vegetation at some sites. Hydrologic integrity is also a concern; as water use increases from neighboring homes and businesses, aquifer drawdowns could impair these water dependent systems and lead to vegetational changes. Invasive plants may be a problem at some sites.

Lakeshore Beaches with abundant sections of exposed sand or sand-gravel may provide valuable nesting habitat for resident painted, snapping, and musk turtles if water levels remain relatively stable during the incubation period of June through October.

Distribution

Extends east, west, and north from Maine; occurs only as scattered areas southward.

Landscape Pattern: Small to large patch, linear.

Characteristic Plants

These plants are frequently found in this community type. Those with an asterisk are often diagnostic of this community.

Sapling/shrub

Gray birch
Red maple
Speckled alder
White pine

Dwarf Shrub

Black chokeberry
Sweetgale

Herb

Beggar-ticks
Boneset
Common horsetail
Common juniper
Creeping spearwort
Flatsedges
Large cranberry
Panic grasses
Pinweed
Poverty oat-grass
Spikerushes
Sweet gale
Switch-grass
Swamp candles
Three square
Tufted hairgrass
Water parsnip

Associated Rare Animals

Black tern

Examples on Conservation Lands You Can Visit

- Fourth Machias Lake – Washington County
- Gero Island – Piscataquis County

Laurentian River Beach

State Rank S2

Community Description

These exposed river beaches feature sparse to extensive cover of low shrubs, forbs, and grasses, with species of northern affinity characteristic. Dominant plants include tufted hairgrass, roses, and mats of sand cherry. Poison ivy may be locally abundant. Canadian tick-trefoil, wild chive, New York aster, and freshwater cordgrass are characteristic herbs. Several rare plants such as Huron tansy, alpine sweet-broom, or alpine milk-vetch may be locally common. Bryoids are virtually absent.

Soil and Site Characteristics

These are rivershores where coarse deposits remain after flooding and ice scour. The substrate is usually cobbly and often dry at the surface. Sites flood in the spring and may be partially underwater for brief periods in the summer. The slope is usually very slight and the sites are exposed to full sun.

Glaucous Rattlesnake-root

Diagnostics

Patchy herbs and low shrubs occur on a rivershore below the high water mark and adjacent trees, with sandy cherry typical and circumneutral indicator plants present (e.g., freshwater cordgrass, wild chive). The cover of grasses and forbs exceeds that of sweetgale and sedges, and the substrate is unconsolidated gravelly to cobbly glaciofluvial deposits.

Similar Types

This type is related to and sometimes contiguous with other rivershore types. Bluejoint Meadows are dominated by bluejoint and occur on finer substrates. Circumneutral Riverside Seeps have more

Location Map

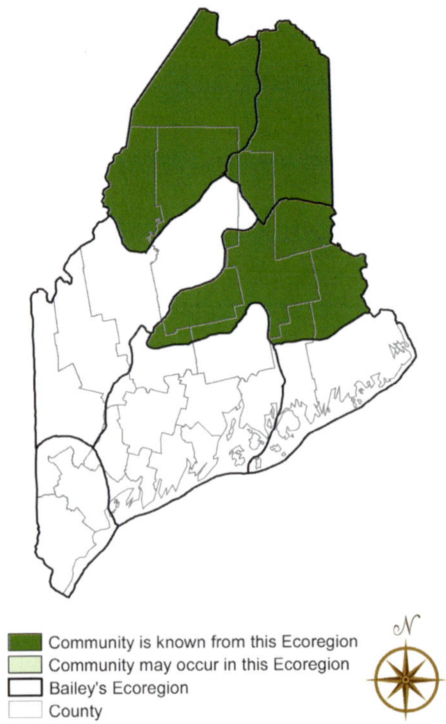

Alpine Sweet-broom

sweetgale and sedges, seepage waters at the surface, and certain indicator species (e.g., grass-of-parnassus and yellow sedge). Dogwood - Willow Shoreline Thickets are dominated by shrubs over 1 m tall. Hudsonia River Beach features beach heather, little bluestem, and other temperate species compared to the northern plants typical of this type.

Conservation, Wildlife, and Management Considerations

This community is tightly linked to naturally fluctuating water levels and occasional ice scour. Most known sites are privately owned and conservation depends on the landowner. Foot traffic has been light and poses no threat, but if off road vehicles are used, the shore and the vegetation could be easily degraded. Exotic agricultural species are common at some sites, and at least one site has the invasive Japanese knotweed.

Distribution

Currently known only from the St. John River, but may be elsewhere in northern Maine; poorly documented. Presumably extends to eastern Canadian rivers. (New England - Adirondack and Laurentian Mixed Forest Provinces.)

Landscape Pattern: Small Patch, linear.

Characteristic Plants

These plants are frequently found in this community type. Those with an asterisk are often diagnostic of this community.

Sapling/shrub

Meadowsweet
Poison-ivy
Red osier dogwood
Smooth rose
Willow

Dwarf Shrub

Sand cherry*

Herb

Bladder campion
Bluebell
Canadian tick-trefoil
Early goldenrod
Freshwater cordgrass
Hemp dogbane
New York aster
Northern blue flag
Reed canarygrass
Silverweed
Tufted hairgrass*
Wild chive

Associated Rare Plants

Alpine milk-vetch
Alpine sweet-broom
Anticosti aster
Canada burnet
Glaucous rattlesnake-root
Huron tansy
Northern gentian
Soft-leaf muhly
St. John oxytrope

Examples on Conservation Lands You Can Visit

- St. John River Preserve – Aroostook Co.

Leatherleaf Bog

State Rank S4

Community Description

This peatland vegetation type is dominated by leatherleaf and other low heath shrubs. Most of the vegetation is usually less than 1 m tall, although taller shrubs including black huckleberry, maleberry, and sweetgale may be sporadic. In the dwarf shrub/herb layer, leatherleaf is always present and usually dominant (30-60% cover at most sites). Other heath shrubs and sedges are mixed in with the leatherleaf. Graminoid cover is usually less than 30%. Typical bog plants including pitcher plant, sundew, and small cranberry are scattered on the peat moss substrate. Trees, if present at all, are <15% total cover.

Soil and Site Characteristics

This type is common in the wetter parts of bogs and acidic, nutrient poor fens (average pH is 4.0) with peat substrate. It usually occurs in settings where groundwater contact is maintained, and so is technically a fen from a hydrologic standpoint, although it is often referred to as "bog" because of the dominance of heath vegetation. This type is often a major constituent of "kettlehole bog" ecosystems, and it may be present in lakeshore peatlands or other sites with a fluctuating water table.

Ringed Boghaunter

Diagnostics

In a peatland setting, dwarf shrub cover exceeds herb cover, but sheep laurel is not dominant because most sites are hydrologically fens. Tufted cotton-grass and/or tawny cotton-grass are common sedges; white beak-rush is frequent but does not form high cover as it can in other types.

Similar Types

This type is intermediate in composition and nutrient regime between a Sheep Laurel Dwarf Shrub Bog (which is drier and has sheep laurel more abundant) and Sedge - Leatherleaf Fen Lawn or other graminoid dominated fen community types (which have graminoids more dominant).

Conservation, Wildlife, and Management Considerations

This type is well represented in Maine and is

Location Map

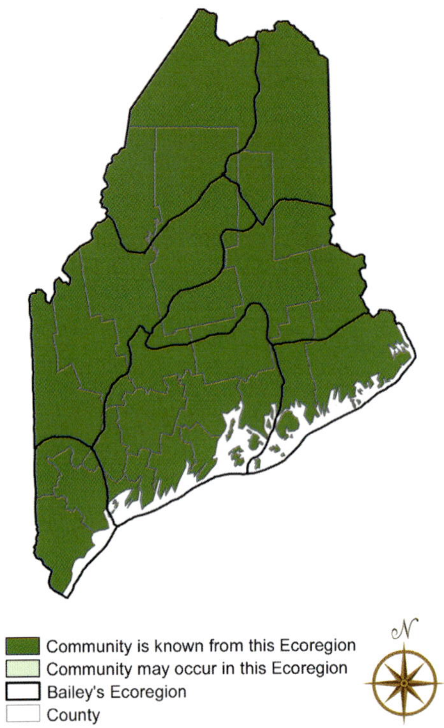

Leatherleaf

fairly stable in extent, with several examples on public lands and private conservation lands. Some sites in kettlehole settings have been degraded by adjacent gravel mining. Changes to bog hydrology through impoundment or draining could lead to vegetation changes. Slow vegetation growth rates, due to the nutrient-poor environment, mean slow recovery from physical disturbances, such as recreational trail use. If disturbance, such as foot traffic, is a necessity, traversing during frozen conditions or using boardwalks can minimize impacts.

The ringed boghaunter, a rare dragonfly restricted to York and southern Oxford counties, is found in this natural community type, especially in very wet locations with abundant inundated peat moss (often suspended in the water column). Occurrences of this community type in northwestern Maine may include the bog fritillary butterfly, which uses small cranberry as its larval host plant. Occurrences in northern Maine may be inhabited by the subarctic bluet, an uncommon damselfly that inhabits open marshes and fens and reaches the southern edge of its range in northern Maine.

Distribution

Statewide; extends in all directions from Maine.

Landscape Pattern: Small Patch

Characteristic Plants

These plants are frequently found in this community type. Those with an asterisk are often diagnostic of this community.

Sapling/shrub

Black huckleberry*
Maleberry*

Dwarf Shrub

Bog rosemary*
Leatherleaf*
Sheep laurel*
Small cranberry
Sweetgale*

Herb

Beaked sedge*
Bog aster*
Few-flowered sedge*
Pitcher plant*
Tawny cotton-grass
Three-leaved false Solomon's seal*
Tufted cotton-grass*
White beak-rush*

Bryoid

Sphagnum mosses*

Associated Rare Plants

Inkberry
Long's bulrush
Screwstem
Swamp birch

Associated Rare Animals

Ringed boghaunter

Examples on Conservation Lands You Can Visit

- Brownfield Bog Wildlife Management Area – Oxford Co.
- Great Heath Public Lands – Washington Co.
- Number Five Bog Public Lands – Somerset Co.
- Salmon Brook Lake Bog Public Lands – Aroostook Co.
- Sunkhaze Meadows National Wildlife Refuge – Penobscot Co.

Low Sedge Fen

State Rank S3

Community Description

This type is characterized by peatland vegetation dominated by low mats of sedges (typically 40-60% cover), sometimes with sparse low heaths, over a continuous and very wet peat moss substrate. White beak-rush, mud sedge, and few-seeded sedge are usually dominant. Podgrass and buckbean are particularly characteristic, and sundews and horned bladderwort are typical in openings among the sedges. Heath shrubs are sparse; the most frequent are leatherleaf or bog rosemary, with other heaths on scattered hummocks. Bog-mat liverwort is an indicator species, although it is not present in all examples of this community type.

Soil and Site Characteristics

This community occurs in very wet portions of peatlands but not in raised portions nor adjacent to open water. It occasionally occurs in mineral soil marshes that grade into peatland areas. The substrate is constantly saturated, often unstable, and acidic (pH 4.0-5.0).

Diagnostics

Graminoids are dominant over a well developed bryoid layer. White beak-rush, mud sedge, and occasionally few-seeded sedge are the dominant sedges. Dwarf shrubs form less than 20% cover and the substrate is saturated to the surface.

Bog Buckbean

Similar Types

Sedge - Leatherleaf Fen Lawns are closely related but have more leatherleaf (usually > 25%) and often have coast sedge as a characteristic species. Leatherleaf Boggy Fens are shrub dominated, rather than sedge dominated. Shrubby Cinquefoil - Sedge Circumneutral Fens are also typically graminoid dominated but by different sedges, and will have the characteristic calciphilic species that are generally absent from this type. Mixed Tall Sedge Fens feature taller and denser graminoids and usually have abundant slender sedge.

Conservation, Wildlife, and Management Considerations

This community type is not particularly

Location Map

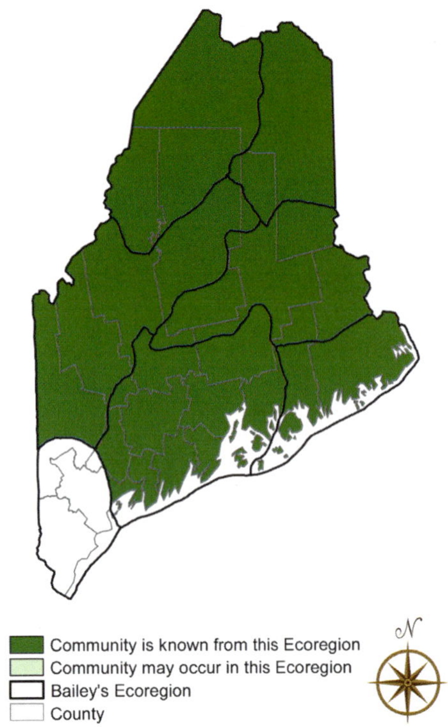

Low Sedge Fen

common, but it has been subject to few threats to date. Some examples occur on public lands and private conservation lands. Impoundment or draining would have negative impacts on hydrology and on vegetation. Slow vegetation growth rates, due to the nutrient poor environment, result in slow recovery from physical disturbances. Degradation from recreational use is unlikely, because of the unstable substrate; but if disturbance, such as foot traffic, is a necessity, traversing during frozen conditions or using boardwalks can minimize impacts.

Several uncommon dragonfly species may be found where bog pools and seasonally inundated depressions occur, including the zigzag darner, subarctic darner, and incurvate emerald. Occurrences in northern Maine may be inhabited by the subarctic bluet, an uncommon damselfly that inhabits open marshes and fens and reaches the southern edge of its range in northern Maine. The delicate emerald, a dragonfly that inhabits bogs and fens covered with a carpet of low sedges, is also a likely associate, as is the rare Quebec emerald.

Distribution

Northern and eastern Maine (New England Adirondack and Laurentian Mixed Forest Provinces), presumably extending into Canada.

Landscape Pattern: Small Patch

Characteristic Plants

These plants are frequently found in this community type. Those with an asterisk are often diagnostic of this community.

Sapling/shrub

Speckled alder

Dwarf Shrub

Bog rosemary
Labrador tea
Leatherleaf
Rhodora
Small cranberry

Herb

Bog goldenrod
Buckbean*
Mud sedge*
Pitcher plant
Podgrass*
Round-leaved sundew
Tawny cotton-grass
Three-leaved false Solomon's seal
White beak-rush*

Bryoid

Sphagnum mosses*

Associated Rare Plants

Livid sedge
Sparse-flowered sedge

Associated Rare Animals

Quebec emerald

Examples on Conservation Lands You Can Visit

- Marble Fen Preserve – Penobscot Co.
- St. John River Preserve – Aroostook Co.
- Telos Lake Public Lands Unit – Piscataquis Co.

Maritime Huckleberry Bog

State Rank S3

Community Description

This peatland type is characterized by low (usually < 60 cm) heath shrubs, such as sheep laurel and leatherleaf, as the dominant layer. Dwarf huckleberry is typically prominent. Black crowberry and/or deer-hair sedge are also present, though not necessarily abundant. In far Downeast examples, black crowberry may entirely replace dwarf huckleberry. Small islands of stunted black spruce may be scattered among the shrubs. Pitcher plants, sundews, bog goldenrod, and other typical bog herbs are mixed in with the shrubs; herb cover is usually <40%. Deer-hair sedge may be locally dominant. A dense layer of peat mosses underlies the plants.

Soil and Site Characteristics

These are coastal or near coastal peatlands, either in raised bogs or in weakly minerotrophic areas transitional to true bogs. They are saturated during the growing season and typically highly acidic (pH < 5.0, occasionally slightly higher).

Diagnostics

Dwarf shrub peatland vegetation is characterized by dwarf huckleberry and/or black crowberry together forming >20% cover. Sheep laurel and leatherleaf are usually present but at lower cover than in other dwarf-shrub peatland types.

Pitcher Plant Leaf with Dragonfly Exuvia

Similar Types

Sheep Laurel Dwarf Shrub Bog and Leatherleaf Boggy Fen communities are similarly found in true bogs and dominated by heath shrubs. They may be contiguous with this type in some bogs. A few of these sites in south-coastal Maine have dwarf huckleberry even though they are not Huckleberry - Crowberry Bogs. The dominance of dwarf huckleberry with either crowberry or deer-hair sedge distinguishes Huckleberry-Crowberry Bogs from other bog vegetation types.

Conservation, Wildlife, and Management Considerations

This community type is not widely

Location Map

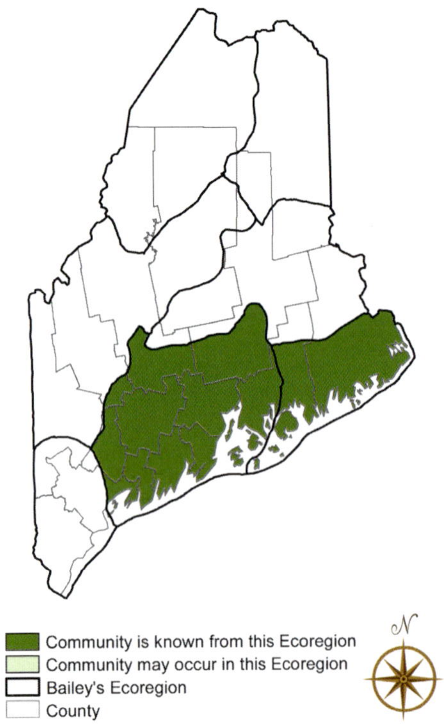

Huckleberry – Crowberry Bog

distributed, but has been subject to few threats to date. Slow vegetation growth rates, due to the nutrient-poor conditions, mean slow recovery from physical disturbances, such as recreational trail use. If disturbance, such as foot traffic, is a necessity, traversing during frozen conditions or using boardwalks can minimize impacts. Peat harvesting could threaten some sites. Changes to bog hydrology, through impoundment or draining, could lead to vegetation changes. Several occurrences are on public lands or private conservation lands.

The rare crowberry blue butterfly is restricted to coastal heaths in east-coastal Maine. It uses black crowberry as a larval host plant. Several uncommon dragonfly species may be found in examples of this community where bog pools occur, including the zigzag darner, subarctic darner, and incurvate emerald.

Distribution

Most typical of Downeast Maine from Penobscot Bay eastward. Always within 15 miles of the shore (Laurentian Mixed Forest Province). Extends east into the Canadian Maritimes.

Landscape Pattern: Small Patch

Characteristic Plants

These plants are frequently found in this community type. Those with an asterisk are often diagnostic of this community.

Sapling/shrub

Black spruce

Dwarf Shrub

Black crowberry*
Dwarf huckleberry*
Labrador tea*
Leatherleaf*
Pale laurel
Sheep laurel*
Small cranberry

Herb

Bog goldenrod
Deer-hair sedge*
Pitcher plant
Round-leaved sundew

Bryoid

Cladonia lichen*
Reindeer lichen*
*Sphagnum fuscum**

Associated Rare Plants

Northern comandra

Associated Rare Animals

Crowberry blue

Examples on Conservation Lands You Can Visit

- Acadia National Park – Hancock Co.
- Great Wass Island Preserve – Washington Co.
- Quoddy Head State Park – Washington Co.

Maritime Slope Bog

State Rank S2

Community Description

A well developed layer of dwarf shrubs is dominated by heath shrubs and black crowberry in a dense carpet in this vegetation type. There may be scattered small conifers, and typically at least a small amount of common juniper. Baked apple-berry is diagnostic and is restricted to this type and other coastal or subalpine peatlands. Herbaceous "bog" species (deer-hair sedge, pitcher plant, etc.) are also common. The bryoid layer is extensive (>70% cover, usually close to 100%) and is dominated by peat mosses and small islands of reindeer lichens.

Soil and Site Characteristics

Sometimes called "blanket bogs," these occur on bedrock or other rocky substrate. Soil is a thin organic layer over rock, and slopes are usually 5-10%. Sites occur on cool microsites near the Downeast coast.

Diagnostics

Tree cover is less than 25% and heath shrubs are dominant. Although the ground layer is composed of peat mosses, peat forms only a thin layer over bedrock or mineral substrate, so this is not a true peatland type.

Baked Apple–berry

Similar Types

Sheep Laurel Dwarf Shrub Bogs and Huckleberry - Crowberry Bogs share species and structure with Heath - Crowberry Maritime Slope Bogs; however, those are found in true peatlands (basins with deep accumulations of saturated peat), not on thin peat over rock. Heath - Lichen Subalpine Slope Bogs are compositionally similar but occur in the mountains.

Location Map

Black Crowberry

Conservation, Wildlife, and Management Considerations

This is an extremely restricted community type, but most documented occurrences are on public lands or private conservation lands. Recreation and climate change are the primary threats; careful planning of trails and ensuring that users stay on trails can help minimize recreational impacts.

The rare crowberry blue butterfly is restricted to coastal heaths in east-coastal Maine. It is uses black crowberry as a larval host plant.

Distribution

Downeast coastal Maine, extending eastward into the Canadian Maritimes (Laurentian Mixed Forest Province).

Landscape Pattern: Small Patch

Round-leaved Sundew

Characteristic Plants

These plants are frequently found in this community type. Those with an asterisk are often diagnostic of this community.

Sapling/shrub

Black chokeberry
Common juniper
Mountain holly
Wild-raisin

Dwarf Shrub

Baked apple-berry
Black crowberry*
Black huckleberry
Labrador tea
Mountain cranberry
Rhodora
Sheep laurel*

Herb

Bog goldenrod
Deer-hair sedge
Pitcher plant
Round-leaved sundew
Starflower
White beak-rush

Bryoid

Bog hair-cap moss
Grey reindeer-lichen
Sphagnum mosses*
Woodland reindeer-lichen

Associated Rare Animals

Crowberry blue

Examples on Conservation Lands You Can Visit

- Great Wass Island Preserve – Washington Co.
- Petit Manan Point, Petit Manan National Wildlife – Washington Co.

Mixed Saltmarsh

State Rank S4

Community Description

These saltmarshes contain a mixture of graminoids and forbs, sometimes with patches of saltmarsh cordgrasses, but saltmeadow cordgrass is not strongly dominant. Chair-maker's rush is almost always present, at least in small amounts. Dominants can vary, but indicator species include creeping bentgrass, freshwater cordgrass, sea lavender, wire rush, saltmarsh bulrush, New England aster, saltmarsh sedge, and narrow-leaved cattail. Sweetgrass is often present, though not abundant. The vegetation occurs as a mosaic of dominants and lacks the strong zonation typical of the larger Spartina Saltmarshes.

Soil and Site Characteristics

These are often fringe marshes in sheltered coastal pockets, estuaries, and tidal creeks; not typically covering large acreages although they may be strung along a fairly long stretch of shoreline. They often occur along tidal creeks, or as a shoreline fringe in coves. Vegetation consists predominantly of low marsh species (saltmarsh regularly inundated twice daily by tides).

Diagnostics

These are tidal marshes in which various saltmarsh plants share dominance with cordgrasses and/or black-grass; chair-maker's rush is typically present and may be dominant; saltmarsh sedge is also characteristic. Vegetation tends to be patchy rather than zoned.

Saltmarsh Bulrush

Similar Types

Spartina Saltmarshes have many of the same species, but have much greater relative cover of saltmeadow cordgrass, smooth cordgrass, and black-grass. They also appear more uniform, and tend to occur at the outer reaches of estuaries (back-barrier marshes and finger marshes). Brackish Tidal Marshes also share many species, but lack the saltmarsh cordgrasses and other strictly saltmarsh species (black-grass, saltmarsh bulrush, saltmarsh false-foxglove, sea lavender, etc.).

Location Map

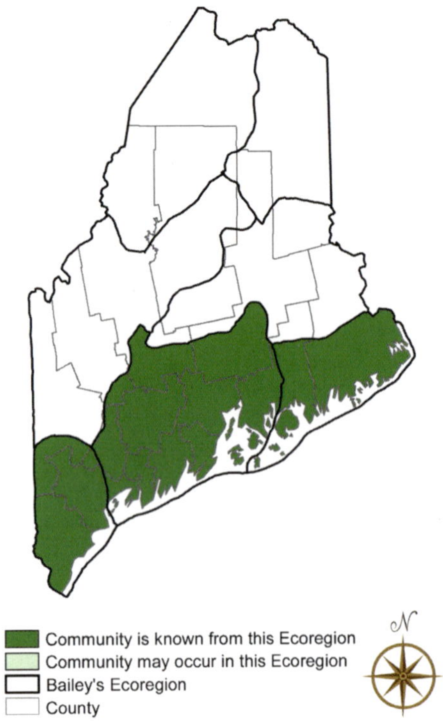

Mixed Graminoid – Forb Saltmarsh

Conservation, Wildlife, and Management Considerations

Saltmarshes have received considerable conservation attention. Many occur on public lands or private conservation lands, only a few of which are listed. With development of the uplands that border these marshes, maintenance of appropriate wetland buffers can help reduce degradation that could result from adjacent land uses.

Saltmarshes are important nesting habitat for Nelson's sharp-tailed sparrow, seaside sparrow, and the rare saltmarsh sharp-tailed sparrow. These wetlands also provide foraging habitat for a large number of wading birds and shorebirds, including the rare least tern. The big bluet, a rare damselfly, inhabits saltmarsh ponds with emergent vegetation in southern Maine.

Distribution

Coastwide; almost all of the east coastal Maine saltmarshes contain this type. Extends eastward into the Canadian Maritimes (Laurentian Mixed Forest Province) and westward into New Hampshire and Massachusetts.

Landscape Pattern: Small Patch

Characteristic Plants

These plants are frequently found in this community type. Those with an asterisk are often diagnostic of this community.

Herb

Alkali bulrush*
Black-grass*
Chaffy sedge*
Chair-maker's rush*
Common arrow-grass
Creeping bentgrass*
Freshwater cordgrass
New York aster*
Salt-loving spikerush*
Saltmeadow cordgrass*
Wire rush*

Associated Rare Plants

Gaspé arrow-grass
Marsh-elder
Saltmarsh false-foxglove
Saltmarsh sedge
Small saltmarsh aster

Associated Rare Animals

Big bluet
Black-crowned night-heron
Laughing gull
Least tern
Saltmarsh sharp-tailed sparrow
Short-eared owl

Examples on Conservation Lands You Can Visit

- Cobscook Bay State Park – Washington Co.
- Great Wass Island Preserve – Washington Co.
- Petit Manan National Wildlife Refuge – Hancock Co.
- Reid State Park – Sagadahoc Co.
- Scarborough Marsh Wildlife Management Area – Cumberland Co.

Mossy Bog Mat

State Rank S4

Community Description

The bryophyte layer is the most obvious component of this peatland type. A dense and usually very wet layer of peat mosses contributes most of the cover. Low herbs and stunted shrubs are often scattered across the moss lawn, but usually form <25% cover overall. Characteristic vascular plant species include leatherleaf, bog rosemary (very dwarfed), horned bladderwort, small cranberry, and white beak-rush. The most typical bryophytes are *Sphagnum cuspidatum* and bog-mat liverwort.

Soil and Site Characteristics

Bog Moss Lawns occur within raised bogs and fens, typically in the wettest areas such as bog pools, boggy pond margins, and water tracks. They may form extensive areas on the higher areas of raised bogs. The substrate is highly acidic (pH ~ 4.0).

Diagnostics

A carpet of bryophytes, mostly Sphagnum mosses, is the main feature in this peatland setting. Vascular plants generally contribute <25% cover.

Small Cranberry

Similar Types

This is the only vegetation type in which bryophytes form the major continuous vegetation layer. Sedge - Leatherleaf Fen Lawns, Low Sedge - Buckbean Fen Lawns, and Leatherleaf Boggy Fens can all have similar vascular plant species, but will have the vascular plants more strongly dominant (typically >50% cover).

Conservation, Wildlife, and Management Considerations

This community type is well represented in Maine and has been subject to few threats to date. Several examples occur

Location Map

Mossy Bog Mat

on public lands and private conservation lands. Impoundment or draining would have negative impacts on bog hydrology and consequently on the vegetation. Slow vegetation growth rates, due to the nutrient-poor environment, result in slow recovery from physical disturbances, such as recreational trail use. If disturbance, such as foot traffic or logging access, is a necessity, traversing during frozen conditions or using boardwalks can minimize impacts.

Several rare dragonflies may be found in this community. Very wet locations with abundant inundated peat moss may host a number of uncommon damselflies and dragonflies including the Quebec emerald, zigzag darner, subarctic darner, incurvate emerald, and delicate emerald. Sites in northwestern Maine may include the bog fritillary butterfly, which uses small cranberry as its larval host plant.

Distribution

Statewide, mostly from central Maine northward (New England - Adirondack and Laurentian Mixed Forest Provinces). Presumably extends westward, northward, and eastward from Maine.

Landscape Pattern: Small Patch

Characteristic Plants

These plants are frequently found in this community type. Those with an asterisk are often diagnostic of this community.

Dwarf Shrub

Black crowberry
Dwarf huckleberry
Leatherleaf*
Pale laurel
Small cranberry

Herb

Horned bladderwort
Pitcher plant
Round-leaved sundew
Tawny cotton-grass
White beak-rush

Bryoid

Bog-mat liverwort
Mylia liverwort
Sphagnum mosses*

Associated Rare Animals

Quebec emerald

Examples on Conservation Lands You Can Visit

- Crystal Bog Preserve – Aroostook Co.
- Great Heath Public Lands – Washington Co.
- Number Five Bog Public Lands – Somerset Co.
- Wiggins Brook, Squaw Mountain Public Lands – Piscataquis Co.

Sphagnum Mosses

Open-water Marsh

State Rank S5

Community Description

Water-lilies or pondweeds are dominant in this floating aquatic vegetation type. It is currently broadly defined and may be divisible into two or more types with additional data. The strict type has white water-lily, yellow water-lily, or more rarely water-shield as the most abundant species. Variants may lack water-lilies altogether and be dominated by floating leaved pondweed species such as bigleaf pondweed or others. Submerged aquatic plants are also common and may include bladderworts and pipewort.

Soil and Site Characteristics

These aquatic beds occur in quiet waters at depths mostly between 0.5 and 2.5 m. The substrate is typically a mixture of silty organic muck.

Diagnostics

Water-lily species are prominent (or in variants, pondweed species are prominent); pickerelweed is basically absent.

Similar Types

Pickerelweed - Macrophyte Aquatic Bed and Bulrush Bed vegetation have greater dominance of emergent versus submerged plants. Pipewort - Water Lobelia Aquatic Bed vegetation has low cover of floating aquatics and is dominated instead by plants with basal rosettes growing on the bottom. The pondweed variant can be similar to Circumneutral - Alkaline Water Macrophyte Suite but lacks the alkaline water indicator species such as water-marigold, common waterweed, Robbins' pondweed, and straight-leaved pondweed.

Scarlet Bluet Damsel Flies Mating

Conservation, Wildlife, and Management Considerations

This aquatic community type is widespread and abundant in Maine. It can be found in the quieter portions of streams and rivers as well as in lakes and ponds, and it provides habitat for many

Location Map

Water-Lily – Macrophyte Aquatic Bed

Characteristic Plants

These plants are frequently found in this community type. Those with an asterisk are often diagnostic of this community.

Herb

Bayonet rush
Bigleaf pondweed*
Pipewort
Spotted bladderwort*
Water-shield*
White water-lily*
Yellow water-lily*

Associated Rare Plants

Pygmy water-lily
Spotted pondweed

Associated Rare Animals

American coot
Common moorhen
Spatterdock darner

water-dependent animals. Many examples occur on public and private conservation lands; however, because the type is so common it is often not documented. The major threats to this natural community are water quality degradation from excess nutrients in runoff and the spread of invasive aquatic plants, such as Eurasian water-milfoil and variable water-milfoil.

In southern and central parts of the state, this community type hosts a number of common reptiles such as northern water snakes, common snapping turtles, and eastern painted turtles. Muskrats, bullfrogs, and green frogs inhabit this community statewide and mink frogs may be found in occurrences from central Maine northward. Northern leopard frogs may utilize these wetlands as breeding grounds. These productive wetlands provide foraging habitat for a number of waterfowl including rare species such as the common moorhen and American coot. Dragonfly and damselfly diversity is typically high in these wetlands and may include species such as the turquoise bluet, little bluet, lilypad forktail, and the rare spatterdock darner.

Distribution

Statewide; extending southward and westward from Maine and presumably into Canada

Landscape Pattern: Small Patch, often linear along lakeshores.

Examples on Conservation Lands You Can Visit

- Acadia National Park - Hancock Co.
- Tyler Pond Wildlife Management Area - Kennebec Co.

Yellow and White Water Lilies

Outwash Plain Pondshore

Three-way Sedge - Goldenrod Outwash Plain Pondshore

State Rank S1

Community Description

This community consists of concentric zones of different herbs around a central pond. A band of shrubs (highbush blueberry, maleberry, buttonbush, leatherleaf) is typical at the upland/pondshore edge. Moving pondward, the next zone is dominated by narrow-leaved goldenrod and three-way sedge, with patches of flat-sedge and brown-fruited rush. In a narrow band at the top of this zone, golden pert and meadow beauty are characteristic and may form dense patches. The next zone, exposed less frequently and for a shorter time, is dominated by pipewort and spikerushes. There is no well developed bryoid layer.

Soil and Site Characteristics

This community forms a band around the perimeter of shallow, sandy bottomed ponds in glacial outwash plains. It occurs on shores that are inundated for the early part of the growing season and exposed later in the growing season, although actual exposure varies from year to year. The substrate is sandy, occasionally mucky, and usually saturated to the surface or nearly so.

Ribbon Snake

Diagnostics

Three-way sedge and usually narrow-leaved goldenrod are dominant in a sandy pondshore setting, with evidence of water level changes through the season. Golden pert and meadow beauty are indicator species.

Similar Types

Mixed Graminoid - Shrub Marshes can also occur on temporarily flooded mineral soils and can share some dominants such as three-way sedge, but they lack the concentric zonation of outwash plain pondshores and typically intermingle shrubs and herbs rather than segregating them into zones. The more variable and widespread Lakeshore Beaches lack three-way sedge, golden pert, and meadow beauty.

Location Map

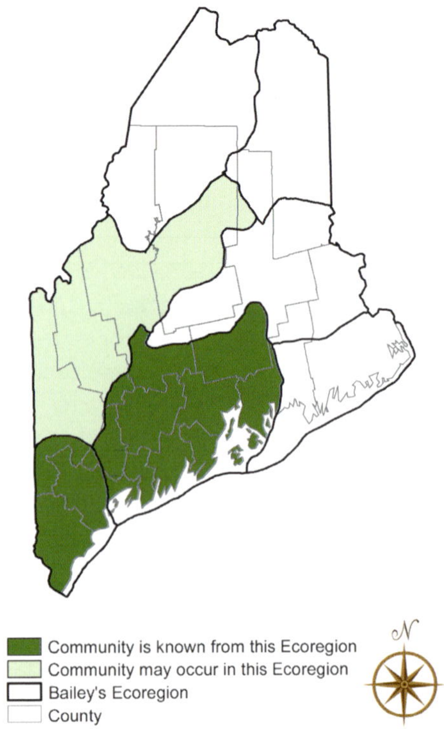

Three-way Sedge

Characteristic Plants

These plants are frequently found in this community type. Those with an asterisk are often diagnostic of this community.

Herb

Bluejoint*
Brown-fruited rush*
Bur-reed*
Canada rush
Fly-away grass
Golden pert*
Narrow-leaved goldenrod*
Pipewort*
Robbin's spikerush*
Three-way sedge
Toothed flat-sedge*
Yellow loosestrife

Associated Rare Plants

Dwarf bulrush
Englemann's spikerush
Fall fimbry
Huron tansy
Long-tubercled spike-rush
Narrow-leaved goldenrod

Associated Rare Animals

Big bluet
Ribbon snake

Conservation, Wildlife, and Management Considerations

This extremely rare natural community is under pressure from adjacent land uses and recreational impacts. The periphery of several sites has been developed or converted to other uses. At the few known sites on conservation lands, the major recreational impact is off-road vehicle use. At low water, ATV use has significantly altered the vegetation at some sites. Hydrologic integrity is also a concern, as water use increases from neighboring homes and businesses and aquifer drawdowns could impair these water dependent systems and lead to vegetational changes.

These outwash plain pondshores provide excellent foraging habitat for the ribbon snake. The pondshores also provide habitat for the big bluet, a rare damselfly. Other more wide-ranging rare insects are likely to be found in this community. This community may also provide feeding habitat for wading birds.

Distribution

Extreme southwestern Maine (Eastern Broadleaf Forest Province), extending southward along the coast to Massachusetts; disjunct in Nova Scotia and Ontario.

Landscape Pattern: Small Patch

Examples on Conservation Lands You Can Visit

- Killick Pond Wildlife Management Area - York Co.
- Waterboro Barrens Preserve - Oxford Co.

Pickerelweed Marsh

State Rank S5

Community Description

This shallow water aquatic type is dominated by a mixture of emergent plants, floating plants, and submerged plants suspended in the water column. Pickerelweed, yellow water-lily, and bladderworts are almost always present, and one or more is typically dominant. A variety of pondweed species, bulrushes, bur-reed species, and other aquatics may be present. Total cover ranges from 30-100% and is typically >50%.

Soil and Site Characteristics

This community occupies quiet waters along the shores of lakes, ponds, and streams. The substrate is usually mucky, and the water is generally <0.7 m deep. Waters are acidic to circumneutral.

Diagnostics

This type is characterized by emergent aquatic vegetation with pickerelweed and/or yellow water-lily prominent. Water bulrush and Oakes' pondweed are also indicator species.

Maine Moose

Similar Types

Water-lily - Macrophyte Aquatic Bed vegetation shares many species but is dominated by floating or submerged plants, not emergent plants. Both types have water-lily species, but they are less abundant in the pickerelweed type than in the water-lily type. Pipewort - Water Lobelia Aquatic Bed has more of its vegetation on the bottom surface and fewer floating leaved plants. Cattail Marshes may be adjacent but feature cattails and shrubs rather than plants that die back below the surface at the end of the growing season.

Location Map

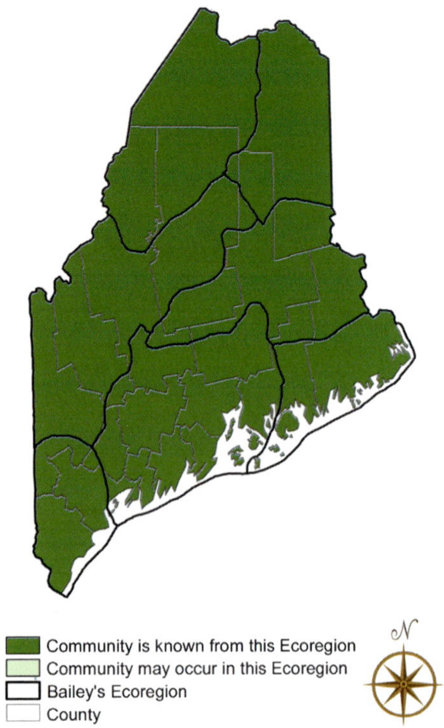

Pickerelweed Marsh

Characteristic Plants

These plants are frequently found in this community type. Those with an asterisk are often diagnostic of this community.

Herb

Pickerelweed*
Pipewort
Spotted bladderwort*
Water bulrush*
White water-lily*
Yellow water-lily*

Associated Rare Animals

American coot
Black-crowned night heron
Common moorhen
Least bittern
Spatterdock darner

Conservation, Wildlife, and Management Considerations

This aquatic vegetation type is widespread and abundant in Maine. It can be found in the quieter portions of streams and rivers as well as in lakes and ponds. It provides habitat for a variety of water dependent animals. Many examples occur on public lands and private conservation lands; however, this common type is often not documented. The major threats to this community are water quality degradation from excess nutrients in runoff and the spread of invasive aquatic plants such as Eurasian water-milfoil and variable water-milfoil.

In southern and central parts of the state, this community type hosts a number of common reptiles such as northern water snakes, common snapping turtles, and eastern painted turtles. Bullfrogs and green frogs inhabit this community statewide, and mink frogs may be found in occurrences from central Maine northward. Northern leopard frogs may utilize these wetlands as breeding grounds. These productive wetlands provide foraging habitat for a number of waterfowl including rare species such as the common moorhen and American coot. Dragonfly and damselfly diversity is typically high in these wetlands.

Distribution

Statewide, extending into lower New England and Canada.

Landscape Pattern: Small Patch

Examples on Conservation Lands You Can Visit

- Merrymeeting Bay Wildlife Management Area – Sagadahoc Co.
- Range Ponds State Park – Androscoggin Co.
- Sebago Lake State Park – Cumberland Co.

Pickerelweed

Rivershore Shrub Thicket

Dogwood - Willow Shoreline Thicket

State Rank S2

Community Description

Dense riparian shrub vegetation (>80% cover) is dominated by a mixture of red osier dogwood and shrub willows. Alders may be present but are not dominant. A band of bush-honeysuckle often forms at the upslope edge, where the shrub vegetation abuts upland forest. Herb richness may be high in openings among the shrubs and may include some calciphiles as well as more widespread species. Bryoids are minor, and consist of bryophytes rather than lichens. More study of alluvial shrub communities is needed.

Soil and Site Characteristics

Sites occupy shores of larger rivers, below the annual high-water line, in areas somewhat protected from the extremes of ice scour and flooding. Riverbanks are moderate to steep, not flat; and the silty to sandy soils are not constantly saturated. Successional dynamics have not been studied, but at least some sites appear to persist through disturbance.

Pussy Willow Leaves

Diagnostics

This type is characterized by 1-3 m tall shrub vegetation, with red osier dogwood and willows prominent, on riverbanks where annual disturbance creates a non-forested zone between the summer water level and adjacent forest. It occurs on soils that are not regularly saturated, and mostly on larger rivers.

Similar Types

Related to and sometimes contiguous with other rivershore communities. Bluejoint Meadows are strongly dominated by bluejoint and occur on flatter substrates. Circumneutral Riverside Seeps have more sweetgale and sedges, often have seepage waters at the surface, and have indicators such as grass-of-parnassus and yellow

Location Map

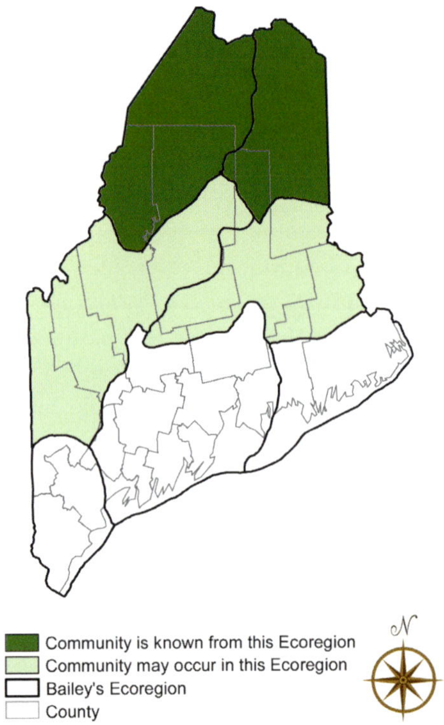

Slender Willow

sedge. Sand Cherry - Tufted Hairgrass River Beaches are dominated by herbs and low shrubs.

Conservation, Wildlife, and Management Considerations

Natural river hydrology, with annual fluctuations, is important in the maintenance of this community. These areas receive little direct use but may be affected by adjacent land use, so an intact buffer of adjoining upland would be helpful. No studies of the dynamics of this community have been done.

These rivershore shrublands provide habitat for bird species that inhabit open shrublands such as common yellowthroat, alder flycatcher, Wilson's warbler, and Lincoln's sparrow. Wood turtles may also use this riparian habitat.

Distribution

Documented from northern Maine along the St. John and Allagash Rivers, and may occur on other fairly large rivers in northern Maine (Laurentian Mixed Forest Province and New England - Adirondack Province). Presumably extends into Quebec and New Brunswick.

Landscape Pattern: Small Patch, linear.

Characteristic Plants

These plants are frequently found in this community type. Those with an asterisk are often diagnostic of this community.

Sapling/shrub

Bush-honeysuckle*
Red osier dogwood*
Red-tipped willow*
Round-leaved dogwood
Shining willow
Speckled alder

Herb

Bluejoint
Flat-topped white aster
Fringed bromegrass
Fringed loosestrife
Spotted joe-pye weed

Associated Rare Plants

Auricled twayblade
Blueleaf willow
Furbish's lousewort
Nantucket shadbush
Northern painted-cup
Sandbar willow

Examples on Conservation Lands You Can Visit

- The Nature Conservancy's St. John River lands – Aroostook Co.

Furbish's Lousewort

Riverside Seep

State Rank S2

Community Description

This type can include graminoid dominated, graminoid/forb, or shrubby vegetation, often with all three forms adjacent to one another. Shrubs include sweetgale, willows, and alders, either low or tall depending on how recently they have been scoured by ice. An array of forbs and graminoids grows among the sparse or low shrubs or downslope on the shore and includes fen indicators such as sticky false asphodel, grass-of-parnassus, various sedges and grasses, and forbs both native (various asters, Canada goldenrod, rose twisted-stalk) and introduced (clovers, ox-eye daisy, cow vetch). The bryoid layer is typically extensive, and features numerous bryophytes other than peat mosses.

Soil and Site Characteristics

These riparian seeps are below the annual high water mark on a substrate of unconsolidated, coarse textured soil and are constantly saturated by groundwater seepage. The substrate varies from fairly steep gravelly banks with locally extensive moss cover to lower shore flats with stabilized cobble. This type is restricted to larger rivers where spring floods and ice scour maintain a semi-open shore, versus a shoreline of dense shrubs. Continued exposure of glacial tills produces comparatively high pH.

Hairy Arnica

Diagnostics

Intermingled short graminoids, forbs, and shrubs grow on an open rivershore. The substrate is usually unconsolidated sandy to gravelly soil, saturated by groundwater discharge. Indicator species include grass-of-parnassus, hidden-scale sedge, and sweetgale.

Similar Types

Shrubby Cinquefoil - Sedge Circumneutral Fens contain many of the same species but occur in a peatland setting, not on a rivershore. Bluejoint Meadows, which may be adjacent to these seeps, are more homogeneous and are dominated by tall graminoids, particularly bluejoint grass. Sand Cherry - Tufted Hairgrass River Beach vegetation occurs on somewhat drier cobble

Location Map

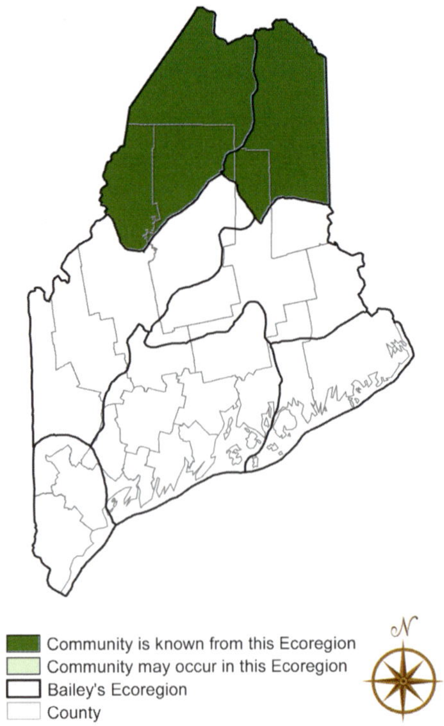

Circumneutral Riverside Seep

or sand substrates and includes few if any sedges.

Conservation, Wildlife, and Management Considerations

The structure and richness of this community is linked to naturally fluctuating water levels and occasional ice scour. Maintenance of an intact forest buffer is important, both for light conditions and for ensuring the flow of seepage waters. Many seeps are protected along the upper portions of the St. John River, but elsewhere conservation depends on the individual landowner. Some sites have been degraded by clearing the adjacent overstory. Foot traffic has been light and poses no threat, but where off road vehicles use the shore, the effects can be devastating.

Distribution

Known primarily from the St. John River and sparsely on other northern Maine rivers and lakes, west to other parts of northern New England (New England - Adirondack and Laurentian Mixed Forest Provinces). Many Canadian affinities, but distribution there unknown.

Landscape Pattern: Small Patch, linear.

Characteristic Plants

These plants are frequently found in this community type. Those with an asterisk are often diagnostic of this community.

Sapling/shrub

Beaked hazelnut*
Bush-honeysuckle*
Meadowsweet
Mountain alder*
Mountain maple*
Red-tipped willow
Shining willow
Speckled alder*

Dwarf Shrub

Shrubby cinquefoil*
Sweetgale*

Herb

Bluejoint
Common horsetail*
Dwarf raspberry
Field mint
Flat-topped white aster
Fowl mannagrass
Grass-of-parnassus
Purple-stemmed aster
Sticky false asphodel
Swamp buttercup
Tall meadow-rue
Wild strawberry*

Associated Rare Plants

Auricled twayblade
Black sedge
Furbish's lousewort
Garber's sedge
Glaucous rattlesnake-root
Mistassini primrose
Mountain timothy
Northern gentian
Northern painted-cup

Examples on Conservation Lands You Can Visit

- Allagash Public Lands – Aroostook Co.
- St. John River Preserve – Aroostook Co.

Riverwash Sand Barren

State Rank S1

Community Description

These riverside barrens support dwarf shrub - graminoid vegetation dominated by little bluestem and other grasses, with beach heather in patches. Big bluestem is characteristic, as are pinweed and goldenrod species. Small white pine trees may be scattered throughout. The vegetation is patchy, with blowouts and other areas of bare sand or gravel. Some occurrences also support populations of the rare silverling. Although individual plants shift over time as their substrates are reworked by floods, the overall composition of the community does not change appreciably over the short term (10-20 years).

Soil and Site Characteristics

Sites occur on gravelly or sandy barrens along river beaches and back beach areas where periodic flooding and xeric soils produce a distinctive disturbance regime. Most are in the two-year floodplain. Soils are very nutrient poor.

Silverling

Diagnostics

Sites occur on sandy floodplains with tree cover less than 25%. Little bluestem is prominent in the herb layer, and beach heather is present and locally dominant.

Similar Types

This type is quite distinct. Little Bluestem - Blueberry Sandplain Grasslands share some species with this type, but are on flat outwash plains, not in rivercourses.

Location Map

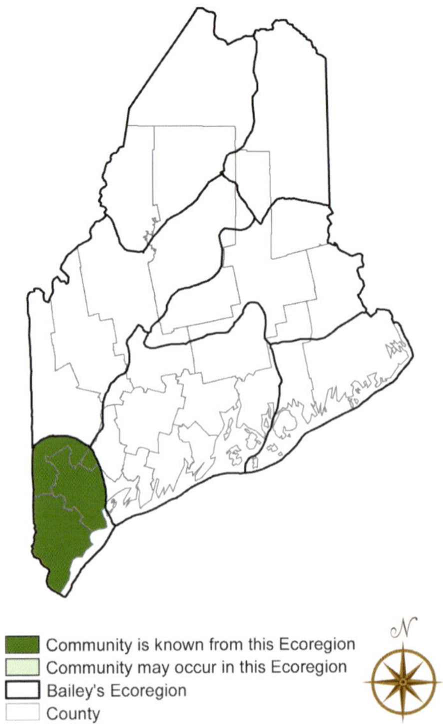

Beach Heather

Conservation, Wildlife, and Management Considerations

All occurrences in Maine are on or adjacent to private land. Heavy recreational use on the Saco River poses a threat to some occurrences, which can be attractive picnicking or camping spots. Although this is a disturbance-adapted community, the impacts of ATVs and heavy foot traffic can be more concentrated and ecologically different from the natural flooding. Because of the dynamic riparian environment required by this community, conservation entails protection of the entire floodplain on particular segments of the river where these communities occur.

Distribution

Known only from the Saco River in southern Maine and adjacent New Hampshire.

Landscape Pattern: Small Patch

Characteristic Plants

These plants are frequently found in this community type. Those with an asterisk are often diagnostic of this community.

Sapling/shrub

Pitch pine
Red Oak
White pine

Dwarf Shrub

Beach heather*

Herb

Big bluestem
Early goldenrod
Flat-topped goldenrod
Gray goldenrod
Jointweed
Little bluestem*
Pinweed
Poverty oatgrass
Rand's goldenrod
Silverling
Silverrod
Wooly panic-grass

Bryoid

Awned hair-cap moss
Reindeer lichen

Associated Rare Plants

Silverling

Examples on Conservation Lands You Can Visit

- None documented on conservation lands.

Salt-hay Saltmarsh

State Rank S3

Community Description

These tidal marshes consist of expanses of saltmeadow cordgrass, smooth cordgrass, and/or black-grass. Shrubs are virtually absent, and the herbaceous cover is usually >85%. Much of the marsh is high marsh, where saltmeadow cordgrass forms meadows, and where black-grass may be dominant at slightly higher elevations. In the low marsh, along creeks or at elevations just below mean high water, smooth cordgrass is abundant. Salt pannes with abundant seashore saltgrass may dot the high marsh; goosetongue may also be locally common. Sea lavender and seaside goldenrod are often found at the upper tidal fringe. The dominant species typically form bands corresponding to tidal inundation zones.

Soil and Site Characteristics

Spartina saltmarshes are typically associated with beach-dune systems (back barrier marshes) or the outer reaches of estuaries (finger marshes). They are extensive along both sides of the tidal river or stream. The extensive high marsh zone is only flooded by above average tides. Salt marsh peat is typically several meters thick. Most are large (>10 acres), but they occasionally occur as smaller pockets along estuaries and coves.

Saltmarsh False-foxglove

Diagnostics

These types are coastal back dune marshes, or near the outer reaches of estuaries, with saltmeadow cordgrass, smooth cordgrass, and black-grass totaling >35% cover, often in bands. The high marsh is well developed.

Similar Types

Mixed Graminoid-Forb Saltmarshes may also have cordgrasses and/or black-grass abundant, but will also have a mix of other co-dominant species, which tend to occur in patches rather than tidal zones; they are typically smaller, often less than 5 acres, and tend to occur farther upstream in estuaries or in smaller, more protected pockets. Brackish Tidal Marshes, which also occur farther upstream in estuaries, lack saltmarsh cordgrasses.

Location Map

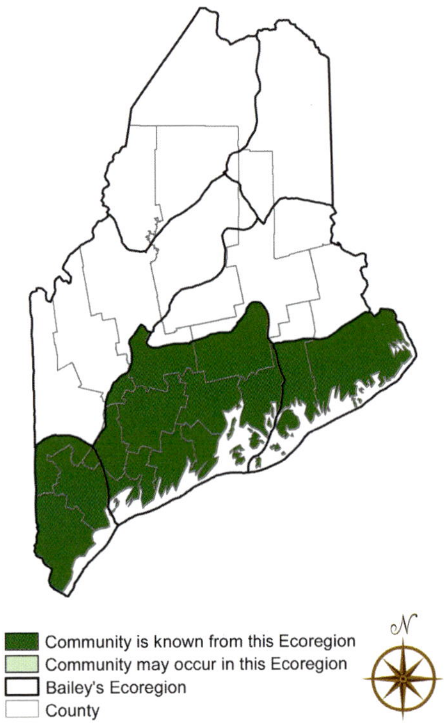

Spartina Saltmarsh – Kinney Shores

Characteristic Plants

These plants are frequently found in this community type. Those with an asterisk are often diagnostic of this community.

Herb

Alkali bulrush
Black-grass*
Common arrow-grass*
Goosetongue*
Purple-stemmed aster
Saltmeadow cordgrass*
Sea milkwort*
Seashore saltgrass*
Seaside goldenrod*
Smooth cordgrass*
Wire rush*

Associated Rare Plants

Dwarf glasswort
Lilaeopsis
Saltmarsh bulrush
Saltmarsh false-foxglove
Slender blue flag

Associated Rare Animals

Big bluet
Black-crowned night-heron
Laughing gull
Least tern
Saltmarsh sharp-tailed sparrow
Short-eared owl

Conservation, Wildlife, and Management Considerations

Few of the larger saltmarshes in Maine are pristine, with some having been filled and nearly all ditched at one time or another. With wetland protection in recent decades many of those that remain are reverting to a more natural hydrologic regime. Many of the remaining high quality Spartina Saltmarshes are on public land or private conservation land. Maintenance of appropriate wetland buffers can help reduce degradation that could result from adjacent land uses.

Saltmarshes are important nesting habitat for Nelson's sharp-tailed sparrow, seaside sparrow, and the rare saltmarsh sharp-tailed sparrow. These wetlands also provide foraging habitat for a large number of wadingbirds and shorebirds, including rare species such as the laughing gull, black-crowned night-heron, and least tern. The big bluet, a rare damselfly, inhabits saltmarsh ponds with emergent vegetation in southern Maine.

Distribution

Coastal Maine, mostly southwest of Merrymeeting Bay (Eastern Broadleaf Forest Province); sporadic and less well developed downeast. Extends southward along the Atlantic coast.

Landscape Pattern: Large Patch

Examples on Conservation Lands You Can Visit

- Bass Harbor Marsh, Acadia National Park – Hancock Co.
- Morse Mountain Preserve – Sagadahoc Co.
- Rachel Carson National Wildlife Refuge – York Co.
- Reid State Park– Sagadahoc Co.
- Scarborough Marsh Wildlife Management Area – Cumberland Co.

Sandy Lake-bottom

Pipewort - Water Lobelia Aquatic Bed

State Rank S5

Community Description

In this shallow water vegetation type almost all of the plant growth is underwater, with only the flowering portions of the plants above water. The most typical species, pipewort and water lobelia, grow as rosettes on the substrate below the water surface. Associated species may be rosette plants or submerged plants growing in the water column (e.g., leafless water-milfoil). Water-shield, a plant with leaves that float on the surface, may be abundant in patches. Water-lilies and pickerelweed may be present, but at low cover. Vegetation cover ranges from sparse to extensive.

Soil and Site Characteristics

This community type can be found in quiet waters of lakes, ponds, streams, and rivers; sites are usually shallow (depths 0.2-1.1 m). The lake bottom substrate almost always has a predominant mineral soil component, rather than muck.

Diagnostics

The most abundant species are those that grow as rosettes on the lake bottom, especially pipewort and water lobelia. Floating leaved plants are absent or very low in cover.

Pipewort

Similar Types

Water-lily - Macrophyte Aquatic Bed vegetation and Pickerelweed - Macrophyte Aquatic Bed vegetation can share many species with this type, but these community types are dominated by floating leaved plants or emergent plants, respectively.

Conservation, Wildlife, and Management Considerations

This aquatic community type is widespread in Maine. It can be found in the quieter portions of streams and rivers as well as in lakeshores and pondshores. It provides habitat for a variety of water-dependent animals. Many examples occur on public lands and private conservation

Location Map

Pipewort – Water Lobelia Aquatic Bed

Characteristic Plants

These plants are frequently found in this community type. Those with an asterisk are often diagnostic of this community.

Herb

Creeping spearwort
Pipewort*
Water lobelia*
Water-shield*

Associated Rare Plants

Acadian quillwort
Prototype quillwort
Shore quillwort
Spotted pondweed
Water stargrass

Associated Rare Animals

Scarlet bluet

lands; however, because the type is so common it is not always documented. The major threats to this community are water quality degradation from excess nutrients, damage from boat wakes, and the spread of invasive aquatic plants, such as Eurasian water-milfoil and variable water-milfoil.

These aquatic beds, especially where they contain floating-leaved plants like water-shield, may provide habitat for damselflies such as the rare scarlet bluet, which seems to prefer acidic sandy-bottomed habitats with water-shield and rushes. Alkaline ponds with aquatic vegetation may support many of the same wildlife species found in more acidic aquatic communities such as Water-lily Macrophyte Aquatic Bed or Pickerelweed - Macrophyte Aquatic Bed. It is unclear whether any wildlife species prefer the more alkaline conditions that this community provides.

Distribution

Statewide; extending southward and westward from Maine, and presumably into Canada.

Landscape Pattern: Small Patch, linear.

Examples on Conservation Lands You Can Visit

- Chemo Pond, Penobscot Experimental Forest – Penobscot Co.
- Kidney Pond, Baxter State Park – Piscataquis Co.
- Pepperpot Pond, Richardson Lake Public Lands – Oxford Co.
- Range Ponds State Park – Androscoggin Co.

Water-shield

Sedge - Heath Fen

Sedge - Leatherleaf Fen Lawn

State Rank S4

Community Description

This open peatland type is dominated by a layer of mixed dwarf heath shrubs and sedges. Small larches, rarely tree sized, are often scattered across the surface but contribute little cover. Leatherleaf, sweetgale, or bog rosemary may be the dominant shrub, and shrub cover is generally 20-40%. Sedges contribute 20-70% cover. Narrow-leaved cotton-grass, few-seeded sedge, and Michaux's sedge are typical dominants. Pitcher plants are usually present. The ground layer is a carpet of peat mosses, often with tracings of large cranberry running across the surface.

Soil and Site Characteristics

This type occurs in open peatlands, often in areas transitional from raised bog (ombrotrophic) to fen (minerotrophic) conditions. Sites are typically acidic (pH 4.0-5.4) but sometimes circumneutral. Peat substrate is saturated to the surface, or nearly so. This type most often occurs at low to moderate elevations.

Diagnostics

Open peatland vegetation consists of sedges and dwarf shrubs (leatherleaf, bog rosemary, sweetgale). Sedge cover exceeds shrub cover. Dominant sedges include few-seeded sedge, coast sedge, Michaux's sedge, white beakrush, and narrow-leaved cotton-grass.

Sheep Laurel

Similar Types

Leatherleaf Boggy Fen is shrubbier, has more leatherleaf than sedge cover, and has tufted cotton-grass or tawny cotton-grass as prominent sedges. Sheep Laurel Dwarf Shrub Bog is drier and shrubbier and features sheep laurel as the dominant shrub. Low Sedge - Buckbean Fen Lawn occurs in similar, although usually somewhat wetter settings. It has a greater dominance of sedges than shrubs and often features mud sedge and podgrass. Mixed Tall Sedge Fen can consist of leatherleaf among sedges, but the sedges will be large and robust species such as slender sedge, beaked sedge, and inflated sedge.

Conservation, Wildlife, and Management Considerations

This community type is well represented

Location Map

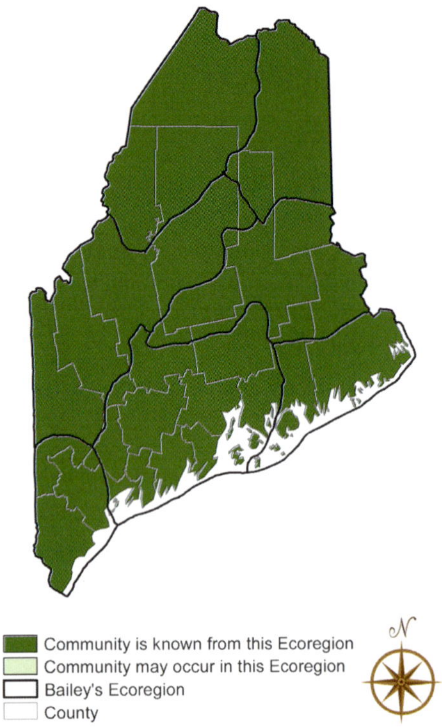

Sedge – Heath Fen

in Maine and is fairly stable in extent, with several examples on public lands and private conservation lands. Impoundment or draining would have negative impacts on bog hydrology and on vegetation. Slow vegetation growth rates, due to the nutrient poor setting, result in slow recovery from physical disturbances, such as recreational trail use. If disturbance, such as trail crossing, is a priority, traversing during frozen conditions or using boardwalks can minimize impacts.

Several rare dragonflies may be found in this community, especially in very wet locations with abundant peat moss (often suspended in the water column). The Quebec emerald is found in northern Maine, and the ringed boghaunter is restricted to the southern part of the state in York and southern Oxford Counties. In northwestern Maine this type may support the bog fritillary butterfly, which uses small cranberry as its larval host plant.

Distribution

Statewide, though more common northward. Throughout northern New England and New York; Canadian distribution unknown.

Landscape Pattern: Small Patch

Characteristic Plants

These plants are frequently found in this community type. Those with an asterisk are often diagnostic of this community.

Sapling/shrub

Larch*
Mountain holly*

Dwarf Shrub

Bog rosemary*
Large cranberry*
Leatherleaf*
Sheep laurel
Small cranberry

Herb

Bog aster*
Bog goldenrod
Coast sedge*
Few-flowered sedge*
Few-seeded sedge*
Narrow-leaved cotton-grass*
Pitcher plant
Round-leaved sundew
Slender sedge
Spatulate-leaved sundew
White beak-rush*
Yellowish sedge

Bryoid

*Sphagnum magellanicum**

Associated Rare Animals

Quebec emerald
Ringed boghaunter

Examples on Conservation Lands You Can Visit

- Acadia National Park – Hancock Co.
- Great Heath Public Lands – Washington Co
- Great Wass Island Preserve – Washington Co
- Number Five Bog Public Lands – Somerset Co.
- Salmon Brook Lake Bog Public Lands – Aroostook Co.
- St. John River Preserve – Aroostook/ Somerset Co.

Sedge Meadow

Tussock Sedge Meadow

State Rank S4

Community Description

These graminoid marshes are dominated by hummocks of tussock sedge interspersed with bluejoint, other graminoids, and a few shrubs. Shrub cover is usually less than 30% but may occasionally be higher; meadowsweet is a characteristic shrub. Other wetland sedges and grasses are scattered in with the tussock sedge and bluejoint, usually in small amounts. Plant species vary from site to site but typically include royal fern, cinnamon fern, sensitive fern, St. Johnswort, flat-topped white aster, or wool-grass. Bryophytes are usually very sparse.

Soil and Site Characteristics

Soils are saturated and usually flooded, sometimes only seasonally. Soils may be entirely organic peat or muck, or a layer of organic matter over mineral soil. Standing water is present through much of the growing season. This type typically occurs in large flat basins that are often associated with drainage streams and may be influenced by beaver activity.

Mulberrywing

Diagnostics

Tussock sedge forms greater than 30% cover (usually >50%); shrub cover usually is less than 30%. Vegetation is strongly hummocked with standing water between hummocks for much of the growing season.

Similar Types

Bluejoint Meadows share many species but have lower abundance of tussock sedge and are usually only temporarily flooded. Sweetgale Mixed Shrub Fens and Alder Shrub Thickets can also share species with this type but are more strongly dominated by shrubs (>25% shrub or dwarf shrub cover, and usually >50%), and with <25% cover of tussock sedge. Mixed Graminoid-Shrub Marshes also have <25% tussock sedge.

Location Map

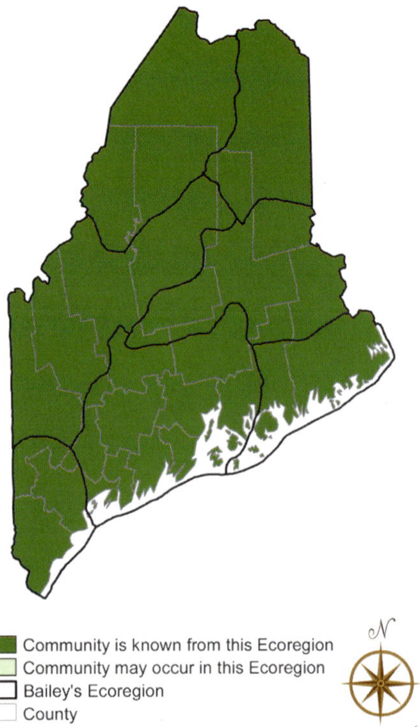

Sedge Meadow

Conservation, Wildlife, and Management Considerations

While graminoid marshes are common throughout the state, the Tussock Sedge Meadow type of graminoid marsh is more restricted in its distribution. Several are known from public lands and private conservation lands, including some very large and intact examples. Maintaining appropriate wetland buffers can help ensure that adjacent land uses do not result in degradation. This wetland type is particularly susceptible to alteration from the non-native purple loosestrife.

The rare Tomah mayfly is found in a few large sedge meadows in central, eastern, and northern Maine. In addition to numerous common bird species, the rare sedge wren nests in graminoid marshes and wet meadows. Northern harriers may also nest and forage in these meadows. Northern leopard frogs forage in large grassy meadows associated with watercourses in mid-summer. Some sites may function as vernal pools, which are important breeding habitat for a variety of amphibians including wood frogs, spotted salamanders, and blue-spotted salamanders. In southern Maine, these wetlands may provide foraging habitat for ribbon snakes, Blanding's turtles, and spotted turtles.

Characteristic Plants

These plants are frequently found in this community type. Those with an asterisk are often diagnostic of this community.

Sapling/shrub
Meadowsweet*
Speckled alder*
Winterberry*

Dwarf Shrub
Leatherleaf

Herb
Bluejoint*
Tussock sedge*

Bryoid
Sphagnum mosses

Associated Rare Plants

Bog bedstraw

Associated Rare Animals

Blanding's turtle
Comet darner
Least bittern
Ribbon snake
Sedge wren
Short-eared owl
Spotted turtle
Tomah mayfly
Yellow rail

Distribution

Statewide, extending throughout the northeastern U.S.; Canadian distribution unknown.

Landscape Pattern: Large Patch

Examples on Conservation Lands You Can Visit

- Cold Stream, Passadumkeag – Penobscot Co.
- Great Heath Public Lands – Washington Co.
- Middle Pond State Park – Oxford Co.
- Roberts Pond Inlet, Massabesic Experimental Forest – York Co.

Sweetgale Fen

Sweetgale Mixed Shrub Fen

State Rank S4

Community Description

A mixture of shrubs, typically about 1 m high (generally taller than bog shrubs, but shorter than most alder thickets), is dominated by sweetgale, leatherleaf, and hardhack or meadowsweet. Alder is usually present but not dominant. Graminoids, typically slender sedge, tussock sedge, and/or bluejoint grass, are usually mixed with the shrubs but are less abundant (averaging around 20% cover). Where shrubs are dense, herb cover is very limited. The bryoid layer is usually very minor; when present it is dominated by peat mosses.

Soil and Site Characteristics

These basin wetlands occur either as part of larger peatlands bordering open water or in impounded areas with peat or muck soils (e.g. beaver flowages). Slow moving open water usually borders this vegetation. The substrate is seasonally to semi-permanently flooded organic material.

Diagnostics

This type has a dominance of medium-height shrubs of sweetgale, meadowsweet, and leatherleaf. Graminoids are present but subordinate to shrubs. Sites occur on saturated or flooded organic soils.

Sweetgale

Similar Types

Sweetgale Mixed Shrub Fens are usually embedded within a mosaic of other peatland types. Mixed Tall Sedge Fens occur in similar settings but have graminoids far more dominant than shrubs. Mountain Holly - Alder Woodland Fens have more alder or mountain holly and usually occur at the peatland/upland interface. Mixed Graminoid - Shrub Marshes have graminoids equaling or exceeding shrub cover and occur on mineral soils or with only a thin organic layer over saturated mineral soil. Alder Shrub Thickets also usually occur on mineral soils rather than peat or muck and have a stronger dominance of alder.

Location Map

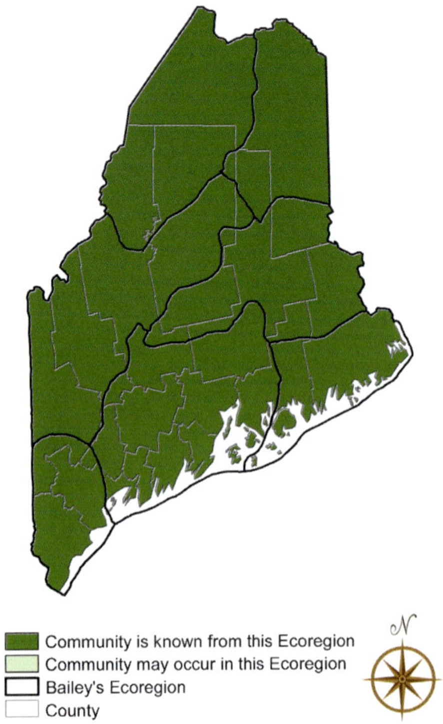

Sweetgale Mixed Shrub Fen

Characteristic Plants

These plants are frequently found in this community type. Those with an asterisk are often diagnostic of this community.

Sapling/shrub

Alder*
Black spruce*
Larch*
Leatherleaf*
Meadowsweet*
Mountain holly*
Red maple*
Sweetgale*
Winterberry holly*

Dwarf Shrub

Leatherleaf*
Rhodora*
Sheep laurel*
Sweetgale*

Herb

Bluejoint
Bog aster*
Few-seeded sedge*
Royal fern*
Slender sedge
Tussock sedge*
White beak-rush*

Bryoid

Sphagnum mosses*

Associated Rare Plants

Long's bulrush

Associated Rare Animals

Rusty blackbird

Conservation, Wildlife, and Management Considerations

This wetland type is well distributed throughout the state and receives little direct use. Maintaining appropriate wetland buffers and water quality are appropriate conservation measures. Public lands and private conservation lands contain many examples of this community.

These shrublands, especially in close proximity to open water, may provide habitat for bird species such as common yellowthroat, alder flycatcher, Wilson's warbler, Lincoln's sparrow, and the rare rusty blackbird. Thaxter's pinion moth uses sweetgale as one of its larval host plants and may be found in this community. The black meadowhawk, a dragonfly of open fens and marshes, may occur here as well. Sites of this community type in northern Maine may be inhabited by the subarctic bluet.

Distribution

Statewide; extends westward and probably eastward and northward as well.

Landscape Pattern: Small Patch

Examples on Conservation Lands You Can Visit

- Mattagodus Wildlife Management Area – Penobscot Co.
- Middle Pond State Park – Oxford Co.
- Moose River – Somerset Co.
- Nahmakanta Public Lands – Piscataquis Co.
- Wiggins Brook, Squaw Mountain Public Lands – Piscataquis Co.

Subalpine Hanging Bog

Heath - Lichen Subalpine Slope Bog

State Rank S1

Community Description

A dwarf shrub bog tilted on its side, this community consists of typical heath bog shrubs growing on a dense peat moss carpet over rocky slopes. Typical species include Labrador tea, sheep laurel, and rhodora. Stunted trees, such as northern white cedar, heart-leaved paper birch, or black spruce, may be scattered among the shrubs, but form <25% cover overall. Herbs are sparse, but may include boreal species such as northern comandra.

Soil and Site Characteristics

Known sites are documented from nearly vertical talus slides in at elevations >2000’, but this type may occur in other subalpine settings.

Diagnostics

Heath shrubs (sheep laurel, Labrador tea, leatherleaf) are dominant on a peat moss substrate, but this type is not a peatland. Vegetation occurs as a thin layer over bedrock, talus, or scree in a subalpine setting.

Larch

Similar Types

Sheep Laurel Dwarf Shrub Bog vegetation is compositionally similar, but occurs in a peatland setting. Heath - Crowberry Maritime Slope Bogs are floristically almost identical but occur in a different setting. Cotton-grass - Heath Alpine Bogs occur in basins rather than on slopes and typically have higher herbaceous cover. Labrador Tea Talus Dwarf-shrublands have less vegetation cover (particularly of peat moss) and occur at lower elevations; but the two types may grade into one another.

Conservation, Wildlife, and Management Considerations

The few documented sites are away from

Location Map

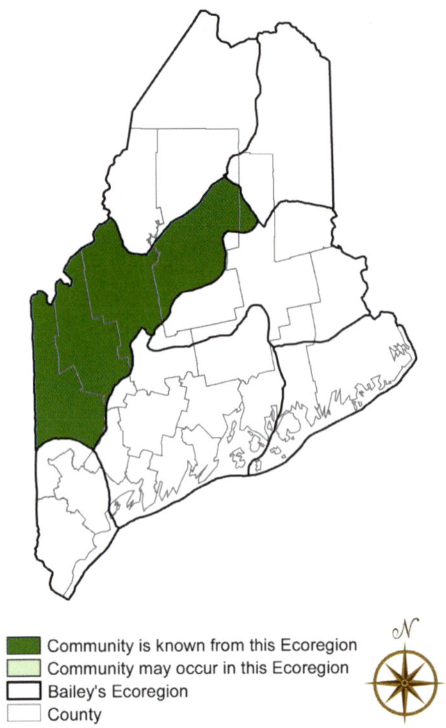

Subalpine Hanging Bog

trails. Access is difficult to dangerous, and foot traffic could damage the vegetation, which grows densely enough that one cannot avoid it by picking one's way over rocks, as can be done with some other alpine/subalpine vegetation types. Other subalpine settings where this might occur would have similar access constraints. Because alpine vegetation grows slowly under harsh conditions, recovery from damage can take a long time.

Distribution

Montane western Maine, extending west into New Hampsire and possibly Vermont and upstate New York (New England - Adirondack Province); likely extends northeasterly to the Gaspé Peninsula.

Landscape Pattern: Small Patch

Characteristic Plants

These plants are frequently found in this community type. Those with an asterisk are often diagnostic of this community.

Sapling/shrub

Black spruce
Heart-leaved paper birch
Larch
Mountain alder

Dwarf Shrub

Alpine bilberry
Labrador tea*
Leatherleaf
Rhodora
Sheep laurel*
Squashberry

Herb

Bunchberry
Goldthread
Northern comandra

Bryoid

Reindeer lichen*
Sphagnum mosses*

Associated Rare Plants

Northern comandra

Examples on Conservation Lands You Can Visit

- Crocker Mountain, Appalachian Trail – Franklin Co.

Tall Grass Meadow

State Rank S3

Community Description

These dense swards of tall grassy vegetation are dominated by bluejoint, with smaller amounts of shrubs (alder, meadowsweet, willow) mixed in. Depending on the disturbance history, the shrubs may be low and not easily visible among the grasses, or taller, in which case the vegetation appears as mixed shrub-graminoid. Other graminoids, such as tussock sedge and other sedges are occasional. Flat-topped white aster and spotted joe-pye weed are common tall forbs. Freshwater cordgrass is often present in small amounts. Bryophytes are very minor (0-10% cover), and lichens are absent. Two principal variants occur; those in alluvial soils of larger rivers and those of more peaty soils along small streams.

Soil and Site Characteristics

Sites occupy mineral soils in temporarily flooded rivershores or streamsides that are flat to slightly sloped. At some sites this type extends onto poorly drained uplands as grassy barrens. Soils are sandy to silty along rivers, with higher clay or organic content in other settings.

Bluejoint

The acidic to neutral soils (pH 5.0-7.0) are saturated or moist not far from the surface. Most sites are disturbance maintained, either by ice scour flooding (larger rivers), by beaver flooding, or by fire (some Downeast grasslands).

Diagnostics

Bluejoint grass exceeds 50% cover and shrubs are less than 25% cover. Sites are flooded for only part of the season. Whether the two variants should be separated as distinct types needs additional study.

Similar Types

Tussock Sedge Meadows share many species, but have less bluejoint grass (usually <30% cover) and more tussock

Location Map

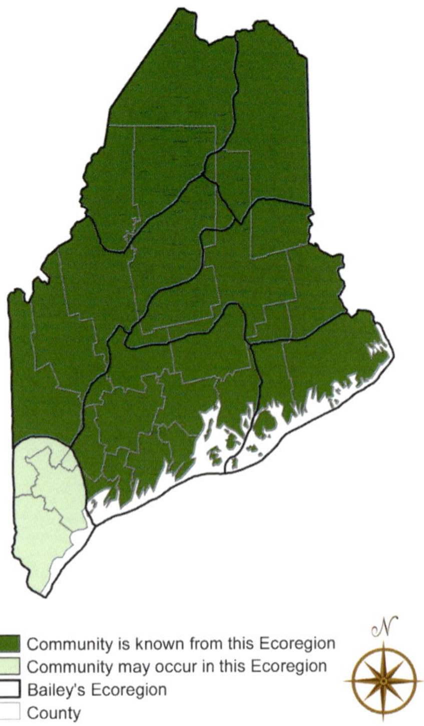

Tall Grass Meadow

sedge (>30%). Mixed Graminoid-Shrub Marshes also can be floristically similar, but are not strongly dominated by bluejoint grass (usually <20%, rarely up to 50%). Other open non-peat wetland types are shrub dominated, not herb dominated.

Conservation, Wildlife, and Management Considerations

In the absence of disturbance (flooding or fire, the latter often human-initiated), this community develops into dense shrublands dominated by alder. Maintaining both the natural disturbance regime and the hydrologic integrity of these systems is key to their conservation. Several high quality examples occur on public lands and private conservation lands.

Northern leopard frogs inhabit large grassy meadows associated with rivers in mid-summer where they forage. Northern harriers, Lincoln's sparrows, and rare short-eared owls may also nest and forage in these meadows. Northern Maine sites may provide grassy nesting sites for very rare breeding waterfowl such as northern pintail, northern shoveler, and gadwall.

Distribution

Larger rivers of far northern Maine, coastal grasslands of Downeast Maine (New England - Adirondack and Laurentian Mixed Forest Provinces), and meandering streams statewide. Distribution outside of Maine is not clear.

Landscape Pattern: Small patch; some Downeast grasslands approaching Large Patch

Examples on Conservation Lands You Can Visit

- Allagash Public Lands – Aroostook Co.
- Cutler Public Lands – Washington Co.
- Narraguagus Wildlife Management Area – Washington Co.
- Petit Manan Point, Petit Manan National Wildlife – Washington Co.
- St. John River Preserve – Aroostook Co.

Characteristic Plants

These plants are frequently found in this community type. Those with an asterisk are often diagnostic of this community.

Sapling/shrub
Meadowsweet
Speckled alder

Herb
Bluejoint*
Flat-topped white aster
Tall meadow-rue
Tussock sedge*

Bryoid
Sphagnum mosses

Associated Rare Plants

Blue-leaf willow

Associated Rare Animals

Short-eared owl
Yellow rail

Tall Sedge Fen

State Rank S4

Community Description

Expanses of tall sedges and grasses grow on saturated peat, with the silvery hue of slender sedge, typically the dominant species, often visible from a distance. Beaked sedge and lake bank sedge are also characteristic; the ubiquitous bluejoint is often present in small amounts. Tussock sedge may be present in small amounts but does not form dense tussocks. The herb layer is often continuous, and whatever shrubs occur are usually below or mixed in with the graminoid cover, except for an occasional alder or meadowsweet protruding above. Dwarf shrubs are usually <50% cover and always less abundant than the herbs. The bryoid layer varies from sparse to almost continuous, inverse to the amount of standing water.

Soil and Site Characteristics

In wetland basins, this fen type occurs as part of a peatland or on peaty deposits adjacent to open water in a minerotrophic setting. The substrate is always saturated and is often flooded at high water. Sites are acidic to circumneutral (pH 4.8-6.8).

Diagnostics

Tall peatland sedges, including slender sedge and beaked sedge, are dominant. Grasses such as bluejoint are present. Sites are often adjacent to open water. Sweetgale, leatherleaf, and meadowsweet, if present, are subdominant to the sedges.

Tall Sedge Fen

Similar Types

Sweetgale Mixed Shrub Fens share many species and can occur in similar settings but are strongly shrub dominated rather than herb dominated. Other graminoid dominated fen types feature different, less robust sedges, and usually have a more developed bryoid layer. Tussock Sedge Meadows can occur in similar settings but are strongly dominated by tussock sedge. Mixed Graminoid - Shrub Marshes can share some species, but occur on mineral substrates or on a thin organic layer over mineral soil, rather than on peat.

Conservation, Wildlife, and Management Considerations

This community type is well represented in Maine and is fairly stable in extent,

Location Map

Tall Sedge Fen

with several examples on public lands and private conservation lands. Impoundment or draining would have negative impacts on hydrology and on vegetation. Slow vegetation growth rates, due to the nutrient poor environment, result in slow recovery from physical disturbances. Degradation from recreational use is unlikely, because of the unstable substrate; but if disturbance, such as foot traffic, is a necessity, traversing during frozen conditions or using boardwalks can minimize impacts.

In southern Maine, these wetlands may provide habitat for several rare reptiles including ribbon snakes, Blanding's turtles, and spotted turtles. The rare ringed boghaunter dragonfly, restricted to the southern part of the state in York and southern Oxford Counties, is found in this community, especially in very wet locations with peat moss often suspended in the water column. The black meadowhawk, an uncommon dragonfly of open fens and marshes, may also be found in this community. Sites in northern Maine may be inhabited by the subarctic bluet, an uncommon damselfly that inhabits open marshes and fens and reaches the southern edge of its range in northern Maine.

Distribution

Statewide. Probably extends throughout northern New England, New York, and adjacent Canada, but not well documented.

Landscape Pattern: Small to Large Patch

Characteristic Plants

These plants are frequently found in this community type. Those with an asterisk are often diagnostic of this community.

Sapling/shrub

Meadowsweet

Dwarf Shrub

Large cranberry*
Leatherleaf
Sweetgale

Herb

Beaked sedge*
Bluejoint
Lake sedge*
Inflated sedge
Silvery sedge
Slender sedge*
Three-way sedge
Tussock sedge
Yellow loosestrife

Bryoid

Sphagnum mosses

Associated Rare Plants

Long's bulrush

Associated Rare Animals

Blanding's turtle
Ribbon snake
Ringed boghaunter
Spotted turtle

Examples on Conservation Lands You Can Visit

- Fourth Machias Lake, Duck Lake Public Lands – Hancock Co.
- Jones Pond, Bigelow Preserve – Franklin Co.
- Killick Pond Wildlife Management Area – York Co.
- Saco Heath Preserve – York Co.

Tall Shrub Fen

State Rank S4

Community Description

This peatland vegetation is characterized by tall shrubs, with spotty tree cover above. Shrub cover is usually >70%, with alder and mountain holly almost always present and with other shrubs locally common. Red maple is typical in the tree layer, along with black spruce. The herb layer is patchy, with 10-50% cover. Labrador tea and pitcher plants are frequent, though not necessarily in high abundance. Cinnamon fern and wild calla are good indicators. The bryoid layer is mostly peat mosses and is patchy, averaging ~50% cover.

Soil and Site Characteristics

Sites occur in peat filled basin wetlands, often at the upland/peatland transition. Areas bordering the upland (the peatland "lagg") have standing water for much of the growing season; however, this community can be found in settings other than the peatland lagg. Substrates are somewhat acidic to circumneutral, with relatively high levels of nitrogen, presumably from the nitrogen fixing alders.

Diagnostics

Vegetation associated with a peatland, with well-developed shrub layer (1-3 m) is dominated by non-heath shrubs and often with scattered trees. Some heaths and other peatland species are present. The substrate is peat or is transitional from peat to mineral soil.

Cinnamon Fern

Similar Types

Alder Shrub Thickets can be similar, but occur on mineral/muck soil and are not associated with peatlands. Sweetgale Mixed Shrub Fens can have similar structure and can be associated with peatlands, but they are more strongly dominated by sweetgale and heath shrubs rather than alder or other shrubs. Spruce - Larch Wooded Bogs and Red Maple Woodland Fens have greater tree cover, less tall shrub cover, and typically occur on deeper peat.

Conservation, Wildlife, and Management Considerations

This community is common around the perimeter of peatlands and in some non-peatland drainages as well. It often receives runoff from adjacent uplands, so

Location Map

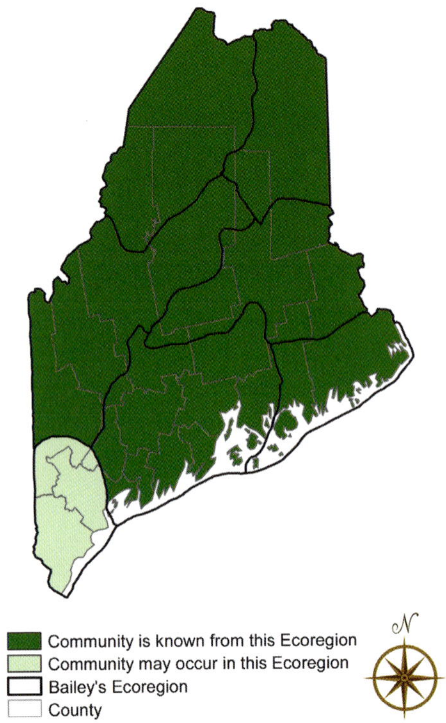

Mountain Holly Flowers

maintenance of appropriate wetland buffers can help retain the quality of individual occurrences.

Wetland species that use this community type include common yellowthroat, Wilson's warbler, Lincoln's sparrow, spotted salamander, wood frog, and northern waterthrush. The black meadowhawk, an uncommon dragonfly of open fens and marshes, may be found in this community. Some occurrences of this community type host vernal pools, which are important breeding habitat for a variety of amphibians including wood frogs, spotted salamanders, and blue-spotted salamanders. The rusty blackbird may occur in this type in northern Maine.

Distribution

Central Maine northward, extending west through New England and New York, and presumably north and east into Canada.

Landscape Pattern: Small Patch

Characteristic Plants

These plants are frequently found in this community type. Those with an asterisk are often diagnostic of this community.

Canopy
Black spruce
Red maple
Sapling/shrub
Balsam fir
Highbush blueberry
Mountain holly*
Red maple
Speckled alder*
Wild-raisin

Dwarf Shrub
Labrador tea
Leatherleaf
Small cranberry

Herb
Cinnamon fern*
Marsh St. Johnswort
Pitcher plant
Three-leaved false Solomon's seal
Three-seeded sedge

Bryoid
Sphagnum mosses

Associated Rare Plants

Bog bedstraw

Associated Rare Animals

Rusty blackbird

Examples on Conservation Lands You Can Visit

- Acadia National Park – Hancock Co.
- Appleton Bog Preserve – Knox Co.
- Great Heath Public Lands – Washington Co.
- Number Five Bog Public Lands – Somerset Co.
- Sunkhaze Meadows National Wildlife Refuge – Penobscot Co.

Appendix A: State and Global Rarity Ranks

Native plants, animals, natural communities, and ecosystems constitute the elements of natural diversity. State and global rarity ranks indicate the biological status, or relative rarity of plants, animals, natural communities, and ecosystems. Rarity ranks are based on the known number of occurrences for each element in Maine, modified by additional considerations such as the number of occurrences that are protected and the ecological fragility of the element -- that is, how sensitive it is to human-caused disturbance. Each state rank begins with an "S" to indicate state (as opposed to national or global) rank. Other states and provinces throughout North America apply these state rarity ranks similarly.

Global rarity ranks follow the same criteria as state ranks but are applied over the entire range of the element. They are preceded by "G" rather than by "S". The majority of Maine's species and communities are not globally rare. Those that are globally rare should be considered especially high priorities for conservation efforts.

The state or global rarity rank is an objective reflection of an element's known distribution and abundance (in Maine or worldwide, respectively). Because continuing inventories bring new knowledge, element ranks sometimes change as our understanding of particular elements becomes more complete.

State Rarity Ranks for Maine Natural Communities

- **S1** Critically imperiled in Maine because of extreme rarity (five or fewer occurrences or very few remaining acres) or because some aspect of its biology makes it especially vulnerable to extirpation from the State.
- **S2** Imperiled in Maine because of rarity (6-20 occurrences or few remaining acres) or because of other factors making it vulnerable to further decline.
- **S3** Rare in Maine (on the order of 20-100 occurrences), though not, to our knowledge, imminently imperiled.
- **S4** Apparently secure in Maine.
- **S5** Demonstrably secure in Maine.
- **SU** Possibly in peril in Maine, but status uncertain; need more information.

Global Rarity Ranks for Maine Natural Communities

Global ranks have been only provisionally assigned for many natural community types, with attention focused on those types considered by NatureServe to be globally rare (Grossman et al. 1994 and the Association for Biodiversity Information 2001). Please check the MNAP web site for current global ranks of Maine's natural community types.

Appendix B:
Cross-walk of Maine Natural Communities, Maine Ecosystems, Maine Widlife Action Plan Habitat Groups, NVC (International Vegetation) Associations, and NVC (International) Ecological Systems

Cross-walk of Maine Natural Communities, Maine Ecosystems, Maine Widlife Action Plan Habitat Groups, NVC (International Vegetation) Associations, and NVC (International) Ecological Systems

Maine Natural Communities: 104 types	Maine Ecosystems: 24 types	Maine Wildlife Action Plan Habitat Types: 21 Types	NVC (International) Vegetation Associations: 182 types	NVC (International) Ecological Systems: 50 types
Acidic cliff - gorge	Rock Outcrop	Cliff Face and Rocky Outcrops (including talus)	Polypodium (virginianum, appalachianum) Cliff Sparse Vegetation	Laurentian-Acadian Acidic Cliff and Talus
Alder floodplain	Streamshore	Rivers and Streams	Alnus incana - Cornus (amomum, sericea) / Clematis virginiana Shrubland	Laurentian-Acadian Wet Meadow-Shrub Swamp; Eastern Boreal Floodplain
Alder shrub thicket	Streamshore	Shrub-scrub Wetland	Alnus incana Swamp Shrublandl Alnus incana ssp. rugosa - Nemopanthus mucronatus / Sphagnum spp. Shrubland	Laurentian-Acadian Wet Meadow-Shrub Swamp
Alpine cliff	Alpine	Alpine	Trichophorum caespitosum - Saxifraga (foliolosa, paniculata, rivularis) Herbaceous Vegetation	Acadian-Appalachian Alpine Tundra
Aspen - birch woodland/forest complex	Spruce – Fir – Northern Hardwood Forest	Shrub / Early Successional and Regenerating Forest	Populus (tremuloides, grandidentata) - Betula (populifolia, papyrifera) Woodland	Acadian Low-Elevation Spruce-Fir-Hardwood Forest; Laurentian-Acadian Pine-Hemlock-Hardwood Forest; Laurentian-Acadian Northern Hardwoods Forest
Atlantic white cedar bog	Unpatterned Fen (most likely) or several other peatland types	Peatlands	Chamaecyparis thyoides / Chamaedaphne calyculata Woodland	North-Central Interior and Appalachian Acidic Peatland
Atlantic white cedar swamp	Appalachian – Acadian Basin Swamp	Forested Wetland	Chamaecyparis thyoides - Picea rubens / Gaylussacia baccata / Gaultheria hispidula Forest	Northern Atlantic Coastal Plain Basin Peat Swamp
Balsam poplar floodplain forest	Streamshore	Rivers and Streams	Populus balsamifera - Fraxinus nigra / Matteuccia struthiopteris Forest	Eastern Boreal Floodplain

Cross-walk of Maine Natural Communities, Maine Ecosystems, Maine Widlife Action Plan Habitat Groups, NVC (International Vegetation) Associations, and NVC (International) Ecological Systems

Maine Natural Communities: 104 types	Maine Ecosystems: 24 types	Maine Wildlife Action Plan Habitat Types: 21 Types	NVC (International) Vegetation Associations: 182 types	NVC (International) Ecological Systems: 50 types
Beach strand	Coastal Dune – Marsh	Unconsolidated Shore	Cakile edentula ssp. edentala - Chamaesyce polygonifolia Sparse Vegetation; Cakile edentula ssp. edentula - Mertensia maritima Sparse Vegetation	Northern Atlantic Coastal Plain Sandy Beach
Beech - birch - maple forest	Spruce – Fir – Northern Hardwood Forest	Deciduous and Mixed Forest	Betula papyrifera / Acer saccharum - Mixed Hardwoods Forest; Acer saccharum - Pinus strobus / Acer pensylvanicum Forest; Acer saccharum - Betula alleghaniensis - Fagus grandifolia / Viburnum lantanoides Forest	Appalachian (Hemlock)-Northern Hardwood Forest; Laurentian-Acadian Northern Hardwoods Forest
Bilberry - mountain-heath alpine snowbank	Alpine	Alpine	Vaccinium uliginosum - Harrimanella hypnoides - Loiseleuria procumbens Dwarf-shrubland; Tricophorum caespitosum - Carex scirpoidea - Carex bigelowii Herbaceous Vegetation	Acadian-Appalachian Alpine Tundra
Birch - oak talus woodland	Central Hardwoods Oak Forest	Deciduous and Mixed Forest	Betula alleghaniensis - Quercus rubra / Polypodium virginianum Woodland; Polypodium (virginianum, appalachianum) / Lichens Nonvascular Vegetation	Laurentian-Acadian Acidic Cliff and Talus
Black ash swamp	Appalachian – Acadian Basin Swamp	Forested Wetland	Populus balsamifera - Fraxinus nigra / Matteuccia struthiopteris Forest	Laurentian-Acadian Alkaline Conifer-Hardwood Swamp
Black spruce woodland	Spruce – Fir – Northern Hardwood Forest	Coniferous Forest	Picea mariana / Kalmia angustifolia Woodland	Acadian Near-Boreal Spruce Barrens; Northern Appalachian-Acadian Rocky Heath Outcrop

Cross-walk of Maine Natural Communities, Maine Ecosystems, Maine Widlife Action Plan Habitat Groups, NVC (International Vegetation) Associations, and NVC (International) Ecological Systems

Maine Natural Communities: 104 types	Maine Ecosystems: 24 types	Maine Wildlife Action Plan Habitat Types: 21 Types	NVC (International) Vegetation Associations: 182 types	NVC (International) Ecological Systems: 50 types
Bluebell - balsam ragwort shoreline outcrop	Streamshore	Rivers and Streams	Andropogon gerardii - Campanula rotundifolia - Solidago simplex Herbaceous Vegetation; Campanula rotundifolia - Packera paupercula - (Aquilegia canadensis) Sparse Vegetation	Boreal Ice-Scour Rivershore; Laurentian-Acadian Floodplain Forest
Blueberry - lichen barren	Spruce – Fir – Northern Hardwood Forest	Coniferous Forest	Vaccinium (angustifolium, myrtilloides, pallidum) - Cladina rangiferina Dwarf-shrubland	Acadian Near-Boreal Spruce Barrens
Bluejoint meadow	Appalachian-Acadian Rivershore; Streamshore	Emergent Marsh and Wet Meadows	Calamagrostis canadensis - Phalaris arundinacea Herbaceous Vegetation; Calamagrostis canadensis - Doellingeria umbellata - Spartina pectinata Herbaceous Vegetation	Laurentian-Acadian Wet Meadow-Shrub Swamp; Boreal Ice-Scour Rivershore
Bog moss lawn	Unpatterned Fen (most likely) or several other peatland ecosystem types	Peatlands	Sphagnum (rubellum, cuspidatum, torreyanum) - Vaccinium macrocarpon Nonvascular Vegetation	Boreal-Laurentian Bog; Acadian Maritime Bog
Boreal circumneutral open outcrop	Rock Outcrop	Dry Woodlands and Barrens	Carex scirpoidea Alkaline Cliff Sparse Vegetation	Laurentian-Acadian Calcareous Rocky Outcrop
Brackish tidal marsh	Tidal Marsh Estuary	Estuaries and Bays	Typha angustifolia - Hibiscus moscheutos Herbaceous Vegetation; Sagittaria subulata - Limosella australis Tidal Herbaceous Vegetation; Schoenoplectus pungens - Eleocharis parvula Herbaceous Vegetation	Acadian Estuary Marsh

Cross-walk of Maine Natural Communities, Maine Ecosystems, Maine Widlife Action Plan Habitat Groups, NVC (International Vegetation) Associations, and NVC (International) Ecological Systems

Maine Natural Communities: 104 types	Maine Ecosystems: 24 types	Maine Wildlife Action Plan Habitat Types: 21 Types	NVC (International) Vegetation Associations: 182 types	NVC (International) Ecological Systems: 50 types
Bulrush bed	Lakeshore; Coastal Plain Pond; Tidal Marsh Estuary	Emergent Marsh and Wet Meadows	Schoenoplectus (tabernaemontani, acutus) Eastern Herbaceous Vegetation	Laurentian-Acadian Freshwater Marsh
Cattail marsh	Lakeshore; Coastal Plain Pond; Tidal Marsh Estuary	Emergent Marsh and Wet Meadows	Typa (angustifolia, latifolia) - (Schoenoplectus spp.) Eastern Herbaceous Vegetation	Laurentian-Acadian Freshwater Marsh
Cedar - spruce seepage forest	Appalachian – Acadian Basin Swamp	Forested Wetland	Thuja occidentalis - (Picea rubens) / Tiarella cordifolia Forest	Acadian-Appalachian Conifer Seepage Forest
Chestnut oak woodland	Central Hardwoods Oak Forest	Dry Woodlands and Barrens	Quercus prinus - Quercus (rubra, velutina) / Vaccinium angustifolium Forest	Central Appalachian Pine-Oak Rocky Woodland
Circumneutral riverside seep	Streamshore	Rivers and Streams	Triantha glutinosa - Carex garberi Herbaceous Vegetation	Boreal Ice-Scour Rivershore
Circumneutral-alkaline water macrophyte suite	Lakeshore; Appalachian - Acadian Rivershore	Peatlands	Elodea canadensis - Potamogeton ssp. Eastern Herbaceous Vegetation [Placeholder]	Laurentian-Acadian Freshwater Marsh
Cotton-grass - heath alpine bog	Alpine	Alpine	Empetrum nigrum - Vaccinium uliginosum - Vaccinium oxycoccos / Rubus chamaemorus Dwarf-shrubland	Acadian-Appalachian Alpine Tundra
Crowberry - bayberry headland	Coastal Headlands	Rocky Coastline and Islands	Morella pensylvanica - Empetrum nigrum Dwarf-shrubland	Acadian-North Atlantic Rocky Coast
Crowberry - bilberry summit bald	Alpine	Alpine	Vaccinium uliginosum / Sibbaldiopsis tridentata Sparse Vegetation	Acadian-Appalachian Subalpine Woodland and Heath-Krummholz
Deer-hair sedge bog lawn	Coastal Plateau Bog	Peatlands	Tricophorum caespitosum - Gaylussacia dumosa / Sphagnum (fuscum, rubellum, magellanicum) Herbaceous Vegetation	Acadian Maritime Bog

Cross-walk of Maine Natural Communities, Maine Ecosystems, Maine Widlife Action Plan Habitat Groups, NVC (International Vegetation) Associations, and NVC (International) Ecological Systems

Maine Natural Communities: 104 types	Maine Ecosystems: 24 types	Maine Wildlife Action Plan Habitat Types: 21 Types	NVC (International) Vegetation Associations: 182 types	NVC (International) Ecological Systems: 50 types
Diapensia alpine ridge	Alpine	Alpine	Diapensia lapponica Dwarf-shrubland	Acadian-Appalachian Alpine Tundra
Dogwood - willow shoreline thicket	Streamshore	Rivers and Streams	Alnus incana - Cornus (amomum, sericea) / Clematis virginiana Shrubland	Boreal Ice-Scour Rivershore
Dune grassland	Coastal Dune – Marsh	Unconsolidated Shore (beaches and mudflats)	Ammophila breviligulata - Lathyrus japonicus Herbaceous Vegetation	Northern Atlantic Coastal Plain Dune and Swale
Dwarf heath - graminoid alpine ridge	Alpine	Alpine	Vaccinium uliginosum - Diapensia lapponica Dwarf-shrub Herbaceous Vegetation; Carex bigelowii Herbaceous Vegetation; Vaccinium uliginosum - Rhododendron lapponicum / Juncus trifidus Dwarf-shrubland; Arctoparmelia centrifuga - Rhizocarpon geographicum Nonvascular Vegetation	Acadian-Appalachian Alpine Tundra
Fir - heart-leaved birch subalpine forest	Alpine	Mountaintop Forest (including krummholz)	Abies balsamea - (Betula papyrifera var. cordifolia) Forest	Acadian-Appalachian Montane Spruce-Fir Forest
Freshwater tidal marsh	Tidal Marsh Estuary	Emergent Marsh and Wet Meadows	Nuphar lutea ssp. advena Tidal Herbaceous Vegetation; Amaranthus cannabinus Tidal Herbaceous Vegetation; Impatiens capensis - Peltandra virginica - Polygonum arifolium - Schoenoplectus fluviatilis - Typha angustifolia Tidal Herbaceous Vegetation; Eriocaulon parkeri - Polygonum punctatum Herbaceous Vegetation	Northern Atlantic Coastal Plain Fresh and Oligohaline Tidal Marsh

Cross-walk of Maine Natural Communities, Maine Ecosystems, Maine Widlife Action Plan Habitat Groups, NVC (International Vegetation) Associations, and NVC (International) Ecological Systems

Maine Natural Communities: 104 types	Maine Ecosystems: 24 types	Maine Wildlife Action Plan Habitat Types: 21 Types	NVC (International) Vegetation Associations: 182 types	NVC (International) Ecological Systems: 50 types
Hardwood river terrace forest	Streamshore	Rivers and Streams	Acer saccharum - Fraxinus spp. - Tilia americana / Matteuccia struthiopteris - Ageratina altissima Forest; Acer saccharum / Ostrya virginiana / Brachyelytrum erectum Floodplain Forest	Laurentian-Acadian Floodplain Forest
Hardwood seepage forest	Central Hardwoods Oak Forest; Appalachian – Acadian Basin Swamp	Deciduous and Mixed Forest; Forested Wetland	Betula alleghaniensis - Acer rubrum - (Tsuga canadensis, Abies balsamea) / Osmunda cinnamomea Forest	Appalachian (Hemlock)-Northern Hardwood Forest; Laurentian-Acadian Northern Hardwoods Forest
Heath - crowberry maritime slope bog	Coastal Plateau Bog	Peatlands	Empetrum nigrum - Gaylussacia dumosa - Rubus chamaemorus / Sphagnum spp. Dwarf-shrubland	Acadian Maritime Bog
Heath - lichen subalpine slope bog	Alpine	Alpine	Kalmia angustifolia - Chamaedaphne calyculata / Rubus chamaemorus / Cladina spp. Dwarf-shrubland	Acadian-Appalachian Subalpine Woodland and Heath-Krummholz
Hemlock - hardwood pocket swamp	Coastal Plain Basin Swamp	Forested Wetland	Acer rubrum - Nyssa sylvatica - Betula alleghaniensis / Sphagnum spp. Forest; Tsuga canadensis - Betula alleghaniensis / Ilex verticillata / Sphagnum spp. Forest	Northern Appalachian-Acadian Conifer-Hardwood Acidic Swamp
Hemlock forest	White Pine – Mixed Hardwood	Conifer Forest	Tsuga canadensis - Betula alleghaniensis - Acer saccharum / Dryopteris intermedia Forest; Tsuga canadensis - (Betula alleghaniensis) - Picea rubens / Cornus canadensis Forest	Appalachian (Hemlock)-Northern Hardwood Forest; Laurentian-Acadian Pine-Hemlock-Hardwood Forest
Huckleberry - crowberry bog	Coastal Plateau Bog	Peatlands	Empetrum nigrum - Gaylussacia dumosa - Rubus chamaemorus / Sphagnum spp. Dwarf-shrubland	Acadian Maritime Bog; Huckleberry - crowberry bog

Cross-walk of Maine Natural Communities, Maine Ecosystems, Maine Widlife Action Plan Habitat Groups, NVC (International Vegetation) Associations, and NVC (International) Ecological Systems

Maine Natural Communities: 104 types	Maine Ecosystems: 24 types	Maine Wildlife Action Plan Habitat Types: 21 Types	NVC (International) Vegetation Associations: 182 types	NVC (International) Ecological Systems: 50 types
Hudsonia river beach	Streamshore	Rivers and Streams	Hudsonia tomentosa - Paronychia argyrocoma Dwarf-shrubland	Laurentian-Acadian Floodplain Forest
Ironwood - oak - ash woodland	Central Hardwoods Oak Forest	Deciduous and Mixed Forest	Acer saccharum - Tilia americana - Fraxinus americana / Ostrya virginiana / Geranium robertianum Woodland	Ironwood - oak - ash woodland
Jack pine forest	Spruce – Fir – Northern Hardwood Forest	Conifer Forest	Pinus banksiana / Vaccinium spp. / Pleurozium schreberi Forest	Boreal Jack Pine-Black Spruce Forest
Jack pine woodland	Rock Outcrop	Dry Woodlands and Barrens	Pinus banksiana / Kalmia angustifolia - Vaccinium spp. Woodland	Northern Appalachian-Acadian Rocky Heath Outcrop
Labrador tea talus dwarf-shrubland	Rock Outcrop	Cliff Face and Rocky Outcrops (including talus)	Picea mariana / Ledum groenlandicum - Empetrum nigrum / Cladina spp. Dwarf-shrubland	Laurentian-Acadian Acidic Cliff and Talus
Lakeshore sand/ cobble beach	Lakeshore	Freshwater Lakes and Ponds	Inland Freshwater Strand Beach Sparse Vegetation	Laurentian-Acadian Lakeshore Beach
Leatherleaf boggy fen	Unpatterned Fen, Patterned Fen, or other peatland types	Peatlands	Chamaedaphne calyculata - (Gaylussacia dumosa) - Decodon verticillatus / Woodwardia virginica Dwarf-shrubland; Chamaedaphne calyculata / Eriophorum virginicum / Sphagnum rubellum Dwarf-shrubland	Boreal-Laurentian-Acadian Acidic Basin Fen; Boreal-Laurentian Bog; North-Central Interior and Appalachian Acidic Peatland
Little bluestem - blueberry sandplain grassland	Pine Barrens	Grassland, Agricultural, Old Field	Vaccinium angustifolium / Schizachyrium scoparium - Carex lucorum Shrub Herbaceous Vegetation	Northern Atlantic Coastal Plain Pitch Pine Barrens

Cross-walk of Maine Natural Communities, Maine Ecosystems, Maine Widlife Action Plan Habitat Groups, NVC (International Vegetation) Associations, and NVC (International) Ecological Systems

Maine Natural Communities: 104 types	Maine Ecosystems: 24 types	Maine Wildlife Action Plan Habitat Types: 21 Types	NVC (International) Vegetation Associations: 182 types	NVC (International) Ecological Systems: 50 types
Low sedge - buckbean fen lawn	Unpatterned Fen, Patterned Fen, or other peatland types	Peatlands	Carex limosa - Rhynchospora alba / Sphagnum pulchrum - Cladopodiella sp. Herbaceous Vegetation	Boreal-Laurentian-Acadian Acidic Basin Fen; Laurentian-Acadian Alkaline Fen; Boreal-Laurentian Bog
Maple - basswood - ash forest	Spruce – Fir – Northern Hardwood Forest	Deciduous and Mixed Forest	Acer saccharum - Fraxinus americana - Tilia americana / Acer spicatum / Caulophyllum thalictroides Forest	Appalachian (Hemlock)-Northern Hardwood Forest; Laurentian-Acadian Northern Hardwood Forest
Maritime spruce - fir forest	Maritime Forest	Deciduous and Mixed Forest	Picea rubens - Picea glauca Forest	Acadian Low-Elevation Spruce-Fir-Hardwood Forest
Mixed graminoid - forb saltmarsh	Tidal Marsh Estuary; Coastal Dune – Marsh	Estuarine Emergent Saltmarsh	Spartina alterniflora / (Ascophyllum nodosum) Acadian/Virginian Zone Herbaceous Vegetation; Salicornia bigelovii - Triglochin maritima Herbaceous Vegetation.	Acadian Coastal Salt Marsh; Northern Atlantic Coastal Plain Tidal Salt Marsh;
Mixed graminoid - shrub marsh	Appalachain-Acadian Basin Swamp; Streamshore	Shrub-scrub Wetland	Cephalanthus occidentalis - Decadon verticillatus Shrubland; Juncus militaris - Eriocaulon aquaticum Herbaceous Vegetation; Scirpus cyperinus Seasonally Flooded Herbaceous Vegetation; Calamagrostis canadensis - Scirpus spp. - Dulichium arundinaceum Herbaceous Vegetation	Laurentian-Acadian Wet Meadow-Shrub Swamp

Cross-walk of Maine Natural Communities, Maine Ecosystems, Maine Widlife Action Plan Habitat Groups, NVC (International Vegetation) Associations, and NVC (International) Ecological Systems

Maine Natural Communities: 104 types	Maine Ecosystems: 24 types	Maine Wildlife Action Plan Habitat Types: 21 Types	NVC (International) Vegetation Associations: 182 types	NVC (International) Ecological Systems: 50 types
Mixed tall sedge fen	Unpatterned Fen (most likely) or several other peatland ecosystem types	Peatlands	Carex (rostrata, utriculata) - Carex lacustris - (Carex vesicaria) Herbaceous Vegetation; Myrica gale - Chamaedaphne calyculata / Carex (lasiocarpa, utriculata) - Utricularia spp. Shrub Herbaceous Vegetation	North-Central Interior and Appalachian Acidic Peatland; Boreal-Laurentian-Acadian Acidic Basin Fen
Mountain alder - bush-honeysuckle subalpine meadow	Alpine	Alpine	Alnus viridis ssp. crispa - Spiraea alba / Solidago macrophylla Shrubland	Acadian-Appalachian Subalpine Woodland and Heath-Krummholz
Mountain holly - alder woodland fen	Unpatterned Fen (most likely) or several other peatland ecosystem types	Peatlands	Alnus incana ssp. rugosa - Nemopanthus mucronatus / Sphagnum spp. Shrubland; Alnus (serrulata, incana) / Osmunda cinnamomea - Sphagnum spp. Shrubland; Vaccinium corymbosum / Sphagnum spp. Shrubland	Boreal-Laurentian-Acadian Acidic Basin Fen
Northern white cedar swamp	Appalachian – Acadian Basin Swamp	Forested Wetland	Thuja occidentalis / Sphagnum (girgensohnii, warnstorfii) Forest	Laurentian-Acadian Alkaline Conifer-Hardwood Swamp
Northern white cedar woodland fen	Appalachian – Acadian Basin Swamp ; Unpatterned Fen	Peatlands	Thuja occidentalis - Abies balsamea / Ledum groenlandicum / Carex trisperma Woodland	Laurentian-Acadian Alkaline Fen
Oak - pine forest	Central Hardwoods Oak Forest	Deciduous and Mixed Forest	Pinus strobus - Quercus (rubra, velutina) - Fagus grandifolia Forest; Quercus rubra - Acer rubrum - Betula spp. - Pinus strobus Forest	Central Appalachian Dry Oak-Pine Forest; Laurentian-Acadian Pine-Hemlock-Hardwood Forest

Cross-walk of Maine Natural Communities, Maine Ecosystems, Maine Widlife Action Plan Habitat Groups, NVC (International Vegetation) Associations, and NVC (International) Ecological Systems

Maine Natural Communities: 104 types	Maine Ecosystems: 24 types	Maine Wildlife Action Plan Habitat Types: 21 Types	NVC (International) Vegetation Associations: 182 types	NVC (International) Ecological Systems: 50 types
Oak - pine woodland	Central Hardwoods Oak Forest	Deciduous and Mixed Forest	Quercus rubra - (Quercus prinus) / Vaccinium spp. / Deschampsia flexuosa Woodland	Central Appalachian Pine-Oak Rocky Woodland; Northern Appalachian-Acadian Rocky Heath Outcrop
Oak Hickory Forest	Central Hardwoods Oak Forest	Deciduous and Mixed Forest	Quercus rubra - Carya (glabra, ovata) / Ostrya virginiana / Carex lucorum Forest; Quercus (alba, rubra, velutina) / Cornus florida / Viburnum acerifolium Forest	Central Appalachian Dry Oak-Pine Forest
Pickerelweed - macrophyte aquatic bed	Lakeshore; Coastal Plain Pond; Appalachian-Acadian Rivershore; Streamshore	Emergent Marsh and Wet Meadows	Pontederia cordata - Sagittaria latifolia Herbaceous Vegetation	Laurentian-Acadian Freshwater Marsh
Pipewort - water lobelia aquatic bed	Lakeshore	Freshwater Lakes and Ponds	Eriocaulon aquaticum - Lobelia dortmanna Herbaceous Vegetation; Eriocaulon aquaticum Sparse Vegetation	Laurentian-Acadian Lakeshore Beach; Laurentian-Acadian Freshwater Marsh
Pitch pine - heath barren	Pine Barrens	Dry Woodlands and Barrens	Pinus rigida / Vaccinium spp. - Gaylussacia baccata Woodland	Northeastern Interior Pine Barrens
Pitch pine - scrub oak barren	Pine Barrens	Dry Woodlands and Barrens	Pinus rigida / Quercus ilicifolia / Piptatherum pungens Woodland	Northeastern Interior Pine Barrens
Pitch pine bog	Pine Barrens	Dry Woodlands and Barrens	Pinus rigida / Chamaedaphne calyculata / Sphagnum spp. Woodland	Atlantic Coastal Plain Northern Bog

Cross-walk of Maine Natural Communities, Maine Ecosystems, Maine Widlife Action Plan Habitat Groups, NVC (International Vegetation) Associations, and NVC (International) Ecological Systems

Maine Natural Communities: 104 types	Maine Ecosystems: 24 types	Maine Wildlife Action Plan Habitat Types: 21 Types	NVC (International) Vegetation Associations: 182 types	NVC (International) Ecological Systems: 50 types
Pitch pine dune woodland	Pine Barrens	Dry Woodlands and Barrens	Pinus rigida / Hudsonia tomentosa Woodland; Hudsonia tomentosa - Arctostaphylos uva-ursi Dwarf-shrubland	Northern Atlantic Coastal Plain Dune and Swale
Pitch pine rocky woodland	Rock Outcrop	Dry Woodlands and Barrens	Pinus rigida / (Quercus ilicifolia) / Photinia melanocarpa / Deschampsia flexuosa Woodland; Pinus rigida / Corema conradii Woodland	Northern Appalachian-Acadian Rocky Heath Outcrop; Central Appalachian Pine-Oak Rocky Woodland
Red maple - sensitive fern swamp	Appalachian – Acadian Basin Swamp	Forested Wetland	Acer rubrum / Carex stricta - Onoclea sensibilis Woodland; Picea rubens - Acer rubrum / Nemopanthus mucronatus Forest; Acer rubrum / Nemopanthus mucronatus - Vaccinium corymbosum Forest; Acer rubrum - Abies balsamea / Viburnum nudum var. cassinoides Floodplain Forest; Acer rubrum - Prunus serotina / Cornus amomum Floodplain Forest	Northern Appalachian-Acadian Conifer-Hardwood Acidic Swamp
Red maple wooded fen	Appalachian – Acadian Basin Swamp	Forested Wetland	Acer rubrum / Alnus incana - Ilex verticillata / Osmunda regalis Woodland	Northern Appalachian-Acadian Conifer-Hardwood Acidic Swamp
Red oak - northern hardwoods - white pine forest	White Pine – Mixed Hardwood	Deciduous and Mixed Forest	Quercus rubra - Acer saccharum - Fagus grandifolia / Viburnum acerifolium Forest	Laurentian-Acadian Pine-Hemlock-Hardwood Forest; Laurentian-Acadian Northern Pine-(Oak) Forest
Red pine - white pine forest	Spruce – Fir – Northern Hardwood Forest	Deciduous and Mixed Forest	Pinus strobus - Pinus resinosa / Cornus canadensis Forest	Laurentian-Acadian Northern Pine-(Oak) Forest

Cross-walk of Maine Natural Communities, Maine Ecosystems, Maine Widlife Action Plan Habitat Groups, NVC (International Vegetation) Associations, and NVC (International) Ecological Systems

Maine Natural Communities: 104 types	Maine Ecosystems: 24 types	Maine Wildlife Action Plan Habitat Types: 21 Types	NVC (International) Vegetation Associations: 182 types	NVC (International) Ecological Systems: 50 types
Red pine woodland	Rock Outcrop	Dry Woodlands and Barrens	Pinus resinosa / Gaylussacia baccata - Vaccinium angustifolium Woodland	Northern Appalachian-Acadian Rocky Heath Outcrop
Red spruce - mixed conifer woodland	Rock Outcrop	Dry Woodlands and Barrens	Pice rubens / Vaccinium angustifolium / Sibbaldiopsis tridentata Woodland	Northern Appalachian-Acadian Rocky Heath Outcrop
Rocky summit heath	Alpine	Alpine; Mountaintop Forest (including krummholz)	Picea mariana / Kalmia angustifolia Dwarf-shrubland	Acadian-Appalachian Subalpine Woodland and Heath-Krummholz
Rose - bayberry maritime shrubland	Coastal Headlands	Rocky Coastline and Islands	Morella pensylvanica - Prunus maritima Shrubland; Prunus serotina - Rhus typhina Shrubland	Acadian-North Atlantic Rocky Coast
Sand cherry - tufted hairgrass river beach	Streamshore	Rivers and Streams	Prunus pumila var. depressa / Deschampsia caespitosa Herbaceous Vegetation	Boreal Ice-Scour Rivershore
Seaside goldenrod - goosetongue open headland	Coastal Headlands	Rocky Coastline and Islands	Solidago sempervirens - (Rhodiola rosea) - Juniperus horizontalis Sparse Vegetatation	Acadian-North Atlantic Rocky Coast
Sedge - leatherleaf fen lawn	Unpatterned Fen (most likely) or several other peatland ecosystem types	Peatlands	Carex oligosperma - Carex pauciflora - Eriophorum vaginatum / Sphagnum spp. Herbaceous Vegetation; Carex (oligosperma, exilis) - Chamaedaphne calyculata Shrub Herbaceous Vegetation	North-Central Interior and Appalachian Acidic Peatland; Boreal-Laurentian-Acadian Acidic Basin Fen
Semi-rich northern hardwood forest	Spruce – Fir – Northern Hardwood Forest	Deciduous and Mixed Forest	Acer saccharum - (Fraxinus americana) / Arisaema triphyllum Forest;	Appalachian (Hemlock)-Northern Hardwood Forest; Laurentian-Acadian Northern Hardwoods Forest

Cross-walk of Maine Natural Communities, Maine Ecosystems, Maine Widlife Action Plan Habitat Groups, NVC (International Vegetation) Associations, and NVC (International) Ecological Systems

Maine Natural Communities: 104 types	**Maine Ecosystems: 24 types**	**Maine Wildlife Action Plan Habitat Types: 21 Types**	**NVC (International) Vegetation Associations: 182 types**	**NVC (International) Ecological Systems: 50 types**
Sheep laurel dwarf shrub bog	Unpatterned Fen (most likely) or several other peatland ecosystem types	Peatlands	Kalmia angustifolia - Chamaedaphne calyculata - (Picea mariana) / Cladina spp. Dwarf-shrubland; Rhododendron canadense - Chamaedaphne calyculata Dwarf-shrubland	Boreal-Laurentian Bog; Acadian Maritime Bog
Shrubby cinquefoil - sedge circumneutral fen	Unpatterned Fen; Patterned Fen	Peatlands	Dasiphora fruticosa ssp. floribunda / Carex lasiocarpa / Campylium stellatum Shrub Herbaceous Vegetation	Laurentian-Acadian Alkaline Fen
Silver maple floodplain forest	Streamshore	Rivers and Streams	Acer saccharinum / Onoclea sensibilis - Boehmeria cylindrica Forest	Laurentian-Acadian Floodplain Forest
Spartina saltmarsh	Coastal Dune – Marsh	Estuarine Emergent Saltmarsh	Salicornia (virginica, bigelovii, maritima) - Spartina alterniflora Herbaceous Vegetation; Spartina patens - Distichlis spicata - (Juncus gerardii) Herbaceous Vegetation; Ruppia maritima Acadian/Virginian Zone Temperate Herbaceous Vegetation	Acadian Coastal Salt Marsh; Northern Atlantic Coastal Plain Tidal Salt Marsh
Spruce - fir - birch krummholz	Spruce – Fir – Northern Hardwood Forest	Deciduous and Mixed Forest	Picea mariana - Abies balsamea / Sibbaldiopsis tridentata Shrubland	Acadian-Appalachian Subalpine Woodland and Heath-Krummholz
Spruce - fir - broom-moss forest	Spruce – Fir – Northern Hardwood Forest	Conifer Forest	Picea rubens - Abies balsamea - Betula papyrifera Forest; Picea rubens - Abies balsamea - Betula spp. - Acer rubrum Forest	Acadian Low-Elevation Spruce-Fir-Hardwood Forest

Cross-walk of Maine Natural Communities, Maine Ecosystems, Maine Widlife Action Plan Habitat Groups, NVC (International Vegetation) Associations, and NVC (International) Ecological Systems

Maine Natural Communities: 104 types	Maine Ecosystems: 24 types	Maine Wildlife Action Plan Habitat Types: 21 Types	NVC (International) Vegetation Associations: 182 types	NVC (International) Ecological Systems: 50 types
Spruce - fir - cinnamon fern forest	Spruce – Fir – Northern Hardwood Forest	Coniferous Forest	Picea rubens - Abies balsamea / Sphagnum magellanicum Forest; Picea rubens - Abies balsamea / Gaultheria hispidula / Osmunda cinnamomea / Sphagnum spp. Forest; Picea mariana - Picea rubens / Pleurozium schreberi Forest	Acadian Near-Boreal Spruce Flat; Northern Appalachian-Acadian Conifer-Hardwood Acidic Swamp
Spruce - fir - wood-sorrel - feather-moss forest	Spruce – Fir – Northern Hardwood Forest	Conifer Forest	Picea rubens - Abies balsamea / Sorbus americana Forest	Acadian-Appalachian Montane Spruce-Fir Forest
Spruce - heath barren	Spruce – Fir – Northern Hardwood Forest	Coniferous Forest	Picea mariana - Picea rubens / Rhododendron canadense / Cladina spp. Woodland	Acadian Near-Boreal Spruce Barrens
Spruce - larch wooded bog	Unpatterned Fen; Domed Bog; Eccentric Bog	Peatlands	Picea mariana / Ledum groenlandicum / Carex trisperma / Sphagnum spp. Forest; Picea mariana - (Larix laricina) / Ledum groenlandicum / Sphagnum spp. Forest; Picea mariana / Rubus chamaemorus / Sphagnum spp. Woodland; Picea mariana / (Vaccinium corymbosum, Gaylussacia baccata) / Sphagnum sp. Woodland	Boreal-Laurentian Bog; North-Central Interior and Appalachian Acidic Peatland
Spruce - northern hardwoods forest	Spruce – Fir – Northern Hardwood Forest	Deciduous and Mixed Forest	Picea rubens - Betula alleghaniensis / Dryopteris campyloptera Forest	Acadian Low-Elevation Spruce-Fir-Hardwood Forest; Acadian-Appalachian Montane Spruce-Fir Fores
Spruce talus woodland	Rock Outcrop	Cliff Face and Rocky Outcrops (including talus)	Picea rubens / Ribes glandulosum Woodland; Polypodium (virginianum, appalachianum) / Lichens Nonvascular Vegetation	Laurentian-Acadian Acidic Cliff and Talus

Cross-walk of Maine Natural Communities, Maine Ecosystems, Maine Widlife Action Plan Habitat Groups, NVC (International Vegetation) Associations, and NVC (International) Ecological Systems

Maine Natural Communities: 104 types	Maine Ecosystems: 24 types	Maine Wildlife Action Plan Habitat Types: 21 Types	NVC (International) Vegetation Associations: 182 types	NVC (International) Ecological Systems: 50 types
Sweetgale mixed shrub fen	Appalachian-Acadian Basin Swamp; Streamshore	Shrub-scrub Wetland	Betula pumila / Chamaedaphne calyculata / Carex lasiocarpa Shrubland; Myrica gale - Spiraea alba - Chamaedaphne calyculata Shrubland	Laurentian-Acadian Wet Meadow-Shrub Swamp; Boreal-Laurentian-Acadian Acidic Basin Fen
Three-toothed cinquefoil - blueberry low summit bald	Rock Outcrop	Dry Woodlands and Barrens	Vaccinium angustifolium - Sorbus americana / Sibbaldiopsis tridentata Dwarf-shrubland	Northern Appalachian-Acadian Rocky Heath Outcrop
Three-way sedge - goldenrod outwash plain pondshore	Lakeshore	Freshwater Lakes and Ponds	Lysimachia terrestris - Dulichium arundinaceum - Rhexia virginica Herbaceous Vegetation; Eleocharis (obtusa, flavescens) - Eriocaulon aquaticum Herbaceous Vegetation; Juncus militaris - Eriocaulon aquaticum Herbaceous Vegetation	Northern Atlantic Coastal Plain Pond
Tussock sedge meadow	Appalachian-Acadian Basin Swamp; Streamshore	Emergent Marsh and Wet Meadows	Carex stricta - Carex vesicaria Herbaceous Vegetation	Laurentian-Acadian Wet Meadow-Shrub Swamp
Twisted sedge cobble rivershore	Appalachian-Acadian Rivershore; Streamshore	Rivers and Streams	Carex torta - Apocynum cannabinum - Cyperus spp. Herbaceous Vegetation; Prunus pumila var. depressa / Deschampsia caespitosa Herbaceous Vegetation	Boreal Ice-Scour Rivershore
Water-lily - macrophyte aquatic bed	Lakeshore; Coastal Plain Pond; Appalachain-Acadian Rivershore	Freshwater Lakes and Ponds	Nuphar lutea ssp. advena - Numphaea odorata Herbaceous Vegetation; Vallisneria americana - Potamogeton perfoliatus Herbaceous Vegetation	Laurentian-Acadian Freshwater Marsh

Cross-walk of Maine Natural Communities, Maine Ecosystems, Maine Widlife Action Plan Habitat Groups, NVC (International Vegetation) Associations, and NVC (International) Ecological Systems

Maine Natural Communities: 104 types	Maine Ecosystems: 24 types	Maine Wildlife Action Plan Habitat Types: 21 Types	NVC (International) Vegetation Associations: 182 types	NVC (International) Ecological Systems: 50 types
White cedar woodland	Rock Outcrop	Cliff Face and Rocky Outcrops (including talus)	Thuja occidentalis / Gaylussacia baccata - Vaccinium angustifolium Woodland; Thuja occidentalis - Fraxinus pennsylvanica / Acer pensylvanicum Woodland	Laurentian-Acadian Calcareous Cliff and Talus
White oak - red oak forest	Central Hardwoods Oak Forest	Deciduous and Mixed Forest	Quercus (alba, rubra, velutina) / Cornus florida / Viburnum acerifolium Forest	Central Appalachian Dry Oak-Pine Forest
White pine - mixed conifer forest	Spruce – Fir – Northern Hardwood Forest	Deciduous and Mixed Forest	Pinus strobus - Tsuga canadensis - Picea rubens Forest; Pinus strobus - Tsuga canadensis Lower New England / Northern Piedmont Forest	Laurentian-Acadian Northern Pine-(Oak) Forest

Appendix C: Plant Common and Scientific Names

Vascular plant nomenclature for scientific names follows *Flora of Maine* (Haines and Vining 1998). Common names are drawn from *Flora of Maine*, *Newcomb's Wildflower Guide* (Newcomb 1977), or from a translation of the scientific name. Moss nomenclature is from the *Checklist of Maine Mosses* (Allen 1999); liverwort names follow Crum 1991; lichen names are drawn from *Hinds and Hinds* (1998), with nomenclature following Esslinger (1997). Visit the Maine Natural Areas Program website for most up-to-date information regarding rare species, *http://www.maine.gov/doc/nrimc/mnap/features/plantlist.htm.*

Common Name	Scientific Name
Acadian quillwort	*Isoetes acadiensis*
Alder	*Alnus* sp./spp.
Alderleaf buckthorn	*Rhamnus alnifolia*
Alkali bulrush	*Bolboschoenus maritimus*
Alpine azalea	*Loiseleuria procumbens*
Alpine bearberry	*Arctostaphylos alpina*
Alpine bilberry	*Vaccinium uliginosum*
Alpine bistort	*Persicaria vivipara*
Alpine bitter-cress	*Cardamine bellidifolia*
Alpine blueberry	*Vaccinium boreale*
Alpine cotton-grass	*Trichophorum alpinum*
Alpine milk-vetch	*Astragalus alpinus*
Alpine pondweed	*Potamogeton filiformis* var. *alpinus*
Alpine sweet-broom	*Hedysarum alpinum*
Alpine sweet-grass	*Hierochloe alpina*
Alternate-leaved dogwood	*Cornus alternifolia*
American beech	*Fagus grandifolia*
American chestnut	*Castanea dentata*
American dunegrass	*Leymus mollis*
American elm	*Ulmus americana*
American ginseng	*Panax quinquefolius*

Common Name	Scientific Name
American shore-grass	*Littorella uniflora*
American stickseed	*Hackelia deflexa*
Anticosti aster	*Symphyotrichum anticostense*
Arctic red fescue	*Festuca prolifera*
Arrow-arum	*Peltandra virginica*
Arrow-leaved violet	*Viola sagittata*
Arrowwood	*Viburnum dentatum* var. *lucidum*
Aspen	*Populus tremuloides* or *P. grandidentata*
Atlantic white cedar	*Chamaecyparis thyoides*
Auricled twayblade	*Listera auriculata*
Autumn coral-root	*Corallorhiza odontorhiza*
Awned hair-cap moss	*Polytrichum piliferum*
Baked apple-berry	*Rubus chamaemorus*
Balsam fir	*Abies balsamea*
Balsam poplar	*Populus balsamifera*
Balsam ragwort	*Senecio pauperculus*
Basswood	*Tilia americana*
Bayberry	*Myrica pensylvanica*
Bayonet rush	*Juncus militaris*
Beach grass	*Ammophila breviligulata*
Beach heather	*Hudsonia tomentosa*
Beach-pea	*Lathyrus japonicus*
Beach plum	*Prunus maritima*
Beachhead iris	*Iris setosa*
Beaked hazelnut	*Corylus cornuta*
Beaked sedge	*Carex utriculata*
Beaked spikerush	*Eleocharis rostellata*
Bearberry willow	*Salix uva-ursi*
Beggar-ticks	*Bidens* spp.
Big bluestem	*Andropogon gerardii*

Common Name	Scientific Name
Bigelow's sedge	*Carex bigelowii*
Bigleaf pondweed	*Potamogeton amplifolius*
Big-leaved aster	*Aster macrophyllus*
Big-toothed aspen	*Populus grandidentata*
Bird's-eye primrose	*Primula laurentiana*
Bitternut hickory	*Carya cordiformis*
Black ash	*Fraxinus nigra*
Black bulrush	*Scirpus atrovirens*
Black cherry	*Prunus serotina*
Black chokeberry	*Photinia melanocarpa*
Black crowberry	*Empetrum nigrum*
Black-grass	*Juncus gerardii*
Black gum	*Nyssa sylvatica*
Black huckleberry	*Gaylussacia baccata*
Black oak	*Quercus velutina*
Black sedge	*Carex atratiformis*
Black spruce	*Picea mariana*
Black willow	*Salix nigra*
Bladder campion	*Silene vulgaris*
Bladderwort	*Utricularia* sp./spp.
Blinks	*Montia fontana*
Bloodroot	*Sanguinaria canadensis*
Blue birch	*Betula* x *caerulea*
Blue cohosh	*Caulophyllum thalictroides*
Bluebead lily	*Clintonia borealis*
Bluebell	*Campanula rotundifolia*
Bluejoint	*Calamagrostis canadensis*
Blue-leaf willow	*Salix myricoides*
Blunt-lobed woodsia	*Woodsia obtusa*
Bog aster	*Oclemena nemoralis*

Common Name	Scientific Name
Bog bedstraw	*Galium labradoricum*
Bog broom-moss	*Dicranum undulatum*
Bog goldenrod	*Solidago uliginosa*
Bog hair-cap moss	*Polytrichum strictum*
Bog-mat liverwort	*Cladopodiella fluitans*
Bog rosemary	*Andromeda polifolia* var. *glaucophylla*
Bog sedge	*Carex magellanica*
Bog willow	*Salix pedicellaris*
Boneset	*Eupatorium perfoliatum*
Boreal bedstraw	*Galium kamtschaticum*
Boreal bentgrass	*Agrostis mertensii*
Bottlebrush grass	*Elymus hystrix*
Bracken fern	*Pteridium aquilinum*
Braun's holly fern	*Polystichum braunii*
Bristly gooseberry	*Ribes hirtellum*
Broad beech fern	*Phegopteris hexagonoptera*
Broad-leaved goldenrod	*Solidago flexicaulis* or *S. macrophylla*
Broom-crowberry	*Corema conradii*
Broom-moss	*Dicranum* sp./spp. or *Dicranella* sp./spp.
Brown-fruited rush	*Juncus pelocarpus*
Brownish sedge	*Carex brunnescens*
Buckbean	*Menyanthes trifoliata*
Bulblet fern	*Cystopteris bulbifera*
Bulrush	*Bolboschoenus* sp/spp. or *Schoenoplectus*
Bulrush sedge	*Carex scirpoidea*
Bunchberry	*Cornus canadensis*
Bur oak	*Quercus macrocarpa*
Bur-reed	*Sparganium* sp./spp.
Bush-honeysuckle	*Diervilla lonicera*
Butterfly weed	*Asclepias tuberosa*

Common Name	Scientific Name
Butternut	*Juglans cinerea*
Butterwort	*Pinguicula vulgaris*
Buttonbush	*Cephalanthus occidentalis*
Campylium fen moss	*Campylium stellatum*
Canada burnet	*Sanguisorba canadensis*
Canada goldenrod	*Solidago canadensis*
Canada mayflower	*Maianthemum canadense*
Canada mountain-ricegrass	*Oryzopsis canadensis*
Canada rush	*Juncus canadensis*
Canadian tick-trefoil	*Desmodium canadense*
Capillary sedge	*Carex capillaris*
Cardinal flower	*Lobelia cardinalis*
Cattail	*Typha* sp./spp.
Cetraria lichen	*Cetraria* sp./spp.
Chaffy sedge	*Carex paleacea*
Chair-maker's rush	*Schoenoplectus pungens*
Chara algae	*Chara* sp./spp.
Chestnut oak	*Quercus montana*
Choke cherry	*Prunus virginiana*
Christmas fern	*Polystichum acrostichoides*
Cinnamon fern	*Osmunda cinnamomea*
Clinton's bulrush	*Tricophorum clintonii*
Clothed sedge	*Carex vestita*
Clover	*Trifolium* sp./spp.
Coast sedge	*Carex exilis*
Coast-blite goosefoot	*Chenopodium rubrum*
Coltsfoot	*Tussilago farfara*
Columbine	*Aquilegia canadensis*
Comb-leaved mermaidweed	*Proserpinaca pectinata*
Common arrow-grass	*Triglochin maritimum*

Common Name	Scientific Name
Common arrowhead	*Sagittaria latifolia*
Common blackberry	*Rubus allegheniensis*
Common broom-moss	*Dicranum scoparium*
Common cattail	*Typha latifolia*
Common grapefern	*Botrychium virginianum*
Common hairgrass	*Deschampsia flexuosa*
Common horsetail	*Equisetum arvense*
Common juniper	*Juniperus communis*
Common pussytoes	*Antennaria howellii*
Common saltwort	*Salsola leali*
Common speedwell	*Veronica officinalis*
Common St. Johnswort	*Hypericum perforatum*
Common waterweed	*Elodea canadensis*
Cotton-grass	*Eriophorum* sp./spp.
Cow vetch	*Vicia cracca*
Cranberry	*Vaccinium oxycoccos* or *V. macrocarpon*
Creeping bentgrass	*Agrostis stolonifera*
Creeping juniper	*Juniperus horizontalis*
Creeping snowberry	*Gaultheria hispidula*
Creeping spearwort	*Ranunculus flammula*
Crinkled hair-grass	*Deschampsia flexuosa*
Crowberry	*Empetrum nigrum* or *E. eamesii atropurpureum*
Cut-leaved anemone	*Anemone multifida*
Cut-leaved toothwort	*Cardamine concatenata*
Cutler's goldenrod	*Solidago multiradiata*
Deer-hair sedge	*Trichophorum cespitosum*
Deertongue	*Dichanthelium clandestinum*
Dewdrop	*Rubus dalibarda*
Diapensia	*Diapensia lapponica*
Dicranum moss	*Dicranum* sp./spp.

Common Name	Scientific Name
Dioecious sedge	*Carex sterilis*
Dogwood	*Cornus* sp./spp.
Doll's eyes	*Actaea pachypoda*
Douglas' knotweed	*Polygonum douglasii*
Dry land sedge	*Carex siccata*
Dutchman's breeches	*Dicentra cucullaria*
Dwarf bilberry	*Vaccinium cespitosum*
Dwarf bulrush	*Lipocarpha micrantha*
Dwarf glasswort	*Salicornia bigelovii*
Dwarf huckleberry	*Gaylussacia dumosa*
Dwarf raspberry	*Rubus pubescens*
Dwarf rattlesnakeroot	*Prenanthes nana*
Early crowfoot	*Ranunculus fascicularis*
Early goldenrod	*Solidago juncea*
Eastern hemlock	*Tsuga canadensis*
Eastern hop-hornbeam	*Ostrya virginiana*
Eastern mannagrass	*Glyceria septentrionalis*
Eaton's bur-marigold	*Bidens eatonii*
Ebony sedge	*Carex eburnea*
Ebony spleenwort	*Asplenium platyneuron*
Englemann's spikerush	*Eleocharis engelmannii*
English sundew	*Drosera anglica*
Eurasian water-milfoil	*Myriophyllum spicatum*
Expanded bulrush	*Scirpus expansus*
Fall fimbry	*Fimbristylis autumnalis*
False nettle	*Boehmeria cylindrical*
False spikenard	*Maianthemum racemosum*
False water-pepper	*Persicaria hydropiperoides*
Featherfoil	*Hottonia inflata*
Fen sickle-moss	*Drepanocladus* sp./spp.

Common Name	Scientific Name
Fern-leaved false foxglove	*Aureolaria pedicularia*
Fern moss	*Thuidium delicatulum*
Few-flowered sedge	*Carex pauciflora*
Few-seeded sedge	*Carex oligosperma*
Fibrous-rooted sedge	*Carex communis*
Field mint	*Mentha arvensis*
Field pussytoes	*Antennaria neglecta*
Flatsedge	*Cyperus* sp./spp.
Flat-topped goldenrod	*Euthamia graminifolia*
Flat-topped white aster	*Doellingeria umbellata*
Flat-tufted feather-moss	*Hypnum imponens*
Floating pondweed	*Potamogeton natans*
Flowering dogwood	*Cornus florida*
Fly-away grass	*Agrostis scabra*
Fly honeysuckle	*Lonicera canadensis*
Foamflower	*Tiarella cordifolia*
Fowl mannagrass	*Glyceria striata*
Fragile fern	*Cystopteris fragilis*
Fragrant wood-fern	*Dryopteris fragrans*
Freshwater cordgrass	*Spartina pectinata*
Fries' pondweed	*Potamogeton friesii*
Fringed bromegrass	*Bromus ciliatus*
Fringed loosestrife	*Lysimachia ciliata*
Fringed Ptilidium liverwort	*Ptilidium ciliare*
Furbish's lousewort	*Pedicularis furbishiae*
Gall of the earth	*Prenanthes trifoliolata*
Garber's sedge	*Carex garberi*
Gaspé arrow-grass	*Triglochin gaspense*
Giant rattlesnake-plantain	*Goodyera oblongifolia*
Glabrous knotted pearlwort	*Sagina nodosa* ssp. *borealis*

Common Name	Scientific Name
Glaucous rattlesnake-root	*Prenanthes racemosa*
Golden heather	*Hudsonia ericoides*
Golden pert	*Gratiola aurea*
Goldie's wood-fern	*Dryopteris goldiana*
Goldthread	*Coptis trifolia*
Gooseberry	*Ribes* sp./spp.
Goosetongue	*Plantago maritima* var. *juncoides*
Grapefern	*Botrychium* sp./spp.
Grass-of-parnassus	*Parnassia glauca*
Gray birch	*Betula populifolia*
Gray goldenrod	*Solidago nemoralis*
Green ash	*Fraxinus pennsylvanica*
Greene's rush	*Juncus greenei*
Grey reindeer-lichen	*Cladina rangiferina*
Grimmia rock-moss	*Grimmia* sp./spp.
Haircap moss	*Polytrichum* sp./spp.
Hairy arnica	*Arnica lanceolata*
Hairy goldenrod	*Solidago hispida*
Hairy wood brome-grass	*Bromus pubescens*
Hardhack	*Spiraea tomentosa*
Hardstem bulrush	*Schoenoplectus acutus*
Hay-scented fern	*Dennstaedtia punctilobula*
Heart-leaved paper birch	*Betula cordifolia*
Hemlock	*Tsuga canadensis*
Hemp dogbane	*Apocynum cannabinum*
Hepatica	*Hepatica nobilis* var. *obtusa*
Herb Robert	*Geranium robertianum*
Hickory	*Carya* sp./spp.
Hidden-scale sedge	*Carex cryptoleptis*
Highbush blueberry	*Vaccinium corymbosum*

Common Name	Scientific Name
Highbush cranberry	*Viburnum opulus*
Highland rush	*Juncus trifidus*
Hoary willow	*Salix candida*
Hobblebush	*Viburnum lantanoides*
Hollow joe-pye weed	*Eupatorium fistulosum*
Horned beak-rush	*Rhynchospora capillacea*
Horned bladderwort	*Utricularia cornuta*
Hornemann's willow-herb	*Epilobium hornemannii*
Huckleberry	*Gaylussacia baccata* or *G. dumosa*
Huron tansy	*Tanacetum bipinnatum* ssp. *huronense*
Hypnum moss	*Hypnum* sp./spp.
Indian cucumber-root	*Medeola virginiana*
Indian grass	*Sorghastrum nutans*
Indian pipe	*Monotropa uniflora*
Inflated sedge	*Carex vesicaria*
Inkberry	*Ilex glabra*
Intermediate sedge	*Carex norvegica*
Interrupted fern	*Osmunda claytoniana*
Ironwood	*Ostrya virginiana*
Jack pine	*Pinus banksiana*
Jack-in-the-pulpit	*Arisaema triphyllum*
Japanese knotweed	*Fallopia japonica*
Joe-pye weed	*Eupatorium* sp./spp.
Jointweed	*Polygonella articulata*
Juniper hair-cap moss	*Polytrichum juniperinum*
Kalm's lobelia	*Lobelia kalmii*
Labrador tea	*Rhododendron groenlandicum*
Lady fern	*Athyrium filix-femina*
Lake sedge	*Carex lacustris*
Lance-leaved draba	*Draba cana*

Common Name	Scientific Name
Lapland buttercup	*Ranunculus lapponicus*
Lapland rosebay	*Rhododendron lapponicum*
Larch	*Larix laricina*
Large cranberry	*Vaccinium macrocarpon*
Large hair-cap moss	*Polytrichum commune*
Large-leaved goldenrod	*Solidago macrophylla*
Large yellow pond-lily	*Nuphar advena*
Leatherleaf	*Chamaedaphne calyculata*
Lesser wintergreen	*Pyrola minor*
Lilaeopsis	*Lilaeopsis chinensis*
Little bluestem	*Schizachyrium scoparium*
Little-tree reindeer-lichen	*Cladina arbuscula*
Livid sedge	*Carex livida*
Long sedge	*Carex folliculata*
Long-stalked sedge	*Carex pedunculata*
Long-tubercled spikerush	*Eleocharis tuberculosa*
Long's bitter-cress	*Cardamine longii*
Long's bulrush	*Scirpus longii*
Low rough aster	*Aster radula*
Low spikemoss	*Selaginella selaginoides*
Lowbush blueberry	*Vaccinium angustifolium*
Mad-dog skullcap	*Scutellaria lateriflora*
Maidenhair fern	*Adiantum pedatum*
Male fern	*Dryopteris filix-mas*
Maleberry	*Lyonia ligustrina*
Map lichen	*Rhizocarpon geographicum*
Maple-leaved viburnum	*Viburnum acerifolium*
Marginal wood-fern	*Dryopteris marginalis*
Marsh-elder	*Iva frutescens*
Marsh-felwort	*Lomatogonium rotatum*

Common Name	Scientific Name
Marsh fern	*Thelypteris palustris*
Marsh muhly	*Muhlenbergia glomerata*
Marsh St. Johnswort	*Triadenum virginicum*
Marsh valerian	*Valeriana uliginosa*
Mayflower	*Epigaea repens*
Meadow beauty	*Rhexia virginica*
Meadowsweet	*Spiraea alba*
Michaux's sedge	*Carex michauxiana*
Missouri rockcress	*Arabis missouriensis*
Mistassini primrose	*Primula mistassinica*
Moss-plant	*Harrimanella hypnoides*
Mountain alder	*Alnus viridis* ssp. *crispa*
Mountain ash	*Sorbus decora* or *S. americana*
Mountain cranberry	*Vaccinium vitis-idaea*
Mountain fern moss	*Hylocomium splendens*
Mountain firmoss	*Huperzia appalachiana*
Mountain-heath	*Phyllodoce caerulea*
Mountain holly	*Nemopanthus mucronatus*
Mountain laurel	*Kalmia latifolia*
Mountain maple	*Acer spicatum*
Mountain sandwort	*Minuartia groenlandica*
Mountain timothy	*Phleum alpinum*
Mountain wood fern	*Dryopteris campyloptera*
Mud sedge	*Carex limosa*
Musclewood	*Carpinus caroliniana*
Mylia liverwort	*Mylia anomala*
Nannyberry	*Viburnum lentago*
Nantucket shadbush	*Amelanchier nantucketensis*
Narrow false oats	*Trisetum spicatum*
Narrow-leaved cattail	*Typha angustifolia*

Common Name	Scientific Name
Narrow-leaved cotton-grass	*Eriophorum angustifolium*
Narrow-leaved goldenrod	*Euthamia tenuifolia*
Neglected reed grass	*Calamagrostis stricta* ssp. *stricta*
New England northern reed	*Calamagrostis stricta* ssp. *inexpansa*
New England violet	*Viola novae-angliae*
New Jersey tea	*Ceanothus americanus*
New York aster	*Symphyotrichum novi-belgii*
New York fern	*Thelypteris novaeboracensis*
Nodding beggar ticks	*Bidens cernua*
Nodding pogonia	*Triphora trianthophora*
Northern blazing star	*Liatris scariosa*
Northern blue flag	*Iris versicolor*
Northern bog aster	*Symphiotrichum boreale*
Northern bog sedge	*Carex gynocrates*
Northern comandra	*Geocaulon lividum*
Northern dewberry	*Rubus flagellaris*
Northern firmoss	*Huperzia selago*
Northern gentian	*Gentianella amarella*
Northern meadow groundsel	*Senecio pauperculus*
Northern painted-cup	*Castilleja septentrionalis*
Northern slender pondweed	*Stuckenia filiformis* ssp. *alpinus*
Northern water-horehound	*Lycopus uniflorus*
Northern water-plaintain	*Alisma triviale*
Northern white cedar	*Thuja occidentalis*
Northern wood sorrel	*Oxalis montana*
Northern woodsia	*Woodsia alpina*
Nova Scotia false-foxglove	*Agalinis neoscotica*
Oak fern	*Gymnocarpion dryopteris*
Ostrich fern	*Matteucia struthiopteris*
Ox-eye daisy	*Leucanthemum vulgare*

Common Name	Scientific Name
Oysterleaf	*Mertensia maritima*
Painted trillium	*Trillium undulatum*
Pale corydalis	*Corydalis sempervirens*
Pale green orchis	*Platanthera flava*
Pale jewel-weed	*Impatiens pallida*
Pale laurel	*Kalmia polifolia*
Panic grass	*Panicum* sp./spp.
Paper birch	*Betula papyrifera*
Parker's pipewort	*Eriocaulon parkeri*
Partridgeberry	*Mitchella repens*
Peat moss	*Sphagnum* sp./spp.
Pickerelweed	*Pontederia cordata*
Pin cherry	*Prunus pensylvanica*
Pincushion moss	*Leucobryum glaucum*
Pinweed	*Lechea intermedia*
Pipewort	*Eriocaulon aquaticum*
Pitch pine	*Pinus rigida*
Pitcher plant	*Sarracenia purpurea*
Plantain-leaved pussytoes	*Antennaria plantaginifolia*
Podgrass	*Scheuchzeria palustris*
Poison ivy	*Toxicodendron radicans*
Polytrichum moss	*Polytrichum* sp./spp.
Pondweed	*Potamogeton* sp./spp.
Poverty oatgrass	*Danthonia spicata*
Prairie sedge	*Carex prairea*
Prototype quillwort	*Isoetes prototypus*
Pubescent sedge	*Carex hirtifolia*
Purple avens	*Geum rivale*
Purple clematis	*Clematis occidentalis*
Purple false-oats	*Trisetum melicoides*

Common Name	Scientific Name
Purple loosestrife	*Lythrum salicaria*
Purple-stemmed aster	*Symphyotrichum puniceum*
Pygmy water-lily	*Nymphaea leibergii*
Pygmyweed	*Crassula aquatica*
Quaking aspen	*Populus tremuloides*
Ram's-head lady's-slipper	*Cypripedium arietinum*
Rand's goldenrod	*Solidago simplex* var. *randii*
Raspberries	*Rubus* sp./spp.
Red elderberry	*Sambucus racemosa*
Red fescue	*Festuca rubra*
Red maple	*Acer rubrum*
Red oak	*Quercus rubra*
Red osier dogwood	*Cornus sericea*
Red pine	*Pinus resinosa*
Red raspberry	*Rubus idaeus*
Red-root flatsedge	*Cyperus erythrorhizos*
Red spruce	*Picea rubens*
Red-stemmed moss	*Pleurozium schreberi*
Red-tipped willow	*Salix eriocephala*
Reed canarygrass	*Phalaris arundinacea*
Reindeer lichen	*Cladina* or *Cladonia* sp./spp.
Rhodora	*Rhododendron canadense*
Robbins' pondweed	*Potamogeton robbinsii*
Robbins' ragwort	*Senecio schweinitzianus*
Robbins' spikerush	*Eleocharis robbinsii*
Rock polypody	*Polypodium virginianum* or *P. appalachianum*
Rock spikemoss	*Selaginella rupestris*
Rock tripe lichen	*Umbilicaria* sp./spp.
Rock whitlow-grass	*Draba arabisans*
Rose	*Rosa* sp./spp.

Common Name	Scientific Name
Rose twisted stalk	*Streptopus roseus*
Roseroot	*Sedum rosea*
Rough-leaved ricegrass	*Oryzopsis asperifolia*
Rough-stemmed goldenrod	*Solidago rugosa*
Round-leaved dogwood	*Cornus rugosa*
Round-leaved pyrola	*Pyrola americana*
Round-leaved sundew	*Drosera rotundifolia*
Round-leaved violet	*Viola rotundifolia*
Round-lobed hepatica	*Hepatica nobilis* var. *obtusa*
Royal fern	*Osmunda regalis*
Rugosa rose	*Rosa rugosa*
Russett sedge	*Carex saxatilis*
Rusty cliff fern	*Woodsia ilvensis*
Salt-loving spikerush	*Eleocharis halophila*
Saltmarsh bulrush	*Bolboschoenus robustus*
Saltmarsh false-foxglove	*Agalinis maritima*
Saltmarsh sedge	*Carex vacillans*
Saltmeadow cordgrass	*Spartina patens*
Sand cherry	*Prunus pumila*
Sandbar willow	*Salix interior*
Sarsparilla	*Aralia nudicaulis*
Sassafras	*Sassafras albidum*
Scarlet oak	*Quercus coccinea*
Screwstem	*Bartonia paniculata*
Scrub oak	*Quercus ilicifolia*
Sea-beach angelica	*Angelica lucida*
Sea-beach sandwort	*Honckenya peploides*
Sea-kale	*Cakile edentula*
Sea lavender	*Limonium carolinianum*
Sea milkwort	*Glaux maritima*

Common Name	Scientific Name
Seashore saltgrass	*Distichilis spicata*
Seaside goldenrod	*Solidago sempervirens*
Sensitive fern	*Onoclea sensibilis*
Shadbush	*Amelanchier* sp./spp.
Shagbark hickory	*Carya ovata*
Sharp-pointed ricegrass	*Oryzopsis pungens*
Sheep fescue	*Festuca ovina*
Sheep laurel	*Kalmia angustifolia*
Shining clubmoss	*Huperzia lucidula*
Shining willow	*Salix lucida*
Shore quillwort	*Isoetes riparia*
Showy lady's-slipper	*Cypripedium reginae*
Shrubby cinquefoil	*Pentaphylloides floribunda*
Silver maple	*Acer saccharinum*
Silverling	*Paronychia argyrocoma*
Silverrod	*Solidago bicolor*
Silverweed	*Argentina anserina*
Silvery sedge	*Carex canescens*
Silvery spleenwort	*Deparia acrostichoides*
Skunk cabbage	*Symplocarpus foetidus*
Slender blue flag	*Iris prismatica*
Slender cliffbrake	*Cryptogramma stelleri*
Slender-leaved sundew	*Drosera linearis*
Slender sedge	*Carex lasiocarpa*
Slender water-milfoil	*Myriophyllum alterniflorum*
Small cranberry	*Vaccinium oxycoccos*
Small enchanter's nightshade	*Circaea lutetiana* ssp. *canadensis*
Small purple bladderwort	*Utricularia resupinata*
Small round-leaved orchis	*Amerorchis rotundifolia*
Small saltmarsh aster	*Symphyotrichum subulatum*

Common Name	Scientific Name
Small St. Johnswort	*Triadenum fraseri*
Small yellow water crowfoot	*Ranunculus gmelinii*
Smartweeds	*Persicaria* spp.
Smooth alder	*Alnus serrulata var. subelliptica*
Smooth cordgrass	*Spartina alterniflora*
Smooth rockcress	*Arabis laevigata*
Smooth rose	*Rosa blanda*
Smooth sandwort	*Minuartia glabra*
Smooth shadbush	*Amelanchier laevis*
Smooth winterberry (holly)	*Ilex laevigata*
Smooth woodsia	*Woodsia glabella*
Soft-leaf muhly	*Muhlenbergia richarsdonis*
Softstem bulrush	*Schoenoplectus tabernaemontanii*
Sparse-flowered sedge	*Carex tenuiflora*
Spatulate-leaved sundew	*Drosera intermedia*
Spearscale	*Atriplex patula*
Speckled alder	*Alnus incana*
Sphagnum moss	*Sphagnum* sp./spp.
Spicebush	*Lindera benzoin*
Spikenard	*Aralia racemosa*
Spikerush	*Eleocharis* sp./spp.
Spinulose wood fern	*Dryopteris carthusiana*
Spongy arrowhead	*Sagittaria calycina* var. *spongiosa*
Spotted bladderwort	*Utricularia purpurea*
Spotted joe-pye weed	*Eupatorium maculatum*
Spotted pondweed	*Potamogeton pulcher*
Spotted touch-me-not	*Impatiens capensis*
Spring beauty	*Claytonia caroliniana*
Squashberry	*Viburnum edule*
Squirrel-corn	*Dicentra canadensis*

Common Name	Scientific Name
St. John oxytrope	*Oxytropis campestris*
St. Johnswort	*Hypericum* sp./spp.
Star saxifrage	*Saxifraga foliolosa*
Starflower	*Trientalis borealis*
Stair-step moss	*Hylocomium splendens*
Sticky false asphodel	*Triantha glutinosa*
Stiff arrowhead	*Sagittaria rigida*
Stiff aster	*Ionactis linariifolius*
Stiff clubmoss	*Lycopodium annotinum*
Straight-leaved pondweed	*Potamogeton strictifolius*
Striped maple	*Acer pensylvanicum*
Sugar maple	*Acer saccharum*
Summer grape	*Vitis aestivalis*
Sundew	*Drosera* sp./spp.
Swamp birch	*Betula pumila*
Swamp buttercup	*Ranunculus hispidus*
Swamp candles	*Lysimachia terrestris*
Swamp dewberry	*Rubus hispidus*
Swamp fly honeysuckle	*Lonicera oblongifolia*
Swamp milkweed	*Asclepias incarnata*
Swamp saxifrage	*Saxifraga pensylvanica*
Swamp white oak	*Quercus bicolor*
Swarthy sedge	*Carex adusta*
Sweet pepper-bush	*Clethra alnifolia*
Sweet-coltsfoot	*Petasites frigidus*
Sweetfern	*Comptonia peregrina*
Sweetflag	*Acorus americanus*
Sweetgale	*Myrica gale*
Sweetgrass	*Hierochloe odorata*
Switchgrass	*Panicum virgatum* var. *spissum*

Common Name	Scientific Name
Tall beak-rush	*Rhynchospora macrostachya*
Tall goldenrod	*Solidago altissima*
Tall meadow-rue	*Thalictrum pubescens*
Tall white violet	*Viola canadensis*
Tamarack	*Larix laricina*
Tapegrass	*Vallisneria americana*
Tawny cotton-grass	*Eriophorum virginicum*
Thicket shadbush	*Amelanchier stolonifera*
Three-leaved false Solomon's seal	*Maianthemum trifolium*
Three-lobed bazzania	*Bazzania trilobata*
Three-seeded sedge	*Carex trisperma*
Three-square	*Schoenoplectus punguns*
Three-toothed cinquefoil	*Sibbaldiopsis tridentata*
Three-way sedge	*Dulichium arundinaceum*
Tidal arrowhead	*Sagittaria calycina*
Trout lily	*Erythronium americanum*
Tufted cotton-grass	*Eriophorum vaginatum* ssp. *spissum*
Tufted hairgrass	*Deschampsia cespitosa*
Tufted reindeer-lichen	*Cladina stellaris*
Tundra dwarf birch	*Betula glandulosa*
Tussock sedge	*Carex stricta*
Twinflower	*Linnaea borealis*
Unicorn root	*Aletris farinosa*
Upright bindweed	*Calystegia spithamea*
Variable sedge	*Carex polymorpha*
Variable water-milfoil	*Myriophyllum heterophyllum*
Vasey's pondweed	*Potamogeton vaseyi*
Velvet-leaf blueberry	*Vaccinium myrtilloides*
Venus' looking-glass	*Triodanis perfoliata*
Violet	*Viola* sp./spp.

Common Name	Scientific Name
Virgin's bower	*Clematis virginiana*
Virginia chain fern	*Woodwardia virginica*
Virginia rose	*Rosa virginiana*
Virginia wild rye	*Elymus virginicus*
Water bulrush	*Schoenoplectus subterminalis*
Water hemlock	*Cicuta maculata*
Water-lily	*Nuphar variegata* or *Nymphaea* sp./spp.
Water lobelia	*Lobelia dortmanna*
Water-marigold	*Megalodonta beckii*
Water-parsnip	*Sium suave*
Water-pimpernel	*Samolus valerandi*
Water-shield	*Brasenia schreberi*
Water stargrass	*Zosterella dubia*
Wavy broom-moss	*Dicranum polysetum*
White adder's-mouth	*Malaxis monophyllos*
White ash	*Fraxinus americana*
White beak-rush	*Rhynchospora alba*
White bluegrass	*Poa glauca*
White oak	*Quercus alba*
White pine	*Pinus strobus*
White sea-blite	*Suaeda maritima*
White spruce	*Picea glauca*
White turtlehead	*Chelone glabra*
White water crowfoot	*Ranunculus aquatilis*
White water-lily	*Nymphaea odorata*
White wood aster	*Aster divaricatus*
White-topped aster	*Sericocarpus asteroides*
Whorled aster	*Oclemena acuminata*
Whorled loosestrife	*Lysimachia quadrifolia*
Wide-leaved sedge	*Carex platyphylla*

Common Name	Scientific Name
Wiegand's sedge	*Carex wiegandii*
Wild calla	*Calla palustris*
Wild chess	*Bromus kalmii*
Wild chive	*Allium schoenoprasum*
Wild garlic	*Allium canadense*
Wild ginger	*Asarum canadense*
Wild indigo	*Baptisia tinctoria*
Wild leek	*Allium tricoccum*
Wild-licorice	*Galium lanceolatum*
Wild lupine	*Lupinus perennis*
Wild-oats	*Uvularia sessilifolia*
Wild raisin	*Viburnum nudum* var. *cassinoides*
Wild rice	*Zizania aquatica* or *Z. palustris*
Wild sarsaparilla	*Aralia nudicaulis*
Wild strawberry	*Fragaria virginiana*
Willow	*Salix* sp./spp.
Winterberry (holly)	*Ilex verticillata*
Wintergreen	*Gaultheria procumbens*
Wire rush	*Juncus arcticus*
Witch grass	*Elytrigia repens*
Witch-hazel	*Hammamelis virginiana*
Witherod	*Viburnum nudum* var. *cassinoides*
Wood fern	*Dryopteris* sp./spp.
Wood lily	*Lilium philadelphicum*
Wood-nettle	*Laportea canadensis*
Wood-sorrel	*Oxalis montana*
Wool-grass	*Scirpus cyperinus*
Woodland horsetail	*Equisetum sylvaticum*
Woodland reindeer-lichen	*Cladina sylvatica*
Woodland sedge	*Carex lucorum*

Common Name	Scientific Name
Wooly panic-grass	*Panicum acuminatum*
Xanthoria lichen	*Xanthoria* sp./spp.
Yarrow	*Achillaea millefolium*
Yellow birch	*Betula alleghaniensis*
Yellow lady's-slipper	*Cypripedium parviflorum* or *C. pubescens*
Yellow loosestrife	*Lysimachia terrestris*
Yellow sedge	*Carex flava*
Yellow violet	*Viola pubescens*
Yellow water-lily	*Nuphar variegata*
Zig-zag goldenrod	*Solidago flexicaulis*

Appendix D: Animal Common and Scientific Names

This list includes rare animals that do or have occurred in Maine. Insects are often less familiar than birds and mammals, and so those mentioned in the natural community profiles are also included in this list. For more information regarding animal rarity ranks and state protection status (Endangered, Threatened, or Special Concern), please contact the Maine Department of Inland Fisheries and Wildlife, *http://www.maine.gov/ifw*.

Common Name	Scientific Name
A moth	*Nepytia pellucidaria*
A noctuid moth	*Chaetaglaea cerata*
American coot	*Fulica americana*
American oystercatcher	*Haematopus palliates*
American pipit	*Anthus rubescens*
Aureolaria seed borer	*Rhodoecia aurantiago*
Bald eagle	*Haliaeetus leucocephalus*
Barrens chaetaglaea	*Chaetaglaea tremula*
Barrens itame	*Itame* sp. 1 nr. *inextricata*
Barrens xylotype	*Xylotype capax*
Bicknell's thrush	*Catharus bicknelli*
Big bluet	*Enallagma durum*
Black-crowned night-heron	*Nycticorax nycticorax*
Black meadowhawk	*Sympetrum danae*
Black tern	*Chlidonias niger*
Blanding's turtle	*Emys blandingii*
Blue-winged warbler	*Vermivora pinus*
Bog elfin	*Callophrys lanoraieensis*
Bog fritillary	*Boloria eunomia*
Chestnut clearwing moth	*Synanthedon casteneae*
Citrine forktail	*Ischnura hastate*
Clayton's copper	*Lycaena dorcas claytoni*

Common Name	Scientific Name
Cobblestone tiger beetle	*Cicendela marginipennis*
Cobweb skipper	*Hesperia metea*
Columbine duskywing	*Erynnis lucilius*
Comet darner	*Anax longipes*
Common moorhen	*Gallinula chloropus*
Common tern	*Sterna hirundo*
Crowberry blue	*Lycaeides idas empetri*
Delicate emerald	*Somatochlora franklini*
Early hairstreak	*Erora laeta*
Edward's hairstreak	*Satyrium edwardsii*
Frigga fritillary	*Boloria frigga*
Frosted elfin	*Callophrys irus*
Golden eagle	*Aquila chrysaetos*
Graceful clearwing	*Hemaris gracilis*
Grasshopper sparrow	*Ammodramus savannarum*
Hemlock woolly adelgid	*Adelges tsugae*
Hessel's hairstreak	*Callophrys hesseli*
Incurvate emerald	*Somatochlora incurvata*
Juniper hairstreak	*Callophrys gryneus*
Katahdin arctic	*Oeneis polixenes katahdin*
Laughing gull	*Larus atricilla*
Least bittern	*Ixobrychus exilis*
Least tern	*Sterna antillarum*
Lilypad forktail	*Ischnura kellicotti*
Little bluet	*Enallagma minusculum*
New England bluet	*Enallagma laterale*
New England siltsnail	*Floridobia winkleyi*
Northern black racer	*Coluber constrictor constrictor*
Northern bog lemming	*Synaptomys borealis*
Oblique zale	*Zale obliqua*

Common Name	Scientific Name
Olive-sided flycatcher	*Contopus cooperi*
Peregrine Falcon	*Falco peregrinus*
Persius dusky wing	*Erynnis persius persius*
Pine barrens bluet	*Enallagma recurvatum*
Pine barrens zale	*Zale* sp. 1 nr. *lunifera*
Pine barrens zanclognatha	*Zanclognatha martha*
Pine devil moth	*Citheronia sepulcralis*
Pine pinion	*Lithophane lepida lepida*
Pine sphinx (Southern pine sphinx)	*Lapara coniferarum*
Pink sallow	*Psectraglaea carnosa*
Piping plover	*Charadrius melodus*
Purple lesser fritillary	*Boloria chariclea grandis*
Quebec emerald	*Somatochlora brevicincta*
Red-winged sallow	*Xystopeplus rufago*
Ribbon snake (Eastern or Northern)	*Thamnophis sauritus*
Ringed boghaunter	*Williamsonia lintneri*
Rock vole	*Microtus chrotorrhinus*
Roseate tern	*Sterna dougallii*
Rusty blackbird	*Euphagus carolinus*
Saltmarsh sharp-tailed sparrow	*Ammodramus caudacutus*
Scarlet bluet	*Enallagma pictum*
Sedge darner	*Aeshna juncea*
Sedge wren	*Cistothorus platensis*
Short-eared owl	*Asio flammeus*
Similar underwing	*Catocala similes*
Sleepy duskywing	*Erynnis brizo brizo*
Spartina borer moth	*Spartiniphaga inops*
Spatterdock darner	*Rhionaeschna mutata*
Spicebush swallowtail	*Papilio troilus*
Spotted turtle	*Clemmys guttata*

Common Name	Scientific Name
Spring salamander (Northern)	*Gyrinophilus porphyriticus*
Subarctic bluet	*Coenagrion interrogatum*
Subarctic darner	*Aeschna subarctica*
Thaxter's pinion moth	*Lithophane thaxteri*
The buckmoth	*Hemileuca maia maia*
Tomah mayfly	*Siphlonisca aerodromia*
Turquoise bluet	*Enallagma divagans*
Twilight moth	*Lycia rachelae*
Upland sandpiper	*Bartramia longicauda*
Western pine elfin	*Callophrys eryphon*
Whimbrel	*Numenius phaeopus*
Whip-poor-will	*Caprimulgus vociferus*
White Mountain tiger beetle	*Cicindela ancocisconensis*
Wood turtle	*Glyptemys insculpta*
Yellow rail	*Coturnicops noveboracensis*
Zigzag darner	*Aeshna sitchensis*

Appendix E: Quantitative Analysis Methods

Quantitative analyses were used as the first approach to evaluating the proposed types, and those analyses were supplemented with qualitative information. The detailed field survey data used for the quantitative analyses came from five major sources:

- MNAP natural community surveys in various regions of the state from 1991 – 1999 (170 samples);
- MNAP Ecological Reserves Inventory data from 1995 – 1997 (330 samples);
- MNAP Acadia National Park vegetation mapping data from 1997 – 1999 (169 samples);
- Peatland data contributed by Dr. Ronald Davis and Dennis Anderson at the University of Maine at Orono (UMO) (441 samples); and
- Alpine data collected and analyzed by Daniel D. Sperduto and Dr. Charles V. Cogbill for the New Hampshire Natural Heritage Inventory (36 samples; Sperduto and Cogbill 1999).

The 705 quantitative vegetation samples, not including the UMO peatland data, were analyzed to help delineate the community types and discern characteristics that make each type unique. The peatland data had already been analyzed and classified (Anderson and Davis 1997, Gawler 1998), and these data were combined with MNAP data on peatland vegetation types for more complete type descriptions.

Once the derivative data sets were assembled and checked, samples-by-species matrices of cover values were analyzed with PC-ORD ordination software (McCune and Mefford 1999). The techniques included ordinations via Detrended Correspondence Analysis (DCA), classifications via Two-way Indicator Species Analyses (TWINSPAN), identification of groups via Multi-Response Permutation Procedures (MRPP), and Indicator Species Analysis. With a project of this scope, where even a fairly large data set does not adequately cover the range of plant community variation across the state, multiple techniques are combined to identify vegetation types. Ordination and classification results are used to identify important gradients or factors in the data, which are then used to develop diagnostics for different vegetation types. Once types have thus been refined, MRPP and Indicator Species Analysis are used, respectively, to look for differences among vegetation types and indicator species for particular types.

The quantitative analyses were supplemented with qualitative information from MNAP files. Hundreds of field forms with notes describing specific examples of natural communities in Maine, as well as journal articles and reports in the primary scientific literature, were reviewed as supplementary materials for the community type delineation.

Appendix F: Glossary

Anastamosing: forming connecting channels.

Barren: a community or ecosystem with irregular tree cover that can range from nearly closed-canopy in some areas to sparse in others. Barrens typically occur on relatively flat sandy or gravelly glacial outwash or till soils. Barrens are prone to fire (given the excessively well-drained soils and predominance of conifers in the canopy), and the time since fire and fire intensity are important determinants in the structure and composition of a particular barren. Most barrens in Maine are pine or spruce dominated.

Brackish: estuary portions with both freshwater and saltwater influences, with salinity intermediate between those extremes.

Broom-mosses: mosses in the genera *Dicranum* and *Dicranella*, which grow in small or large tufts on the forest floor (upright growth form) and whose leaves often appear swept in one direction (hence the name). The growth form is very different from that of feather-mosses (see definition).

Bryoid: the ground layer of mosses, liverworts, and lichens. Those on trees are not included.

Bryophyte: mosses and liverworts; these are non-flowering and non-vascular plants that can be important components of the ground-level flora.

Calciphilic: literally, "calcium-loving", a plant that tends to grow in areas of comparatively high calcium content (often high in pH), which are uncommon in Maine.

Canopy: the uppermost layer of a forest or woodland, with trees >10 cm dbh. This includes trees emerging above the canopy (often called "supercanopy" trees); in some vegetation types, certain species will be characteristic as scattered supercanopy trees. It also may include trees that are shorter than the main canopy layer but are still > 10 cm dbh.

Caribou moss: see reindeer lichen.

Circumneutral: having a pH of around 7 (neutral). Usually circumneutral refers to pH 5.5 – 7.5 (higher values rarely seen in Maine ecosystems).

Cirque: a steep-sided rounded basin left in a mountainside after glacial activity. The Chimney Pond basin (South Basin) on Katahdin is Maine's best-known cirque.

Co-dominant: two species that share dominance of a particular layer.

Coniferous, Deciduous, Mixed: describes the proportion of conifers to deciduous tree cover in the canopy. The cutoff used here and in the National Vegetation Classification System (NVC) (Grossman et al. 1998) is 25%. Using this convention, if a forest has 0 - 25% deciduous tree cover it is classed as coniferous, and vice versa, and if both conifers and deciduous trees contribute 26 – 74% of the canopy, the types is mixed. It is important to note, however, that this should be taken as a fairly general indicator rather than a precise percentage.

Cover: the percentage of the ground that would be covered by a vertical projection of a given layer, group of plants, or plant species. Cover can be visually estimated with enough accuracy for this classification system to work. Canopy cover of each tree species can be estimated visually, but may also be determined by using basal area as a surrogate if tree measurements are being collected. The proportional basal area of each species is taken as that species' canopy proportion, and if the total canopy cover is noted then the cover for each species is easily derived.

DBH: diameter at breast height; the diameter of a tree at 4.5 off the ground.

Dominant: a species that is the most abundant, or nearly so, in its layer.

Emergent: coming out of; referring to aquatic plants whose leaves grow mainly above the water, not floating or submerged. Pickerelweed is an emergent plant.

Ericaceous/Ericad: shrubs of the family Ericaceae, or heath family, which are an ecologically important group that includes blueberry, cranberry, sheep laurel, and many bog shrubs. Some are evergreen and others deciduous.

Estuary: the lower portion of a river that mixes with salt water.

Feather-mosses: a group of mosses from several genera that creep laterally over the forest floor (spreading growth form), often covering relatively large areas, with leaves that resemble small feathers. Common species include red-stemmed moss, stair-step moss, hypnum moss, and fern moss. Feather-mosses have a different growth form than broom-mosses.

Forb: a non-woody plant that is not a graminoid (see that definition). Most plants that are commonly called "wildflowers" are forbs.

Glaciofluvial: pertaining to streams fed by melting glaciers, or the deposits and landforms produced by such streams.

Graminoid: grass-like in growth form, referring collectively to grasses (family *Poaceae*), sedges (family *Cyperaceae*), and rushes (family *Juncaceae*).

Heath shrub: see "Ericaceous" above.

Herb: the ground layer of vascular plants, including all herbaceous plants plus any dwarf shrubs (those that typically grow less than 1 m tall and often < 0.5 m tall). The herb layer may have both an herbaceous component and a woody component.

Hydric: characterized by considerable moisture.

Hydromorphic: a plant specially adapted to live in water. Hydromorphic characteristics include presence of large internal air spaces, lack of support tissue (since the water supports the plant), and ability to float on top of the water or be suspended in the water column.

Krummholz: stunted trees, typically at high altitudes (above treeline).

Lepidoptera: order of insects including moths and butterflies

Lichen: non-vascular plants consisting of a symbiotic relationship of algae and fungus. Occurs in three growth forms-fruticose (upright), foliose (leafy), and crustose (thin crusts).

Liverwort: small non-vascular plants related to mosses, but differing from them in certain structural and reproductive characteristics. See "bryophyte" and "bryoid" above. The most commonly seen liverwort in Maine forests is three-lobed bazzania, whose tiny leaves split into three teeth at the end, easily seen with a hand-lens.

Matrix forming: natural communities that are dominant in the landscape and are of a large landscape scale, covering hundreds to thousands of acres.

Microsite (microhabitat): small areas within a community that differ in some character (e.g., exposure, soil) and may have slightly different flora.

Minerotropic: wetlands (typically peatlands) that receive all or most of their nutrients from contact with surface or ground water sources.

Morainal: pertaining to boulders, stones, and other debris carried and deposited by a glacier.

Muskeg: general term for peatbog or peatland

Ombrotrophic: wetlands (typically peatlands) that receive all or most of their water and nutrients from precipitation.

Outwash: gravelly to sandy deposits that were carried in glacial meltwater and left behind after the glaciers receded. Outwash includes steep-sided eskers as well as more flat outwash plains and deltas.

Palustrine: of or pertaining to non-tidal wetlands.

Peatland: a wetland that develops where water drainage is blocked, precipitation is retained, and decomposition of organic matter is slow. Peatlands include bogs and fens and are typically characterized by accumulations of peat (*Sphagnum*) moss.

Podzolic: refers to soils that develop in temperate to cold moist climates under conifer or heath vegetation.

Reindeer lichen (reindeer moss): a group of lichen species in the genera *Cladina* and *Cladonia* that are like miniature shrubs or loose cushions in their growth form and often cover extensive areas. Though commonly called reindeer mosses, they are not mosses. Also known as "caribou moss" or "caribou lichens".

Relative cover: the proportion of a layer's cover that is composed of a certain species or group of plants. If the herb layer has 40% total cover, made up of 30% lowbush blueberry and 10% woodland sedge, the relative cover of shrubs in the herb layer is 75%.

Saturated: constantly filled with water, referring to poorly drained soils that are always wet. As the term is used here, it excludes soils that are seasonally flooded.

Sedge: a large family of plants (*Cyperaceae*) related to grasses, with narrow, grass-like leaves.

Senesce: to reach maturity and grow old

Serotinous: cones that remain closed until they are released by heat or fire.

Shrub: woody plants > 1 m but < 3 m tall. Note that the cutoff for shrubs vs. dwarf shrubs is about 1 m but is not intended to be constrained to that exactly. Dwarf shrubs are those that mostly grow under 1 m tall at the particular site, even though some individuals may be somewhat over.

***Sphagnum* mosses:** also known as peat mosses, species in the genus *Sphagnum*. This is a tremendously important group of mosses, both ecologically and economically, forming the substrate of most peatlands in Maine. Peat mosses differ from other mosses in their growth form (among other characters), with upright stems bearing groups (fascicles) of leaves spaced along them. At the top of the stem, the fascicles form a dense mass called a capitulum, which can look like a button, a star, or something in between, depending on the species. *Sphagnum* leaves have special cells that hold large amounts of water, making a *Sphagnum* carpet much like a sponge.

Strata: the vertical arrangement of layers of plants within a community. The strata are defined for this classification as canopy, sub-canopy, shrub, and herb.

Subcanopy: trees > 3 m tall but < 10 cm dbh.

Succession: the natural changes in species composition within a community over time.

Talus: large blocks of rock that have come loose from a hillside or cliff and accumulate as a talus slope.

Xeric: of, characterized by, or adapted to an extremely dry conditions.

References

Allen, B. 1999. Checklist of Maine mosses. Evansia 16:28-43.

Anderson, M.G., M.D. Merrill, and F.B. Biasi. 1998. Connecticut River watershed analysis: ecological communities and neotropical migratory birds. Final Report Summary. The Nature Conservancy, Eastern Conservation Science, Boston, MA. 56 pp.

Anderson, M.G. 1999. Viability and Spatial Assessment of Ecological Communities in the Northern Appalachian Ecoregion. Ph.D. Dissertation, University of New Hampshire. 224 pp.

Anderson, D.S. and R.B. Davis. 1997. The vegetation and its environment in Maine peatlands. Can. J. Bot. 75:1785-1805.

Association for Biodiversity Information. 2001. International Classification of Ecological Communities: Terrestrial Vegetation. Natural Heritage Central Databases (data current as of 2 July 2001). The Association for Biodiversity Information, Arlington, VA.

Bailey, R.G. 1995. Description of the Ecoregions of the United States. Miscellaneous Publication No. 1391, U.S. Dept. of Agriculture Forest Service, Washington, DC. 108 pp.

Baldwin, H.I. 1979. The distribution of *Pinus banksiana* Lamb. in New England and New York. Rhodora 81:549-565.

Bard, G.E. 1967. The woody vegetation of the mature forest of the Mianus River Gorge Preserve. Bull. Torr. Bot. Club 94:336-344.

Beatty, S.W. 1984. Influence of microtopography and canopy species on spatial patterns of forest understory plants. Ecology 65:1406-1419.

Bliss. L.C. 1963. Alpine plant communities of the Presidential Range, New Hampshire. Ecology 44:678-697.

Bormann, F.H. and M.F. Buell. 1964. Old-age stand of hemlock-northern hardwood forest in central Vermont. Bull. Torr. Bot. Club 91:451-465.

Bromley, S.W. 1935. The original forest types of southern New England. Ecol. Monographs 5:61-89.

Brown, J.H. Jr., C.A. Castaneda and R.J. Hindle. 1982. Floristic relationships and dynamics of hemlock (*Tsuga canadensis*) communities in Rhode Island. Bull. Torr. Bot. Club 109:385-391.

Bryan, R.R., M. Dionne, R.A. Cook, J. Jones, and A Goodspeed. 1997. Maine Citizen's Guide to Evaluating, Restoring, and Managing Tidal Marshes. Maine Audubon Society. 87 pp.

Buell, M.F. and J.E. Cantlon. 1950. A study of two communities of the New Jersey pine barrens and a comparison of methods. Ecology 31:567-586.

Burke, I. 1982. The Mahoosuc Mountains, Oxford County, Maine: A Natural Areas Inventory and Management Statement. Maine Critical Areas Program, Dept. of Conservation and State Planning Office, Augusta, ME.

Calhoun, A.J.K., J.E. Cormier, R.B. Owen, Jr., A.F. O'Connell Jr., C.T. Roman, and R.W. Tiner, Jr. 1994. The Wetlands of Acadia National Park and Vicinity. Maine Agricultural and Forest Experiment Station Miscellaneous Publication 721. 108 pp.

Caljouw, C. and S. Roeske. 1981. A Natural Resource Inventory and Critical Areas Survey of Bigelow Preserve. Bureau of Public Lands, Dept. of Conservation and Maine State Planning Office, Augusta, ME. 104 pp.

Cameron, D.C. 2000. Aquatic Vegetation Survey for Selected Maine Lakes. Maine Natural Areas Program, Dept. of Conservation, Augusta, ME. 113 pp.

Chapman, V.J. 1937. A note on the salt marshes of Nova Scotia. Rhodora 39:53-57.

Chokkalingam, U. 1998. Spatial and Temporal Patterns and Dynamics in Old-growth Northern Hardwood and Mixed Forests of Northern Maine. Ph.D. Dissertation, University of Maine. 227 pp.

Clayden, S. and A. Bouchard. 1983. Structure and dynamics of conifer-lichen stands on rock outcrops south of Lake Abitibi, Quebec. Can. J. Bot. 61:850-871.

Coffman, M.S. & G.L. Willis. 1977. The use of indicator species to classify climax sugar maple and eastern hemlock forests in Upper Michigan. Forest Ecology and Management 1:149-168.

Cogbill, C.V. 1985. Evaluation of the forest history and old-growth nature of the Big Reed Pond Preserve, T8R10 and T8R11 WELS, Maine. Report to The Nature Conservancy, Brunswick, ME.

Cogbill, C.V. and W.D. Hudson, Jr. 1990. The baseline characterization of the alpine area of Katahdin. Final unpublished report to the Appalachian Mountain Club Murphy Fund committee. 13 pp. + appendices.

Comer, P., D. Faber-Langendoen, R. Evans, S. Gawler, C. Josse, G. Kittel, S. Menard, M. Pyne, M. Reid, K. Schulz, K. Snow, and J. Teague. 2003. Ecological Systems of the United States: A Working Classification of U.S. Terrestrial Systems. NatureServe, Arlington, VA.

Cowardin, L.M., V. Carter, F.C. Golet, and E.T. LaRoe. 1979. Classification of Wetlands and Deepwater Habitats of the United States. U.S. Fish and Wildlife Service FWS/OBS 79/31. 103 pp.

Critical Areas Program. 1983. Jack pine (*Pinus banksiana* Lamb.) in Maine and its relevance to the Critical Areas Program. Planning Report # 77, State Planning Office, Augusta, ME. 79 pp.

Crum, H. 1991. Liverworts and Hornworts of Southern Michigan. University of Michigan Herbarium, Ann Arbor, MI. 233 pp.

Damman, A.W.H. 1977. Geographical changes in the vegetation patterns of raised bogs in the Bay of Fundy region of Maine and New Brunswick. Vegetatio 35:137-151.

Damman, A.W.H. and T.W. French. 1987. The ecology of peat bogs of the glaciated northeastern United States: a community profile. U.S. Fish and Wildlife Service Biological Report 85(7.16). 100 pp.

Davis, M.B., R.W. Spear, and L.C.K. Shane. 1980. Holocene Climate of New England. Quarternary Res. 14:240-250.

Davis, R.B. 1964. Bryophytes and lichens of the spruce-fir forests of the coast of Maine. I. The ground cover. The Bryologist 76:190-194.

Davis, R.B. 1966. Spruce-fir forests of the coast of Maine. Ecol. Monogr. 36:79-94.

Davis, R.B. and D.S. Anderson. 1991. The Eccentric Bogs of Maine: a Rare Wetland Type in the United States. Maine Agricultural Experiment Station Technical Bulletin #146, University of Maine, Orono. 144 pp.

Davis, R.B. and D.S. Anderson. 1999. A Numerical Method and Supporting Database for Evaluation of Maine Peatlands as Candidate Natural Areas. Maine Agricultural and Experiment Station Technical Bulletin #175, University of Maine, Orono. 166 pp.

Davis, R.B. and D.S. Anderson. 2001. Classification and distribution of freshwater peatlands in Maine. Northeastern Naturalist 8:1-50.

Dieffenbacher-Krall, A. 1994. Paleo- and Historical-ecology of the Cutler Grasslands, Cutler, Maine. M.S. Thesis, University of Maine, Orono. 58 pp. + illus.

Doyle, K.M., T.J. Fahey, and R.D. Paratley. 1987. Subalpine heathlands of the Mahoosuc range, Maine. Bull. Torr. Bot. Club 114:429-436.

Dunlop, D.A. and G.E. Crow. 1985. The vegetation and flora of the Seabrook dunes with special reference to rare plants. Rhodora 87:471-486.

Eastman, L.M. 1976. Chestnut oak, *Quercus prinus* L., in Maine and its relevance to the Critical Areas Program. Planning Report #14, State Planning Office, Augusta, ME. 11 pp.

Engstrom, B.E. 1997. Inventory and Classification of Natural Communities along the Upper Saco River, New Hampshire. New Hampshire Natural Heritage Inventory, Dept. of Resources and Economic Development, Concord, NH.

Esslinger, T.L. 1997. A cumulative checklist for the lichen-forming, lichenicolous and allied fungi of the continental United States and Canada. North Dakota State University: http://www.ndsu.nodak.edu/instruct/esslinge/chcklst/chcklst7.htm (First Posted 1 December 1997, Last Accessed 27 August 2001), Fargo, ND.

Fahey, T.J. 1976. The vegetation of a heath bald in Maine. Bull. Torr. Bot. Club 103:23-29.

Fahey, T.J. and W.A. Reiners. 1981. Fire in the forests of Maine and New Hampshire. Bull. Torr. Bot. Club 108:362-373.

Famous, N.C. 1998. Township 40 MD Landscape Analysis, Hancock County, Maine. Prepared for the Maine Bureau of Parks and Lands, Dept. of Conservation, Augusta, ME.

Famous, N.C. and M. Spencer. 1992. An Evaluation of the Maine Bureau of Public Lands Cutler Management Unit using landscape analysis and field verification. Final Report. Spencer/Famous Environmental Consulting, Augusta, ME.

Fernald, M.L. and K.M. Wiegand. 1910. A summer's botanizing in eastern Maine and western New Brunswick. Rhodora 12:101-146.

Flaccus, E. 1959. Revegetation of landslides in the White Mountains of New Hampshire. Ecology 40:692-703.

Foster, J.R. and W.A. Reiners. 1983. Vegetation patterns in a virgin subalpine forest at Crawford Notch, New Hampshire. Bull. Torrey Bot. Club 110:141-153.

Frye, R.F., and J.A. Quinn. 1979. Forest development in relation to topography and soils on a floodplain of the Raritan River, New Jersey. Bull. Torrey Bot. Club 106:334- 345.

Gawler, S.C. 1988. Disturbance-mediated Population Dynamics of *Pedicularis furbishiae* S. Wats., a Rare Riparian Endemic. Ph.D. Dissertation, University of Wisconsin, Madison. 195 pp.

Gawler, S.C. 1991. Natural Landscapes of Maine: A Classification of Ecosystems and Natural Communities. Maine Natural Areas Program, Dept. of Conservation, Augusta, ME. 77 pp.

Gawler, S.C. 1998. Priorities and tools for protecting peatlands in Maine - a report submitted to the U.S. Environmental Protection Agency. Maine Natural Areas Program, Dept. of Conservation, Augusta, ME.

Gawler, S.C. 2000. Vegetation mapping of Acadia National Park: classification, keys, and vegetation types. A report from the Maine Natural Areas Program to The Nature Conservancy, Arlington, VA. 156 pp.

Gawler, S.C. 2001. Natural Landscapes of Maine: A Classification of Vegetated Natural Communities and Ecosystems. Maine Natural Areas Program, Dept. of Conservation, Augusta, ME. 70 pp.

Gawler, S.C. and C. Jessee. 1997. Conservation plan for the Hollis Training Site, Maine Army National Guard. Maine Natural Areas Program, Dept. of Conservation, Augusta, ME.

Gawler, S.C., J.J. Albright, P.D. Vickery, and F.C. Smith. 1996. Biological Diversity in Maine: an Assessment of Status and Trends in the Terrestrial and Freshwater Landscape. Maine Natural Areas Program, Dept. of Conservation, Augusta, ME. 79 pp. + appendices.

Golet, F.C., A.J.K. Calhoun, W.R. DeRagon, D.J. Lowry, and A.J. Gold. 1993. Ecology of Red Maple Swamps in the Glaciated Northeast: A Community Profile. U. S. Fish and Wildlife Service, National Wetlands Research Center, Lafayette, LA. 151 pp.

Grossman, D.H., D. Faber-Langendoen, A.S. Weakley, M. Anderson, P. Bourgeron, R. Crawford, K. Goodin, S. Landaal, K. Metzler, K.D. Patterson, M. Pyne, M. Reid, and L. Sneddon. 1998. International Classification of Ecological Communities: Terrestrial Vegetation of the United States, Volume I. The National Vegetation Classification System: development, status and applications. The Nature Conservancy, Arlington, VA. 126 pp.

Grossman, D.H., K.L. Goodin, and C.L. Reuss (Editors). 1994. Rare Plant Communities of the Conterminous United States: An Initial Survey. The Nature Conservancy and The Association of Natural Heritage Programs and Conservation Data Centers, Arlington, VA. 620 pp.

Haines, A. and T.F. Vining. 1998. Flora of Maine: A Manual for Identification of Native and Naturalized Vascular Plants of Maine. V. F. Thomas Co., Bar Harbor, ME. 847 pp.

Harris, P. 1991. Waterboro Barrens Report. Unpublished report, The Nature Conservancy, Brunswick, ME.

Heinselman, M.L. 1981. Fire intensity and frequency as factors in the distribution and structure of northern ecosystems. pp. 7-57 in Mooney, H. A., T. M. Bonnicksen, N.L. Christensen, J.E. Lotan, and W.A. Reiners (technical coordinators). Fire regimes and ecosystem properties. Proceedings of the conference, December 1978, Honolulu, HI. U.S.D.A. Forest Service General Technical Report WO-26.

Henault, R.L. 1995. The establishment of a permanent plot in a Pitch Pine Dune Woodland at Morse Mountain Conservation Area. Unpublished report, Bates College, Lewiston, ME.

Hill, A.F. 1923. The vegetation of the Penobscot Bay region, Maine. Proc. Portland Soc. Nat. Hist. Vol. 3. Pt. 3:305-438.

Hinds, J.W. and P.L. Hinds. 1998. An annotated checklist of Maine macrolichens. pp. 345-376 in Glenn, M.G., R.C. Harris, R. Dirig, and M.S. Cole (Editors), Lichenographia Thomsoniana: North American Lichenology in Honor of John W. Thomson. Mycotaxon Ltd., Ithaca, NY. 445 pp.

Hoffman, C. 1995. Vegetation Patterns in the Saco River Floodplain Forest in Relation to Elevation and Flooding. M.S. Thesis, Antioch New England Graduate School, Keene, NH. 82 pp.

Holland, M.M. and C.J. Burk. 1984. The herb strata of three Connecticut River oxbow swamp forests. Rhodora 86:397-415.

Jacobson, G.L, I.J. Fernandez, P.A. Mayewski, and C.V. Schmitt (Editors). 2009. Maine's Climate Future: An Initial Assessment. University of Maine, Orono, ME.

Jacobson, G.L., Jr. and H.A. Jacobson. 1989. An Inventory of Distribution and Variation in Salt Marshes from Different Settings along the Maine Coast in Anderson, W.A. and H.W. Borns, Jr. (Editors). Neotectonics of Maine: studies in seismicity, crustal warping, and sea level change. Bulletin 40, pp. 69-83, Maine Geological Survey, Dept. of Conservation, Augusta, ME.

Jones, J.J. 1989. Inland Beach and Shoreline Character Evaluation within the Unorganized Towns in Maine. Maine Dept. of Conservation and Maine State Planning Office, Augusta, ME. 28 pp.

Keddy, P.A. 1983. Shoreline vegetation in Axe Lake, Ontario: effects of exposure on zonation patterns. Ecology 64:331-344.

Keddy, P.A. and I.C. Wisheu. 1989. Ecology, biogeography, and conservation of coastal plain plants: some general principles for the study of Nova Scotia wetlands. Rhodora 91(865):72-94.

Kern, M.J. 1985. T15R9 Aroostook County: A Natural Resources Inventory. Bureau of Public Lands, Dept. of Conservation and State Planning Office, Augusta, ME. 169 pp.

Keys, J., Jr., C. Carpenter, G. Hooks, F. Koenig, W.H. McKlab, W. Russell, and M.L. Smith. 1995. Ecological Units of the Eastern United States – first approximation (map and booklet of map unit tables). U.S. Dept. of Agriculture, Forest Service, Atlanta, GA.

Laderman, A.D. 1989. The Ecology of Atlantic White Cedar Wetlands: A Community Profile. U.S. Fish and Wildlife Service. Biol. Rept. 85(7.21). 114 pp.

Langdon, R., J. Andrews, K. Cox, S. Fiske, N. Kamman and S. Warren. 1998. A Classification of the Aquatic Communities of Vermont. Vermont Office of The Nature Conservancy and Vermont Biodiversity Project.

Leak, W.B. 1987. Characteristics of five climax stands in New Hampshire. Research note NE-336. USDA Forest Service, Northeast Forest Experiment Station, Broomall, PA. 5 pp.

Lorimer, C.G. 1977. The presettlement forest and natural disturbance cycle of northeastern Maine. Ecology 58:139-148.

MacDougall, A. 2001. Conservation status of Saint John River Valley hardwood forest in western New Brunswick. Rhodora 103:47-70.

Marchand, P.J. 1977. Subalpine bogs of the Mahoosuc Range, Maine: Physical characteristics and vegetation development. Contribution No. 11, The Center for Northern Studies, Wolcott, VT. 19 pp.

May, D.E. and R.B. Davis. 1978. Alpine tundra vegetation on Maine mountains and its relevance to the Critical Areas Program. Planning Report #36, State Planning Office. Augusta, ME. 66 pp.

Maycock, P. F. 1961. The spruce-fir forests of the Keweenaw Peninsula, Northern Michigan. Ecology 42:357-365.

McCune, B. and M.J. Mefford. 1999. Multivariate Analysis of Ecological Data, version 4.0. MjM Software, Gleneden Beach, OR.

McIntosh, R.P. 1959. Presence and cover in pine-oak stands in the Shawangunk Mountains, New York. Ecology 40:482-485.

McMahon, J.S. 1993. Saving All the Pieces – An Ecological Reserves Proposal from Maine. Maine Naturalist 1(4): 213-222.

McMahon, J. 1998. An Ecological Reserves System Inventory: Potential Ecological Reserves on Maine's Existing Public Lands and Private Conservation Lands. Maine Forest Biodiversity Project and Maine State Planning Office, Augusta, ME. 122 pp.

McMahon, J.S. 1990. The biophysical regions of Maine: patterns in the landscape and vegetation. M.S. Thesis, University of Maine, Orono. 120 pp.

McPhedran, J. 1998. Natural Community Mapping within a Proposed Ecological Reserve in Central Maine. M.S. Thesis, Dept. of Botany, Field Naturalist Program, University of Vermont, Burlington, VT. 84 pp.

Metzler, K.J. and A.W.H. Damman. 1985. Vegetation patterns in the Connecticut River flood plain in relation to frequency and duration of flooding. Naturaliste Can. (Rev. Ecol. Syst.) 112:535-547.

Mittlehauser, G.H., J.H. Connery, and J. Jacobs. 1996. Inventories of selected flora and fauna on 10 islands of Acadia National Park, Maine. U.S. Dept. of the Interior, National Park Service, Natural Resources Technical Report NPS/NESO-RNR/NRTR/96-01. 118 pp.

Moloney, K.A. 1986. Wave and nonwave regeneration processes in a subalpine *Abies balsamea* forest. Can. J. Bot. 64:341-349.

Motzkin, G. 1991. Atlantic White Cedar Wetlands of Massachusetts. Research Bulletin No. 731, Massachusetts Agricultural Expt. Sta. 53 pp.

Mulligan, A.D. 1980. The flora, vegetation, and phytogeographic relationships of Whaleboat Island, Casco Bay, Maine. Rhodora 82:441-459.

Nelson, B.W. and L.K. Fink. 1980. Geological and Botanical Features of Sand Beach Systems in Maine. Bulletin No. 14, Maine Sea Grant Publications. 163 pp.

Newcomb, L. 1977. Newcomb's Wildflower Guide. illus. by Gordon Morrison. Little, Brown and Co., Boston, MA. 490 pp.

Nichols, G.E. 1935. The hemlock-white pine-northern hardwood region of eastern North America. Ecology 16:403-422.

Nichols, W.F., J.M. Hoy, and D.D. Sperduto. 2001. Open Riparian Communities and Riparian Complexes in New Hampshire. New Hampshire Natural Heritage Inventory, Dept. of Economic and Community Development, Concord, NH.

Niering, W.A. and R.S. Warren. 1980. Vegetation patterns and processes in New England salt marshes. Bioscience 30:301- 307.

Nixon, S.W. 1982. The ecology of New England high salt marshes: a community profile. U.S. Fish and Wildlife Service Office of Biological Services, Washington DC. FWS/OBS-81/5570.

Norton, A.H. 1927. Botanical notes from the Millbridge region. Maine Naturalist 7:148-151.

Odum, W.T., T.J. Smith III, J.K. Hoover, and C.C. McIvor. 1984. The ecology of tidal freshwater marshes of the United States east coast: a community profile. U.S. Fish and Wildlife Service FWS/OBS-83/17. 177 pp.

Olday, F.C., S. C.Gawler, and B. St. John Vickery. 1982. Seven unusual sub-Arctic plants of the Maine coast. Planning Report #78, Critical Areas Program, State Planning Office, Augusta, ME.

Olmsted, C.E. 1937. Vegetation of certain sand plains of Connecticut. Bot. Gaz. 99:209-300.

Oosting, H.J. and J.F. Reed. 1944. Ecological composition of pulpwood forests in northwestern Maine. Am. Midl. Nat. 31:182-210.

Poiani, K. and B. Richter. 1999. Functional Landscapes and the Conservation of Biodiversity. Working Papers in Conservation Science No. 1. The Nature Conservancy, Conservation Science Division, Arlington, VA. 8 pp.

Rappaport, N.R. and F.R. Wesley. 1985. A vegetation study of nine bird islands off the coast of Maine. Unpublished report prepared for the National Audubon Society. 63 pp.

Reiners, W.A . and G.E. Lang. 1979. Vegetational patterns and processes in the balsam fir zone, White Mountains, New Hampshire. Ecology 60:403-417.

Roberts, B.A. and A. Robertson. 1986. Salt marshes of Atlantic Canada: their ecology and distribution. Can. J. Bot. 64 455-467.

Rogers, R.S. 1978. Forests dominated by hemlock (*Tsuga canadensis*): distribution as related to site and postsettlement history. Can. J. Bot. 56:843-854.

Rogers, R.S. 1980. Hemlock stands from Wisconsin to Nova Scotia: transitions in understory composition. Ecology 61:178-193.

Russell, E.W.B. 1983. Indian-set fires in the forests of Northeastern United States. Ecology 64:78-88.

St. John, H. 1929. Plants of the headwaters of the St. John River, Maine. Research Studies of the State College of Washington. Vol 1:28-58.

Siccama, T.G., F.H. Bormann, and G.E. Likens. 1970. The Hubbard Brook ecosystem study: Productivity, nutrients, and phytosociology of the herbaceous layer. Ecol. Monogr. 40:389-402.

Singleton, J., J. Loo and J. Foley. 2000. Conservation guidelines for ecologically sensitive forested sites on private woodlots within the Fundy Model Forest. Information Report M-X-207E. Canadian Forest Service - Atlantic Forestry Centre, Fredericton, NB, Canada. 57 pp.

Society of American Foresters. 1980. Forest Cover Types of the United States and Canada. Society of American Foresters, Bethesda, MD. 148 pp.

Sorenson, E.R. 1986. The ecology and distribution of ribbed fens in Maine and their relevance to the Critical Areas Program. Planning Report #81. State Planning Office. Augusta, ME. 171 pp.

Sorenson E.R., B. Engstrom, M. Lapin, R. Popp and S. Parren. 1998a. Northern white cedar swamps and red maple – northern white cedar swamps of Vermont: Some sites of ecological significance. Nongame and Natural Heritage Program, Vermont Fish and Wildlife Dept., Waterbury, VT. 45 pp.

Sorenson, E., M. Lapin, B. Engstrom, and R. Popp. 1998b. Floodplain forests of Vermont: some sites of ecological significance. Nongame and Natural Heritage Program, Vermont Fish and Wildlife Dept., Waterbury, VT. 18 pp.

Sperduto, D.D. 1994. Coastal Plain Pondshores and Basin Marshes in New Hampshire. Report submitted to the Environmental Protection Agency, Wetlands Protection Division, Region 1. New Hampshire Natural Heritage Program, Dept. of Economic and Community Development, Concord, NH. 97 pp.

Sperduto, D.D. 2000. A classification of wetland natural communities in New Hampshire. New Hampshire Natural Heritage Inventory, Dept. of Economic and Community Development, Concord, NH. 156 pp.

Sperduto, D.D. and N. Ritter. 1994. Atlantic White Cedar Wetlands of New Hampshire. Report submitted to the Environmental Protection Agency Wetlands Protection Section, Region 1. New Hampshire Natural Heritage Program, Dept. of Resources and Economic Development, Concord, NH. 97 pp.

Sperduto, D.D. and C.V. Cogbill. 1999. Alpine and subalpine vegetation of the White Mountains, New Hampshire. A report prepared for the USDA Forest Service, Laconia, NH. New Hampshire Natural Heritage Inventory, Dept. of Economic and Community Development, Concord, NH. 25 pp. + appendices.

Sperduto, D.D. and K.F. Crowley. 2001. Key to Upland Forest Communities in New Hampshire. New Hampshire Natural Heritage Inventory, Dept. of Economic and Community Development, Concord, NH.

Sprugel, D.G. 1976. Dynamic structure of wave-regenerated *Abies balsamea* forests in the northeastern United States. J. Ecol. 64:889-911.

Stern, R.J. 1979a. Old growth white oak (*Quercus alba*) stands and their relevance to the Critical Areas Program. Planning Report #65, State Planning Office, Augusta, ME. 35 pp.

Stern, R.J. 1979b. Old growth shagbark hickory (*Carya ovata*) stands in Maine and their relevance to the Critical Areas Program. Planning Report #66, State Planning Office, Augusta, ME. 26 pp.

Teal, J.M. 1986. The ecology of regularly flooded salt marshes of New England: a community profile. U.S. Fish and Wildlife Service, Biol. Rep. 85(7.4). 61 pp.

The Nature Conservancy and the Environmental Systems Research Institute. 1994. NBS/NPS Vegetation Mapping Program: Standardized National Vegetation Classification System. Prepared for the U.S. Dept. of the Interior, National Biological Survey and National Park Service.

Therres, G.D. 1999. Wildlife species of regional conservation concern in the northeastern United States. Northeast Wildlife 54:93-100.

Thompson, E.H. 1980. The role of fire in the Great Wass Island jack pine stand. Unpublished report submitted to The Nature Conservancy, Maine Chapter, Brunswick, ME. 19 pp.

Tyler, H.R. Jr., and C.V. Davis. 1982. Evaluation of No. 5 Bog and jack pine stand, Somerset, Maine, as a potential National Natural Landmark. Report prepared for the National Park Service, U.S. Dept. of the Interior. Maine Critical Areas Program, State Planning Office, Augusta, ME. 41 pp.

Vaux, P. 2001. Maine Aquatics: Maine Aquatic Biodiversity Project Newsletter. MABP, Room 215, 650 State Street, Bangor, ME. 5 pp.

U.S. Forest Service. 2009. Northeastern Climate Change Atlas. Available at: http://www.nrs.fs.fed.us/atlas.

Vickery, P.D. 1990. Report on grassland habitats in relation to the presence of Grasshopper Sparrows (*Ammodramus savannarum*) and other rare vertebrates in Maine. A report prepared for the Maine Natural Heritage Program, Augusta, ME. 29 pp.

Vogelmann, H.W. 1976. An unusual Black Gum Swamp in Maine. Rhodora 78:326-327.

Walker, S. 2009. Planning for Habitat Resilience in a Changing Climate. Beginning with Habitat Program, Maine Department of Inland Fisheries and Wildlife, Winter 2009 newsletter, Augusta, ME.

Wheeler, A.G. 1991. Plant bugs of *Quercus ilicifolia*: myriads of mirids (Heteroptera) in pitch pine-scrub oak barrens. J. New York Entomol. Soc. 99:405-440.

Widoff, L. 1987. Pitch pine/scrub oak barrens in Maine. 1988. Planning Report #86. Critical Areas Program, Maine State Planning Office, Augusta, ME and Maine Natural Heritage Program, The Nature Conservancy. Topsham, ME. 104 pp.

Widoff, L. 1988. Ecologically significant peatlands in Maine: recommendations for protection. Unpublished report to the Natural Resources Council of Maine.

Index to Natural Community Profiles

Community Name	Page #